Image Analysis and Integration in Remote Sensing

Image Analysis and Integration in Remote Sensing

Editor: Catherine Williamson

Callisto Reference
www.callistoreference.com

Callisto Reference,
118-35 Queens Blvd., Suite 400,
Forest Hills, NY 11375, USA

Visit us on the World Wide Web at:
www.callistoreference.com

ISBN: 978-1-64116-804-5 (Hardback)

Cataloging-in-Publication Data

Image analysis and integration in remote sensing / edited by Catherine Williamson.
p. cm.
Includes bibliographical references and index.
ISBN 978-1-64116-804-5
1. Remote sensing. 2. Remote sensing--Analysis. 3. Remote-sensing images. I. Williamson, Catherine.
G70.4 .R46 2023
621.367 8--dc23

Table of Contents

Preface

Remote sensing is the science of obtaining information about objects or areas from a distance, usually from satellites and aircrafts. It has diverse applications in a number of fields including hazard assessment and natural resource management. Remote sensing data plays a crucial role in global environmental analysis and contributes towards economic development. However, there are some challenges in real life applications of remote sensing including object detection, image classification, image correction, image segmentation, and change detection. Research is underway for developing advanced techniques for interpreting and integrating remote sensing images. Statistical computing methods and machine learning techniques such as graphical models, deep learning, kernel machines, and sparse coding are used to analyze remote sensing data. This data can be combined with other layers of information by using GIS. This book provides significant information to help develop a good understanding of remote sensing image analysis and integration. Through it, we attempt to further enlighten the readers about the new concepts in this area of study.

This book is a comprehensive compilation of works of different researchers from varied parts of the world. It includes valuable experiences of the researchers with the sole objective of providing the readers (learners) with a proper knowledge of the concerned field. This book will be beneficial in evoking inspiration and enhancing the knowledge of the interested readers.

In the end, I would like to extend my heartiest thanks to the authors who worked with great determination on their chapters. I also appreciate the publisher's support in the course of the book. I would also like to deeply acknowledge my family who stood by me as a source of inspiration during the project.

Editor

1

Assessment of the Segmentation of RGB Remote Sensing Images: A Subjective Approach

Giruta Kazakeviciute-Januskeviciene [1,2,*], Edgaras Janusonis [1], Romualdas Bausys [1], Tadas Limba [2] and Mindaugas Kiskis [2]

1 Department of Graphical Systems, Vilnius Gediminas Technical University, Sauletekio al. 11, LT-10223 Vilnius, Lithuania; edgaras.janusonis@stud.vgtu.lt (E.J.); romualdas.bausys@vgtu.lt (R.B.)

2 Faculty of Public Governance and Business, Mykolas Romeris University, Ateities str. 20, LT-08303 Vilnius, Lithuania; tlimba@mruni.eu (T.L.); mkiskis@mruni.eu (M.K.)

* Correspondence: giruta.kazakeviciute-januskeviciene@vgtu.lt

Abstract: The evaluation of remote sensing imagery segmentation results plays an important role in the further image analysis and decision-making. The search for the optimal segmentation method for a particular data set and the suitability of segmentation results for the use in satellite image classification are examples where the proper image segmentation quality assessment can affect the quality of the final result. There is no extensive research related to the assessment of the segmentation effectiveness of the images. The designed objective quality assessment metrics that can be used to assess the quality of the obtained segmentation results usually take into account the subjective features of the human visual system (HVS). A novel approach is used in the article to estimate the effectiveness of satellite image segmentation by relating and determining the correlation between subjective and objective segmentation quality metrics. Pearson's and Spearman's correlation was used for satellite images after applying a k-means++ clustering algorithm based on colour information. Simultaneously, the dataset of the satellite images with ground truth (GT) based on the "DeepGlobe Land Cover Classification Challenge" dataset was constructed for testing three classes of quality metrics for satellite image segmentation.

Keywords: satellite image segmentation; segmentation quality assessment; correlation analysis; objective quality metrics; subjective evaluation

---

1. Introduction

Satellite imagery classification is one of the main tasks in remote sensing applications such as city planning, climate change research [1], earth observations, geographical maps improvement [2], topographic surveys, or for the military purposes. Colour segmentation may also be used to track land cover changes over time [3]. The most prominent goal of remote sensing data analysis is object detection, based on image segmentation [2]. The principal goal of the image segmentation process is to partition the image into a set of segments that are homogeneous, according to specific criteria such as colour spectrum, shape, intensity, or texture [3] and map the individual regions to the corresponding real-world objects, like rivers, fields, roads, and other. Image segmentation is a broad term dependent on the goals of the specific application. Image segmentation is utilized for a wide variety of applications and is also a part of the object detection process.

Remote sensing images (RSI) usually represent various parts of the Earth surface and are characterized by clear details along with diverse spatial and textural information. When compared to most natural images, the satellite images stand out as being highly structured and uniform [1].

While improving current state-of-the-art segmentation methods, it is equally important to have proper methods for segmentation quality evaluation. The performance of colour segmentation greatly impacts the quality of an image understanding system [4].

Since the subjective quality assessment is expensive, resource-intensive, and time-consuming, the aim is to investigate the correlation between objective quality assessment methods (QAM) and the human visual system (HVS) so that appropriate objective QAM can be applied to assess the quality of satellite image segmentation.

A wide variety of objective metrics for examining segmentation quality are proposed. Fewer studies were conducted to compare the correlation of widely used objective segmentation metrics to the subjective quality assessment. Usually, only a small subset of available quality metrics is being compared against new quality measures during their development process [5–7].

The authors proposed to relate various objective metrics to the subjective scores by using metric classes, which is a novel approach in the image segmentation. The defined metrics classes were tested with the constructed dataset of the satellite images with ground truth (GT). The DeepGlobe Land Cover dataset was used as the base for the construction of our dataset. The original results of this study can be applied for the development of image segmentation quality analysis or combined with the related knowledge for the derived results, such as selecting a quality metric for a particular dataset or application that demands both tight correlations with human perception and low computational complexity. Results can also help suggest a possible combination of different metrics for achieving a more accurate relation with human perception of segmented images.

This article has the following structure. Section 2 provides a summary on published papers in the context of subjective and/or objective segmentation quality assessment of the satellite or natural images. Section 3 provides an overview of the k-means++ clustering method. Section 4 describes the characteristics of dataset and dataset preparation for the segmentation, objective, and subjective quality assessment. Internal, external, and full-reference image quality assessment (FR-IQA) metric groups considered in this experiment are explained in Section 5. Section 6 provides information on the subjective quality assessment procedure. Experimental results, including the roadmap of technology and used correlation coefficients, are presented in Section 7. Discussion and directions for future research are presented in Section 8. Finally, conclusions in Section 9 summarize the main observations from the discussion.

2. Related Works

Over many years, the development of image quality assessment (IQA) has drawn extensive and constant attention. Relatively less research was conducted from the perspective of evaluating image segmentation quality, especially the way humans understand and estimate the quality of segmented images.

Authors of Reference [7] systematized human visual properties, important for designing an objective segmentation metric: (1) human visual tolerance (e.g., for border distortions), (2) human visual saturation (difficulty for the human to evaluate similarity when distortions become large), (3) different perception of false negative (FN) and false positive (FP) pixels, and (4) the strongest distortions determine overall segmentation quality.

For natural images, the authors of Reference [8] suggested a contour-based score, combining Jaccard index (*JI*) and *BF* (Boundary F1) score for increased correlation with human rankings, as *JI* does not take into account quality of segmentation boundaries. It was observed that, while the accurate contours are definitely less important than correct classification, the proposed measure can further improve correlation with human rankings for high-quality segmentation. Authors used PASCAL VOC 2007 and 2011 datasets. However, very imbalanced data in PASCAL Visual Object Classes (PASCAL VOC) dataset can create biased results for some metrics like Accuracy (*ACC*). Ground truths contain instances of individual objects labelled separately in addition to the auxiliary label for object contours.

Similar to Reference [8], authors in Reference [9] proposed a subjective quality metric for single object segmentation combining region and boundary-based terms that relate to human visual tolerance and saturation properties. A subjective test showed the improved results, achieving a Spearman Rank Order Correlation Coefficient (*SROCC*) value of 0.88 and then compared to *JI*, *F*1 score (*F*1), Fuzzy Contour (*FC*), *BF* score, and Mixed Measure (*MM*) metrics. For testing the new metric, the dataset was created by selecting images from other object segmentation databases like Microsoft research Asia salient object database, PASCAL VOC 2012, and Microsoft research Cambridges grabcut database.

The effectiveness of Peak signal-to-noise ratio (*PSNR*) as a quality measure for image segmentation algorithms was explored in Reference [10]. In this experiment, GT data was created by modifying images from the Berkeley BSR300 database (intended for evaluating edge detection algorithms) to resemble threshold segmentation. *PSNR* was used to measure the similarity between the GT image and segmentation result obtained by applying salt and pepper noise on the created GT image. It was noted that quality of the edge detection can be more effectively evaluated by *PSNR*. The tests performed by the authors, however, have not considered multi-region segmentation, which is a more practical approach for real-world objects.

Evaluation of segmentation quality for satellite images is regarded as a difficult task since there is no universal standard for evaluating segmentation results of satellite images. The most common evaluation methods are based on external or full-reference quality measures, and the need of sufficient amount of reference data poses a problem. Authors of Reference [11] suggested a Synthetic Image Testing Framework (SITEF) for the evaluation of multispectral satellite image segmentation by using synthetic images. This method provides different evaluation perspectives such as parcel size, shape, and land cover type. The framework was tested using images obtained by SPOT HRG satellite consisting of six land cover classes. The Hammoude metric, Rand coefficient, Corrected Rand coefficient, and *JI* were used for segmentation evaluation.

Authors in Reference [12] proposed attention dilation-LinkNet (AD-LinkNet) neural network that displays a significant improvement on the segmentation accuracy of satellite images. Three satellite segmentation datasets were used: DeepGlobe's road extraction and land cover datasets [1] as well as Inner Mongolia's land classification dataset. For different data sets, different model optimizations are needed. *JI* was used for evaluating both road and land segmentation results.

There are apparent differences between satellite images [1] and natural images, as seen in PASCAL VOC 2012 dataset. In satellite imagery, every object has a semantic meaning, while natural images are more chaotic than satellite, and often include large areas of background, which is less important when compared to foreground objects.

Colour image segmentation can provide more information for detecting objects than using grayscale images. Satellite images consist of homogenous colour regions, that define image data and grouping of pixels can be performed on distinct colour features. Many papers have been published in the past, focusing on using colour features for segmenting satellite images [3,13]. Automatic detection of road segments from RSI for road detection applications [14] suggested an approach that can also be applied to detect other objects in RSI. Authors in Reference [15] proposed a method for land cover classification based on colour moments (mean, standard deviation, and skewness) used in extracting colour features. The segmentation method in Reference [16] used wavelet transform for extracting colour and texture features from satellite images in the YCbCr colour space.

3. Image Segmentation

A wide variety of clustering solutions have been proposed for solving various problems depending on application-specific goals and/or characteristics of a particular dataset. A cluster is a group of pixels, which are similar and also dissimilar to pixels in other clusters, according to specific image features, that can be used for further image analysis. If the segmentation uses colour as a feature, then pixels are assigned to the clusters based on their colour similarity. Additionally, segmentation can be performed by using a combination of different features [16]. Using segmentation by colour, we can also avoid feature calculation for every pixel, thus, improving overall performance.

For this research, the k-means [17] clustering algorithm was selected. Researchers tend to employ clustering methods with well-known advantages and disadvantages, avoiding undiscovered limitations that could impact results [18]. K-means popularity and its simple implementation make it easier for replicating the same experiment independently. However, other clustering algorithms can be used, which might provide a different perspective and/or better results in certain situations (e.g., finding more complex cluster shapes).

The k-means algorithm (Table 1) groups the N data points or pixels $\{x_1, x_2, \cdots, x_i, \cdots, x_N\}$ into K clusters by minimizing its objective function SSE (sum of squared error) within a cluster for all clusters. Minimizing SSE also maximizes SSB (the sum of squares between clusters). For this reason, SSE, SSB, and metrics combining these criteria return higher scores for clusters constructed by k-means.

Table 1. The k-means clustering method.

Step	Description
1	An optimal number of clusters K is selected by the Silhouette method [19].
2	Instead of random initialization, cluster centroids are initialized using k-means++ procedure (Table 2).
3	The Euclidean distance $D = d(x_i, c_i)$ is calculated between the cluster centroids and each pixel of an image.
4	Based on the calculated D, all pixels are assigned to the nearest centroid.
5	Recalculate all cluster centroid positions c_i by computing the mean of currently assigned pixels. $c_i = 1/\lvert C_i \rvert \cdot \sum_{x \in C_i} x_i$.
6	The cycle is repeated (from the third step) until the position of the cluster centroids no longer changes [1] (i.e., no pixels are reassigned).

[1] Alternatively, another criterion, e.g., (1) a limited number of iterations, (2) insignificant changes of cluster centroids positions, and (3) SSE falls below the set limit—possible combinations.

Table 2. The k-means++ cluster centroid initialization method [20].

Step	Description
1	First cluster centroid c_1 is selected evenly randomly from the existing set of pixels X.
2	Calculate distances $D(x)$ from each pixel to the nearest centroid (in the first iteration, this is c_1).
3	Each subsequent cluster centroid c_i is selected from the remaining set of pixels $x \in X$ with probability: $\frac{D^2(x)}{\sum_{x \in X} D^2(x)}$.
4	Go back to step 2 and repeat K-1 times until K cluster centroids have been added.
5	Proceed with step 3 from Table 1.

Before the segmentation stage, no preprocessing is necessary, although, according to Reference [8], the quality metric scores might have less meaning when the quality of segmentation is too low, suggesting that middle-quality segmentation is the best choice for the subjective quality evaluation. As such, we selected the optimal number of clusters by the Silhouette method [19] using a k-means++ algorithm to find their initial centroids (Table 2).

Results produced by k-means are dependent on the initialization procedure of cluster centroids. It is important to configure parameters for more deterministic behavior, not to impact the calculations of the metrics severely. The prepared images from our satellite dataset (described in Section 4) have lower resolution (324 × 220 px) and fewer clusters (2–4). We observed segmentation results returned by MATLAB implementation of k-means seem to be stable even with the default value of 100 iterations and only three replications. Alternatively, it is possible to (1) lock cluster centroids by selecting them manually or (2) use Pseudo-Random Number Generator (*PRNG*) with constant seed [18]. In our case, k-means++ initialization (Table 2) was used, which also can improve segmentation results.

Satellite images are segmented in the CIE 1976 L*a*b* colour space, as suggested in Reference [14], which also more accurately represents the human perception of the colour [21] and have better performance than RGB space in many colour image applications. The Euclidean distance is used to measure colour similarities between pixels in the a*b* plane [3].

4. Satellite Images Dataset

There are many datasets designed for image segmentation problems. The selection of an appropriate dataset is a crucial decision that impacts subsequent choices, such as the selection of the optimal clustering method. It is important to understand the limitations and possible problems of the particular dataset before using it for any research project.

The majority of quality metrics for the assessment of the segmented image quality requires the GT image. For the evaluation of image quality after such distortions as image compression, the GT is treated as the original image, and the GT preparation process is unnecessary. Obtaining GT is always a barrier for automated image segmentation. Human skills are often required for manual labelling, and it is a time-consuming process. The manual labelling itself is subjective, and the different GT versions of the same image may be produced. Widely used datasets like Berkeley segmentation datasets (BSDS) [22] often contain natural images very diverse in their content, but they do not necessarily serve as a target for the specific application or provide GT, which match specific segmentation goals.

Figure 1 depicts images from our constructed dataset based on the "DeepGlobe Land Cover Classification Challenge" dataset [1] intended to solve a mentioned problem by providing satellite imagery with GT data for improving state-of-the-art satellite image processing methods.

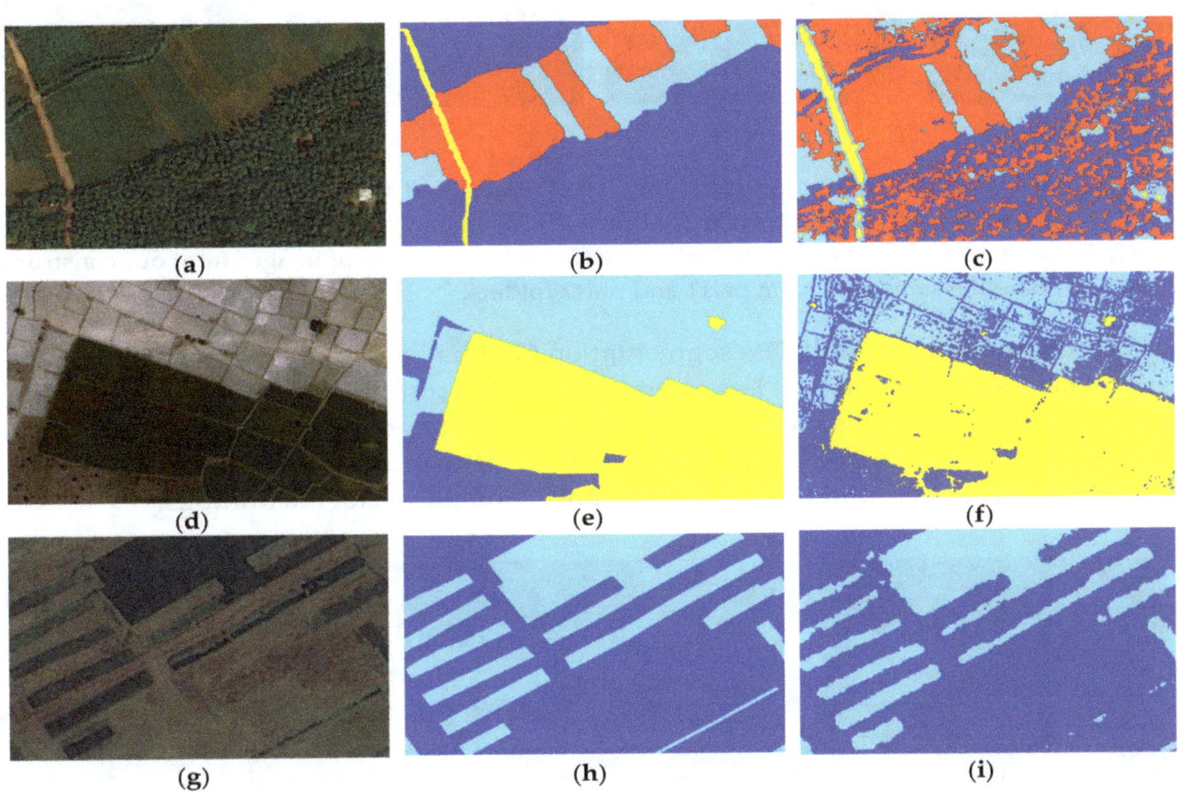

Figure 1. Segmentation results for satellite images from our dataset part 1 [23] (different clusters for each image are marked with selected colours): (**a**) satellite image "71619_sat," (**b**) corrected GT image "71619_gt" with 4 clusters, (**c**) k-means++ segmentation results "71619_seg" of "71619_sat," (**d**) satellite image "161109_sat," (**e**) corrected GT image "161109_gt" with three clusters, (**f**) k-means++ segmentation results "161109_seg" of "161109_sat," (**g**) satellite image "676758_sat," (**h**) corrected GT image "676758_gt" with two clusters, (**i**) k-means++ segmentation results "676758_seg" of "676758_sat.".

The original dataset is divided into three parts: test dataset (172 images), validation dataset (171 images), and training dataset, which consists of satellite images paired with corresponding GT images. Training dataset includes 1606 images in total collected by DigitalGlobe's satellite: 803 satellite images in RGB, a size of 2448 × 2448 px 24-bit JPEG image format and 803 GT images in RGB, and a size of 2448 × 2448 px 24-bit PNG image format. Files are named using the following format: <ID>_sat.jpg for satellite images and <ID>_mask.png for GT images. GT images contain seven land cover types: (1) urban land, (2) agriculture land, (3) forest land, (4) water, (5) barren land, and (6) rangeland and unknown (e.g., clouds). Each cover type is described in (R, G, B) colour codes. Dataset was sampled for land cover classes to have enough representation with agriculture class having 56.76% of total pixel count and water class having at least 3.74%.

The computation time of quality metrics depends on the size of input images. High computational complexity metrics like Silhouette [24] are expensive when calculating distances. For this reason, 324 × 220 px regions and corresponding GT regions were cropped from the existing 2448 × 2448 px images and corresponding GT images, provided by a DeepGlobe Land Cover dataset (Figure 2). This also limits the number of possible clusters, making the land cover objects evaluated less distracting for human observers.

(a) (b) (c) (d)

Figure 2. Sampled satellite images from the DeepGlobe Land Cover dataset with two selected 324 × 220 regions that are highlighted in red and are used for our dataset: (**a**) "676758_sat.jpg," (**b**) "762359_sat.jpg," (**c**) "941237_sat.jpg," (**d**) "668465_sat.jpg." Names of images from our constructed dataset [23] share the same <ID's> in part1 and part2 folders.

The cost for annotating multi-class segmentation GT images provided in the DigitalGlobe Land Cover dataset have segments that are less accurate, missing prominent land cover portions (Figure 3). Therefore, semi-automatic adjustments to the GT images were performed inside MATLAB using an Image Labeler, which allows us to label image data manually or use automation, and to export to the MATLAB workspace as a ground truth object variable containing label definitions.

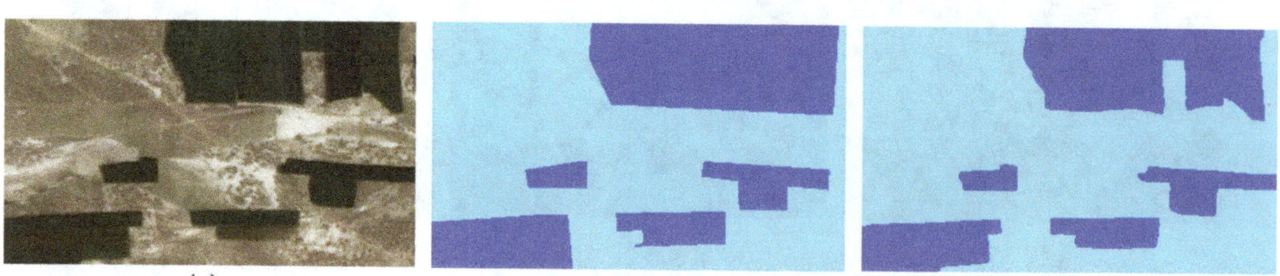

(a) (b) (c)

Figure 3. Example of visible structural inaccuracies in agriculture land (different image clusters are marked with selected colours): (**a**) satellite image "965977_sat_00000.png" from our constructed dataset, (**b**) not corrected GT image of "965977_sat_00000.png," and (**c**) corrected GT image of "965977_sat_00000.png " from our constructed dataset.

The constructed dataset, including average mean opinion score (*MOS*) scores for each image used in our survey, can be accessed at Reference [23].

5. Objective Quality Assessment Methods for Image Segmentation

Depending on whether segmentation results are evaluated by a human or an algorithm, quality assessment is divided into two main branches: objective and subjective [25,26]. The main intention of objective quality models is to approximate properties of HVS in order to avoid slow and impractical subjective testing procedures. For this reason, the design process of new objective quality metrics often includes correlation tests with an obtained *MOS* [5,6].

The objective metrics may require initial information to evaluate the image segmentation quality. Such information is known as a reference image or GT. It could be prepared manually so that a comparison could be made with segmentation results achieved by a particular algorithm. This group of metrics is called external metrics (or supervised evaluation measures). The external information may not always be available, and metrics that do not depend on external information are used. This group of metrics is called internal metrics (or unsupervised evaluation measures).

5.1. External Metrics

The evaluation of segmented image quality using external metrics is equivalent to comparing two segmentation versions (GT image and segmentation result), where each pixel has a unique class label (or index) assigned to it. The GT image is often created (labelled) by an expert in a particular field (e.g., medical image segmentation) or sometimes can be generated [11] from an input image in the form of synthetic information. In our case, the segmentation result is obtained from the clustering algorithm, which returns an array containing cluster indices of each pixel corresponding to which cluster that pixel was assigned. Then the quality of the segmentation result can be evaluated by an external metric taking GT image and segmentation result as an input to determine the level to which two segmentations match. The evaluation process of the external metric is based on the analysis of cluster indices assigned to all pixel pairs between the GT image and segmentation result.

External quality metrics are calculated from the confusion matrix. The confusion matrix is a summary of the results of the segmentation problem. The number of correct and incorrect assignments is summed for a specific class. The confusion matrix is defined as a square matrix ($K \times K$) where K is a number of classes (or clusters) and consists of four parameters (Table 3). These parameters are then used to derive combined external metrics that are presented in Table 4.

It is worth noting that external metrics can also serve as a mean of comparing segmentation results of two different algorithms or segmentation results of a single algorithm, but with different parameters [24].

Table 3. Notation for the external comparison metrics.

Notation	Name	Description
TP	true positive pixels	pixels that belong to cluster C_i in the GT image and are correctly assigned to cluster C_i in the segmented image (i.e., common pixels between GT image and segmented image)
TN	true negative pixels	pixels that are assigned to different clusters both in the segmented and GT images
FP	false positive pixels	pixels that are incorrectly assigned to cluster C_i in the segmented image compared to the GT image
FN	false negative pixels	pixels that are assigned to cluster C_i in the GT image, but assigned to different cluster in the segmented image

Table 4. List of external metrics.

Symbols/Notations Commonly Used	Names Commonly Used	Definition	
ACC	Accuracy	$\frac{TP+TN}{TP+TN+FP+FN}$	(1)
PPV, P	Positive Prediction Value, Precision	$\frac{TP}{TP+FP}$	(2)
TPR, R	True Positive Rate, Recall, Sensitivity	$\frac{TP}{TP+FN}$	(3)
TNR	Specificity, True Negative Rate	$\frac{TN}{TN+FP}$	(4)
MCC	Matthews correlation coefficient	$\frac{TP \cdot TN - FP \cdot FN}{\sqrt{(TP+FP)(TP+FN)(TN+FP)(TN+FN)}}$	(5)
JI, IoU	Jaccard index, Intersection over Union	$\frac{TP}{TP+FP+FN}$	(6)
F_1, *DSC*, Dice	Sørensen–Dice coefficient	$2 \cdot \frac{P \cdot R}{P+R} = \frac{2TP}{2TP+FP+TP}$	(7)
F_2	-	$5 \cdot \frac{P \cdot R}{4P+R}$	(8)
$F_{1/2}$	-	$5 \cdot \frac{P \cdot R}{P+4R}$	(9)
KI	Kulczynski index [27]	$\frac{1}{2}(P+R) = \frac{1}{2}\left(\frac{TP}{TP+FP} + \frac{TP}{TP+FN}\right)$	(10)
FMI	Folkes–Mallows index [27]	$\sqrt{P \cdot R} = \frac{TP}{\sqrt{(TP+FP) \cdot (TP+FN)}}$	(11)

The F_β measure is defined as a weighted harmonic mean of P and R:

$$F_\beta = \left(1 + \beta^2\right) \frac{P \cdot R}{(\beta^2 \cdot P) + R} \tag{12}$$

In Equation (12), the parameter β is any real positive number ($0 \le \beta \le +\infty$) and determines the weighting between P and R. If $\beta > 1$, higher weight is applied to R. If $\beta < 1$, higher weight is applied to P. Depending on the value of β, expression (12) can lead to several possible scenarios/metrics:

- if $\beta \to +\infty$, $\Rightarrow F_\beta = R$,
- if $\beta = 2$, $\Rightarrow F_\beta$ is equal to F_2, where R has double weight compared to P,
- if $\beta = 1$, $\Rightarrow F_\beta$ is equal to the unweighted harmonic mean of P and R, and is equal to F_1. In this case, P and R are equally important,
- if $\beta = 0.5$, $\Rightarrow F_\beta$ is equal to $F_{1/2}$ where R has half weight compared to P,
- if $\beta \to 0$, $\Rightarrow F_\beta = P$.

KI is defined as the arithmetic mean of P and R, while the *FMI* is defined as the geometric mean of P and R. Since the geometric mean is always in-between of arithmetic mean and harmonic mean for any positive number (in this case $0 \le P,\ R \le 1$), then it is also true that $K \ge FMI \ge F_1$.

All external metrics listed in Table 4 have a range of [0; 1], except *MCC*, ranging [−1; 1], where 1 represents a perfect segmentation result. However, various research studies suggest [8,9] that region-based metrics like F_1 or *JI* alone cannot fully reflect the human perception of image segmentation quality.

The definitions of metrics in Table 4 can only be applied to the binary (two-class) segmentation case. For multiclass segmentation, the overall scores for external measures were obtained by finding the score for each cluster C_i (function *confusionmatStats* [28]), and then calculating the unweighted mean among all clusters K.

$$overall\_external\_measure = \frac{1}{K} \sum_{C=1}^{C_K} external\_measure_C \tag{13}$$

5.2. Internal Metrics

Internal metrics require no external information, i.e., reference image, to evaluate the segmentation quality. The segmented result is evaluated based on a particular set of characteristics (criteria), derived from the initial dataset. This feature is important, as generating or creating reference images is time-consuming or sometimes impossible.

The internal quality metrics are usually employed for (1) solving an optimal number of clusters, (2) determining the quality of clustering results without depending on external information, and (3) determining if data have any structure [24].

Internal methods evaluate clustering by examining the separation and the compactness of the clusters.

- Cluster cohesion (or compactness) measures how closely related objects are in a cluster or how close the data points are from the cluster centroid. Better clustering results have pixel values close to their respective cluster centroids.
- Cluster separation measures how a cluster differs or is separated from the other clusters. Better clustering results have centroids of different clusters far from each other.

The primary measures of cohesion (14) and separation (15) are calculated from the image under investigation, while (16), (17), (19), and (21) measures combine both cohesion and separation. Basic notation for internal metrics is provided in Table 5. Clustering quality is considered good when the clusters are well separated and compact.

Table 5. Basic notations and definitions for the internal metrics.

Notation	Description
K	the number of clusters
N	the number of objects (pixels) in image X (i.e., pixel count or resolution)
$\{C_1, C_2 \cdots, C_i, \cdots C_K\}$	the set of K clusters
$D = d(\cdot, \cdot)$	the Euclidean distance between two objects (pixels)
$\lvert C_i \rvert$	the total number of data points (pixels) in a cluster i
$X_i = \left\{x_{i1}, x_{i2}, \cdots, x_{ij}, \cdots, x_{i\lvert C_i \rvert}\right\}$	the set of pixels in C_i
$\{c_1, c_2 \cdots, c_i, \cdots c_K\}$	the set of cluster means (centroids)

The Sum of Squared Errors Within Cluster *(SSW)* is alternatively known as the Sum of Squared Errors *(SSE)* (14) [29]. Lower values indicate higher cluster cohesion. *SSE* decreases with the increase of the number of clusters. *SSE* is defined as:

$$SSE_K = \sum_{i=1}^{i=K} \sum_{j=1}^{j=|C_i|} d\left(x_{ij}, c_i\right)^2 \tag{14}$$

The Sum of Squares Between Clusters *(SSB)* (15) [29] is a measure of separation. Higher *SSB* values indicate more separated clusters. *SSB* is defined as the sum of squared distances from c (known as overall centroid, i.e., centroid of all cluster centroids) to other cluster centroids c_i each time multiplied by the number of pixels in the cluster C_i.

$$SSB_K = \sum_{i=1}^{i=K} |C_i| \cdot d(c_i, c)^2 \tag{15}$$

Using *SSE* and *SSB*, some other combined internal metrics can be calculated. The Calinski–Harabasz index (*CHI*) (16) [30] is alternatively known as the variance ratio criterion (*VRC*).

$$CH = \frac{SSB_K}{SSE_K} \cdot \frac{N-K}{K-1} \tag{16}$$

The larger the SSB_K/SSE_K ratio is, the better the clustering quality is.

The Hartigan index (*HI*) (17) is defined as the logarithmic relationship between *SSB* and *SSE* [31].

$$HI = \log\left(\frac{SSB_K}{SSE_K}\right) \tag{17}$$

Numbers in the range [0; 1] have negative logarithms. Therefore, if $SSB > SSE$, then $HI > 0$. If $SSE > SSB$, then $HI < 0$. $HI = 0$ only if *SSB* is equal to *SSE*.

The *Xu* coefficient (18) combines SSE_K, number of clusters *K*, a total number of pixels, and dimensionality of input data [32].

$$X_u = D \cdot \log_2\left(\sqrt{\frac{SSE_K}{DN^2}}\right) + \log K \tag{18}$$

The Silhouette coefficient (*SH*) (19) [19] is another popular way of combining cohesion and separation [27].

$$S(x_{ij}) = \frac{b(x_{ij}) - a(x_{ij})}{max\left\{b(x_{ij}),\ a(x_{ij})\right\}} \text{ or } S(x_{ij}) = \begin{cases} 1 - \frac{a(x_{ij})}{b(x_{ij})}, & if \quad a(x_{ij}) < b(x_{ij}), \\ 0, & if \quad a(x_{ij}) = b(x_{ij}), \\ \frac{b(x_{ij})}{a(x_{ij})} - 1 & if \quad a(x_{ij}) > b(x_{ij}). \end{cases} \tag{19}$$

The coefficient $\left(-1 \geq S(x_{ij}) \geq 1\right)$ can be calculated for an individual pixel x_{ij}.

- $a(x_{ij})$ average distance from the pixel x_{ij} to other pixels in C_i (cohesion),
- $b(x_{ij})$ is the minimum (worst case) of the all-average distances, each of them computed among the same pixel x_{ij} and all the pixels inside another cluster (separation).

SH value for an individual pixel x_{ij} represents how similar x_{ij} is to pixels inside its own cluster, compared to pixels in other clusters. The Silhouette coefficient for a single cluster C_i.

$$S(C_i) = \frac{\sum_{j=1}^{j=|C_i|} S(x_{ij})}{|C_i|} \tag{20}$$

Finally, the overall *SH* for an image can be calculated similarly to Equation (13). However, alternative ways of calculating the overall score are possible (e.g., by averaging *SH* values for all pixels). Higher *SH* values indicate a better clustering result. More detailed interpretations of *SH* values are described in Reference [19].

The Davies-Bouldin index (*DBI*) [33,34] calculation is based on the ratio of within-cluster distances to between-cluster distances.

$$DBI = \frac{1}{K}\sum_{i=1}^{i=K} R_i,\ were\ R_{ij} = \frac{\overline{S}_i + \overline{S}_j}{d_{ij}},\ and\ S_i = \frac{1}{|C_i|}\sum_{h=1}^{h=|C_i|} d(x_{ih},\ c_i) \tag{21}$$

$R_i = \max_{i \neq j}\left\{R_{ij}\right\}$ —maximum of R between the cluster C_i and each other cluster C_j,

$d_{ij} = d(c_i, c_j)$ —distance between the centroid c_i of cluster C_i and centroid c_j of the cluster C_j,

$\overline{S}_i$ —the average distance between every pixel (within the cluster C_i) and its centroid c_i,

$\overline{S}_j$ —the average distance between every pixel (within the cluster C_j) and its centroid c_j.

Here, cohesion is defined in the form of sum $S_i + S_j$, while d_{ij} defines separation. Lower *DBI* values indicate a better clustering result. If clusters are close to each other (small d_{ij}) and are dispersed (big $\overline{S}_i + \overline{S}_j$), then *DBI* value will be high, indicating less optimal clustering.

5.3. IQA Metrics

IQA metrics can be divided into three groups, depending on the need of the reference information: (1) Full-Reference (FR)—metrics, that require reference/ground-truth image (quality of the segmented image is measured in comparison to the ground-truth image); (2) No Reference (NR)—metrics, that do not require reference/ground truth image for measuring quality; (3) reduced reference (RR) metrics measure quality by comparing distorted/segmented image with a reference/ground truth image, composed of specific extracted features (such as edge information). The reference image is named ground truth from the perspective of image segmentation. In this experiment, we concentrated on commonly used FR metrics that are listed in Table 6.

Table 6. Full-reference image quality assessment (FR-IQA) image quality metrics and their main features.

Notation	Ref [1]	Name	Metric Features [2]
PSNR	-	Peak Signal to Noise Ratio	pixel difference-based, inversely proportional to the *MSE*
SR-SIM	[35]	Spectral Residual based similarity	visual saliency map, gradient modulus
UQI	[36]	Universal quality index	structural distortion, luminance distortion, loss of contrast
MS-SSIM	[37]	Multi-scale Structural Similarity index	structural distortion
IW-SSIM	[38]	Information Content Weighted *SSIM*	NSS
NQM	[39]	Noise Quality Measure	CSF
FSIMc	[40]	Feature Similarity Index	structural distortion
SSIM	[41]	Structural Similarity	luminance, contrast, and structural distortions
MAD	[42]	Most Apparent Distortion	local luminance and contrast masking, spatial-frequency components changes, CSF
GSM	[43]	Gradient Similarity	luminance, structural and contrast changes
WSNR	[44]	Weighted signal to noise ratio	CSF
SFF	[45]	Sparse Feature Fidelity	based on sparse features similarity (structure differences) and luminance correlation (brightness distortions)
VIF_p	[46]	pixel-based Visual Information Fidelity	NSS
VSNR	[47]	Visual Signal-to-Noise Ratio	visual masking, perceived contrast, global precedence
PSNR-HVS-M	[48]	Extension of *PSNR*	incorporates CSF and between-coefficient contrast masking of DCT basis functions
VSI	[49]	Visual Saliency-induced index	visual saliency map, gradient modulus, colour distortions
SNR	-	Signal to Noise Ratio	Pixel difference-based

[1] All references contain research papers provided by the original authors of the specific quality method.
[2] NSS—Natural Scene Statistic models. CSF—Contrast Sensitivity Function. DCT—Discrete Cosine Transform. *MSE*—Mean Squared Error.

IQA metrics can be applied to the image segmentation evaluation [10]. As input, the segmented images have different characteristics compared to the loosely compressed natural images and can be treated more like a synthetic (i.e., artificially generated) image. Segmented images have crisp contours, uniform regions, and are generally less complex. Majority of the proposed IQA metrics are designed for correlation with HVS, which is very sensitive to the contrast [50] or structural information changes. To achieve correlation with human perceived quality, most IQA metrics employ multiple strategies. However, some of the features (Table 6), like contrast changes, contrast masking or luminance masking, may be less or not important in evaluating the satellite image segmentation result. Various research studies emphasize the importance of precise contours for improved perceived segmentation quality, after combining *JI* and *BF* quality metrics [8] in the novel FR-IQA quality index, intended for colour images [5].

All the selected FR-IQA measures were calculated using MATLAB R2019b. *VSNR*, VIF_P, *UQI*, *NQM*, and *WSNR* calculated from MeTriX MuX Visual Quality Assessment Package (v. 1.1) [51] using default settings.

6. Subjective Quality Assessment of Segmented Satellite Images

Subjective evaluation by a human is the most reliable method for determining image quality in various applications (such as image editing, image retargeting [52], and others) as well as in the image segmentation [7]. However, segmentation quality requirements are also application-dependent [8,9].

Subjective quality evaluation tests require careful preparation, human, and time resources. In contrast to the long-established subjective evaluation of the distorted image and video quality, image segmentation lacks dedicated quality evaluation methodologies. For this reason, due to similarities with IQA, many strategies for subjective quality assessment for segmentation are adopted from the existing standards [7].

For the assessment of image quality, there are many test methods and rating scales provided by International Telecommunication Union (ITU) standards, describing acceptable modifications and recommendations. The method describes how a stimulus (in this case, a sequence of segmented images) is presented, and the rating scale describes the way that subjects evaluate their opinion of the stimulus. For this subjective quality assessment test, the simultaneous double stimulus for continuous evaluation (SDSCE) method described in ITU-R BT.500-14 [53] was combined with an absolute category rating (ACR) scale described in T-REC-P.913 [54]. ACR consists of a five-level rating scale: (5—Excellent, 4—Good, 3—Fair, 2—Poor and 1—Bad). T-REC-P.913 also does not recommend increasing the number of levels, since the accuracy of the results does not improve [55], and the evaluation process becomes more complicated for a human.

Before performing the *MOS* experiment, it is recommended to ensure that there are enough segmentation results of bad, average, and good quality between all segmented images. Otherwise, the data points will be concentrated in a single corner of the scatter plot and will not fully cover the ACR scale.

For subjects to fully understand their task, the first part of the test is the training phase and includes examples covering the full range of segmentation quality results combined with verbal instructions given by the administrator explaining voting procedure by following practices described in Reference (Section 11.5 in [54]).

During the second phase, subjects were presented with the electronic form depicted in Figure 4 and were requested to evaluate the differences between the GT segmentation and segmentation result obtained by k-means++. Subjects were aware of which image is the GT and which image is the segmentation result obtained by the selected algorithm. There was no time limit for evaluating a single pair of images. However, each experiment session lasted no more than 20 min.

Similar to the approach in Reference [9], original satellite images were also placed on the left, but only for context purposes and should not impact the judgement of segmentation quality. The segmented images have not been scaled to avoid introducing possible distortions.

In order to avoid fatigue and/or boredom, the test was divided into two parts with each consisting of 45 segmentations (90 segmentations in total). The first part was evaluated by 95 and the second part was evaluated by 92 subjects, which is more than enough for stable average ratings.

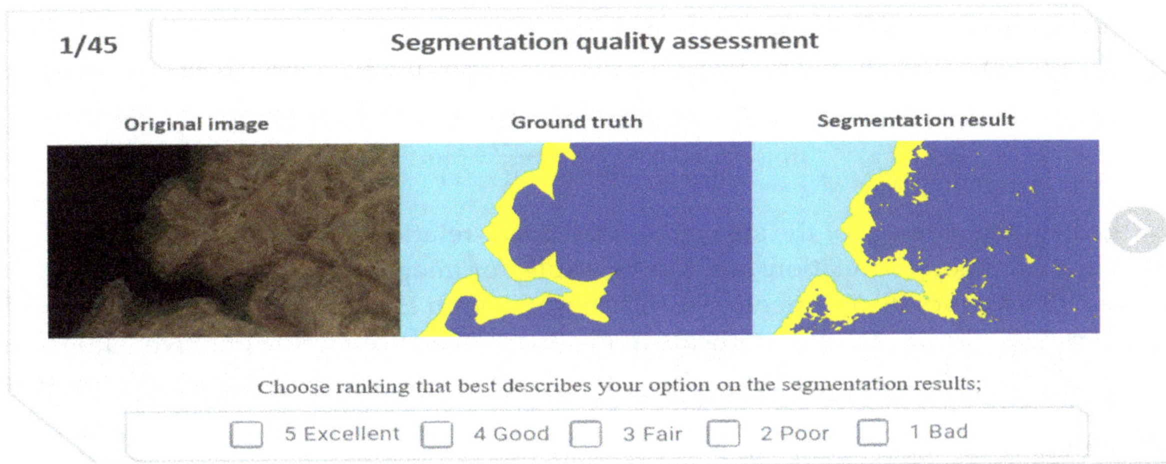

Figure 4. Example of the electronic form for segmentation quality assessment using the absolute category rating (ACR) scale. Different image clusters are marked with selected colours in "Ground truth" and "Segmentation result" images.

After collecting experimental results, ratings were converted to numerical values (1, 2, 3, 4, and 5) and the total *MOS* for the single-segmented image was calculated as the arithmetic mean of the individual assessments that the subjects assigned to the segmentation result [53].

$$MOS = \frac{1}{N} \cdot \sum_{n=1}^{N} o_n \tag{22}$$

Here, o_n is the observed rating for subject n, and N is the total number of subjects (participants) in the experiment. We observed that most segmented images received *MOS* scores from 2.5 to 3.5. This distribution is likely due to the human tendency of trying to avoid giving extreme scores when evaluating images [56].

7. Results

The workflow presented in Figure 5 was performed to collect all necessary data required for calculating the correlation between the subjective and objective scores.

As seen in the general framework, the whole workflow can be divided into three main branches. On the left section are presented steps that deal with constructing our dataset using the original dataset of satellite images and their GT. In the middle section are the described steps related to the selection of the segmentation method and performing segmentation of the satellite images based on the colour feature. Presented steps that are used for obtaining subjective scores are on the right section. The goal is to obtain the objective and subjective scores for the calculation of the Pearson Linear Correlation Coefficient (*PLCC*) and Spearman Rank Order Correlation Coefficient (*SROCC*), which is the final step and is further described in this section. The calculation of the objective scores is presented before the final step.

Correlation between the subjective (*sub*) and objective (*obj*) scores was determined by *PLCC* (23) and *SROCC* (24) [57]. Here, n is the total number of images in the dataset.

$$PLCC = \frac{n(\sum sub \cdot obj) - (\sum sub)(\sum obj)}{\sqrt{\left[n \sum sub^2 - (\sum sub)^2\right] \cdot \left[n \sum obj^2 - (\sum obj)^2\right]}}. \tag{23}$$

SROCC is equal to *PLCC* applied on the ranks, in this case, of subjective and objective scores. For the image i, d_i denotes the difference between the ranks. In order to compute *SROCC*, data have to be converted into ranks. Assuming no tied ranks exist, a simplified *SROCC* formula can be applied.

$$SROCC = 1 - \frac{6 \cdot \sum d_i^2}{n(n^2 - 1)}. \tag{24}$$

PLCC is the measurement of the strength of the linear relationship, while *SROCC* also measures the strength of the monotonic relationship. For example, for images distorted by various compression artifacts, most IQA metrics display nonlinear relationships with HVS [58]. Both correlation coefficients have a [−1; 1] range. Negative values indicate a negative correlation, while positive values indicate a positive correlation. If the values of both correlation coefficients are approaching zero, this suggests that relationship most likely does not exist.

The results of the experiment are presented in the two groups of Tables: Tables 7–9 for the overall correlation between subjective *MOS* scores and objective metric scores and Tables 10–12 for the correlation between different quality groups of *MOS* scores and objective metric scores.

Tables 7–9 show the overall correlation using all images from the dataset. In Tables 8 and 9, metrics are sorted from highest to lowest, according to their *PLCC* and *SROCC* scores. Table 7 shows that the *SH* has the strongest positive correlation, while the *DBI* has the strongest negative correlation for both *PLCC* and *SROCC*. Table 8 presents that the JI has the highest *PLCC*, while *ACC* has the highest *SROCC* closely followed by *JI*. Table 9 shows that the PSNR has the highest *PLCC*, while *UQI* has the highest *SROCC* closely followed by *PSNR*.

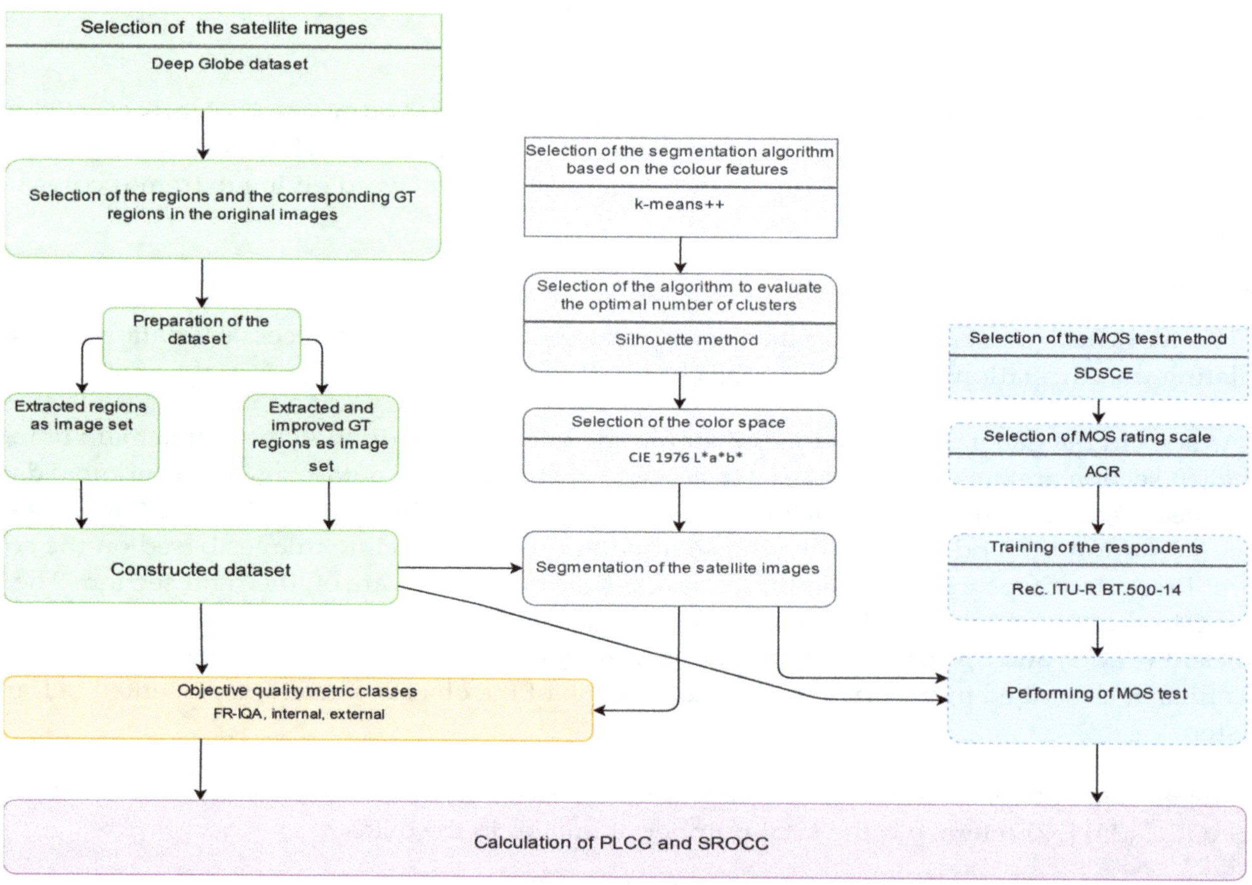

Figure 5. Framework illustrating the workflow for obtaining final results (different colours present different branches of the workflow, including the appropriate steps).

Table 7. The overall correlation between subjective *MOS* scores and internal validation scores.

Metric [1]	*PLCC*	*SROCC*	Remarks
Silhouette coefficient (*SH*)	0.4684	0.4252	Higher is better
Calinski–Harabasz index (*CHI*)	0.3902	0.3426	
SSB	0.1861	0.1636	
Hartigan index (*HI*)	0.0443	0.0428	
Xu coefficient (*Xu*)	0.0173	0.0153	Lower is better
SSE	0.1477	0.1698	
Davies-Bouldin index (*DBI*)	−0.4562	−0.4355	

[1] Internal metrics were calculated using the optimal number of clusters selected by the Silhouette method.

Table 8. The overall correlation between subjective *MOS* scores and external validation scores.

Metric	*PLCC*	Metric	*SROCC*
Jaccard index (*JI*)	0.7497	Accuracy (*ACC*)	0.8147
Fscore (F_1)	0.681	Jaccard index (*JI*)	0.8105
$F_{1/2}$	0.6764	Fscore (F_1)	0.802
Fowlkes–Mallows index (*FMI*)	0.6673	F_2	0.7981
Precision (*P*)	0.6623	Fowlkes–Mallows index (*FMI*)	0.7961
F_2	0.6613	$F_{1/2}$	0.7947
Kulczynski index (*KI*)	0.6379	Kulczynski index (*KI*)	0.7928
Accuracy (*ACC*)	0.6224	Precision (*P*)	0.7782
MCC	0.6214	*MCC*	0.7665
Recall/Sensitivity (*R*)	0.5668	Recall/Sensitivity (*R*)	0.7464
Specificity (*SPEC*)	0.4096	Specificity (*SPEC*)	0.5998

Table 9. The overall correlation between subjective *MOS* scores and FR-IQA measures.

Metric	*PLCC*	Metric	*SROCC*
PSNR	0.8647	*UQI*	0.8632
SR-SIM	0.8083	*PSNR*	0.8608
UQI	0.805	*SSIM*	0.8423
MS-SSIM	0.7881	*MS-SSIM*	0.8264
NQM	0.7677	*SR-SIM*	0.8174
IW-SSIM	0.7511	*SNR*	0.8087
FSIMc	0.7381	*GSM*	0.7774
SSIM	0.7226	*IW-SSIM*	0.761
MAD	0.7118	*NQM*	0.7586
GSM	0.7036	*FSIMc*	0.7534
WSNR	0.6351	*SFF*	0.753
SFF	0.632	*VSI*	0.7001
VIF_P	0.6214	*MAD*	0.6932
VSNR	0.5501	*PSNR-HVS-M*	0.6802
PSNR-HVS-M	0.5357	*WSNR*	0.6333
VSI	0.1365	*VIFP*	0.6276
SNR	−0.082	*VSNR*	0.5609

Information from Tables 7–9 is also represented as the scatterplots in Figure 6a–d, respectively, showing a comparison between *PLCC* and *SROCC* values. Vertical and horizontal axes correspond to *SROCC* and *PLCC*, respectively. Here, metrics close to the dotted red line ($y = x$) have similar *PLCC* and *SROCC* values. Due to very similar values for external metrics, the scatterplot scale and position in Figure 6b were adjusted. The closer metric is to the (0, 0) point, the weaker correlation with *MOS* for that metric is. Best results are for metrics, which are closer to the upper right corner (positive correlation with *MOS*) or lower-left corner (negative correlation with *MOS*). From the visual inspection, it can also be observed that *SROCC* and *PLCC* values themselves display strong positive linear correlation, meaning both correlation coefficients are equally important.

Tables 10–12 show the correlation between each metric and different quality groups based on *MOS*: low-quality (1.0–2.5), middle-quality (2.5–3.5), and high-quality (3.5–5]. For each quality group, the top three best results are highlighted except the results of the correlation in Table 10. Additionally, the information in Tables 10–12 are also presented as the bar charts in Figures 7–9. For each quality group in Tables 11 and 12, the top three best results are highlighted in light green.

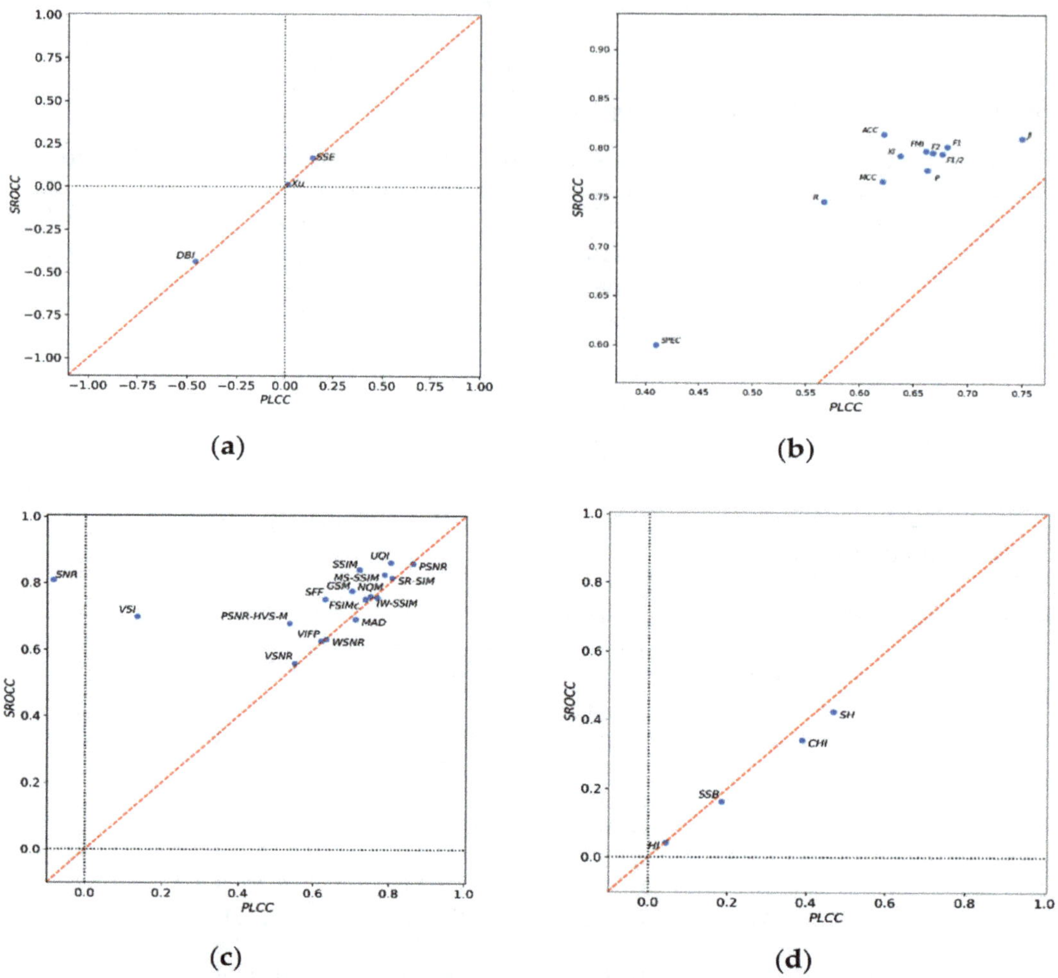

Figure 6. Comparison of *SROCC* and *PLCC* values (obtained from Tables 7–9) for: (**a**) internal validation scores (*DBI*, *Xu*, *SSE*), (**b**) external validation scores, (**c**) FR-IQA measures, and (**d**) internal validation scores (*HI*, *SSB*, *CHI*, *SH*). Red $y=x$ line denotes where SROCC and PLCC values are equal.

Table 10. Correlation between different quality groups of MOS scores and internal validation scores.

Metric [1]	*PLCC*			*SROCC*			Remarks
MOS	[1.0–2.5)	(2.5–3.5)	(3.5–5]	[1.0–2.5)	(2.5–3.5)	(3.5–5]	
Calinski–Harabasz index (*CHI*)	0.1340	−0.0716	0.4339	0.2549	−0.1051	0.4615	Higher is better
SSB	0.1840	−0.1139	0.1004	0.2794	−0.3146	0.1275	
Hartigan index (*HI*)	0.0834	−0.2247	0.2864	0.0270	−0.2550	0.2777	
Silhouette coefficient (*SH*)	0.2659	−0.0967	0.4795	0:2255	−0.0700	0.5237	
Davies–Bouldin index (*DBI*)	−0.1798	0.0126	−0.5035	−0.1104	−0.0057	−0.5198	Lower is better
SSE	0.2734	−0.0988	−0.3966	0.5319	−0.0757	−0.3706	
Xu coefficient (*Xu*)	0.2881	−0.1576	−0.4714	0.4583	−0.1436	−0.4684	

[1] Internal metrics were calculated using the optimal number of clusters selected by the Silhouette method.

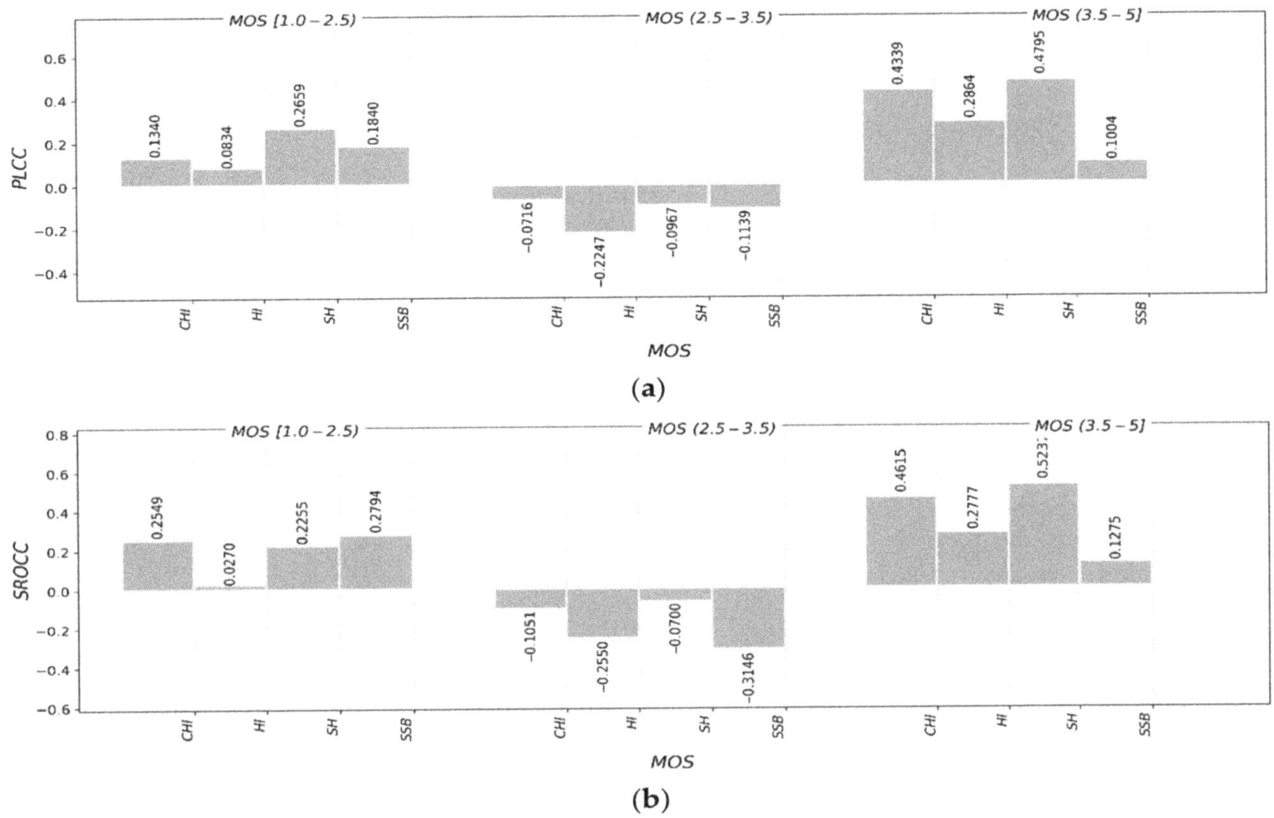

Figure 7. Comparison of the correlation values between different quality groups of *MOS* scores and internal validation scores (*CHI*, *HI*, *SH*, *SSB*): (**a**) *PLCC*, and (**b**) *SROCC*.

Table 11. Correlation between different quality groups of MOS scores and external validation scores (the top three best results are highlighted for each quality group).

Metric		*PLCC*			*SROCC*	
MOS	**[1.0–2.5)**	**(2.5–3.5)**	**(3.5–5]**	**[1.0–2.5)**	**(2.5–3.5)**	**(3.5–5]**
Accuracy (*ACC*)	0.5632	0.3607	0.1949	0.4559	**0.5360**	0.1996
$F_{1/2}$	**0.7142**	**0.4080**	**0.3671**	0.5343	**0.4395**	**0.4832**
F_2	0.5815	0.3884	0.1998	0.5123	0.3848	0.2994
Fowlkes–Mallows index (*FMI*)	0.6675	0.4001	0.3044	0.5417	0.3948	**0.3982**
Fscore (F_1)	0.6578	0.4040	0.3050	0.5392	0.4073	0.3845
Jaccard index (*JI*)	**0.7099**	**0.4215**	**0.3224**	**0.6544**	0.4077	0.3785
Kulczynski index (*KI*)	0.6150	0.3952	0.3040	**0.6152**	0.3911	0.3923
MCC	0.6992	0.3390	0.2577	**0.6176**	0.3603	0.3103
Precision (*P*)	**0.7284**	**0.4053**	**0.3919**	0.5686	**0.4381**	**0.4917**
Recall/Sensitivity (*R*)	0.2563	0.3643	0.1076	0.3088	0.2885	0.1561
Specificity (*SPEC*)	0.4267	0.1054	0.0177	0.4069	0.1367	−0.0460

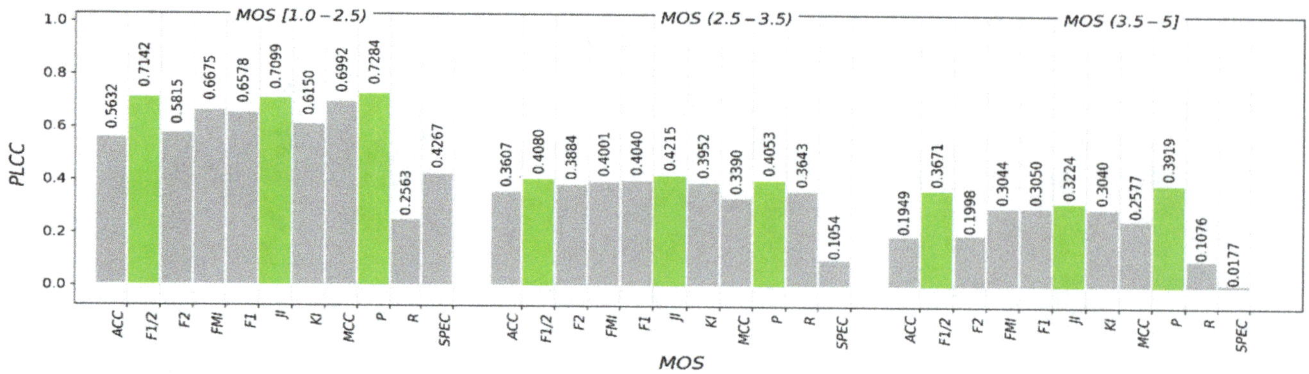

(a)

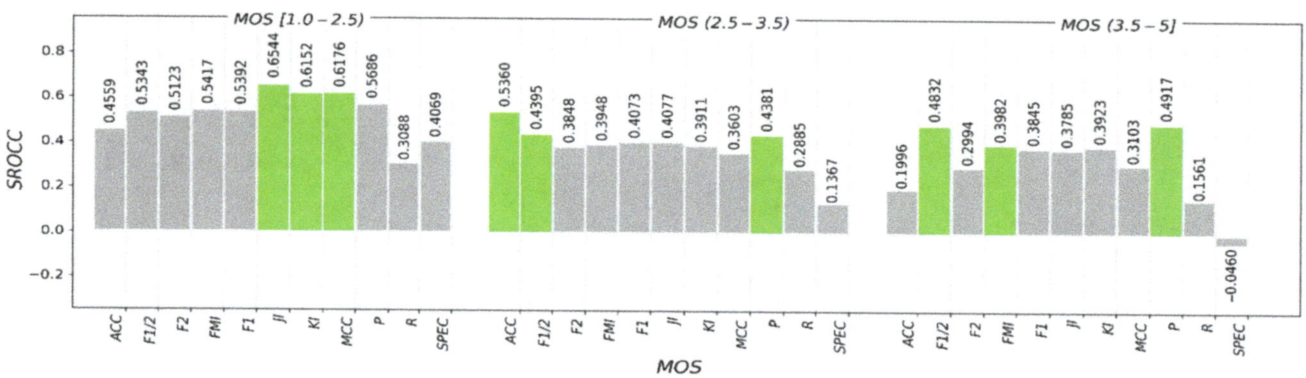

(b)

Figure 8. Comparison of the correlation values between different quality groups of *MOS* scores and external validation scores (the top three best results are highlighted in light green for each quality group): (**a**) *PLCC*, and (**b**) *SROCC*.

Table 12. Correlation between different quality groups of *MOS* scores and FR-IQA measures (the top three best results are highlighted for each quality group).

Metric		*PLCC*			*SROCC*	
MOS	**[1.0–2.5)**	**(2.5–3.5)**	**(3.5–5]**	**[1.0–2.5)**	**(2.5–3.5)**	**(3.5–5]**
$FSIM_c$	0.3949	0.3353	0.2944	0.3235	0.4014	0.3636
GSM	0.5001	0.2497	0.3489	0.4020	0.4166	0.3786
IW-SSIM	0.4898	0.4844	0.4320	0.5686	0.4927	0.4190
MAD	0.2537	0.3174	**0.5544**	0.2745	0.3452	**0.5573**
MS-SSIM	0.4320	**0.6077**	0.3278	0.5025	**0.6208**	0.3804
NQM	0.5617	0.4950	**0.4512**	**0.6275**	0.4894	0.4338
PSNR	**0.5806**	**0.6533**	0.4199	**0.6029**	**0.6239**	0.3982
PSNR-HVS-M	0.3721	0.2432	0.2001	0.4975	0.5109	0.2072
SFF	0.3489	0.2645	0.2283	0.3260	0.3872	0.2945
SNR	**0.6290**	−0.2113	0.3831	0.6005	0.4852	0.4061
SR-SIM	0.5148	0.3783	**0.8001**	0.4706	0.4584	**0.8053**
SSIM	0.5095	**0.6124**	0.3386	0.5662	**0.6044**	0.4093
UQI	**0.6288**	0.4667	0.4493	0.5368	0.5735	**0.5229**
VIF_P	0.4729	0.2645	0.2310	**0.6716**	0.2659	0.2530
VSI	0.3506	0.0847	−0.1368	0.2794	0.2637	0.2866
VSNR	0.3802	0.3706	0.2615	0.3676	0.3506	0.2500
WSNR	0.5580	0.1429	0.1860	0.4877	0.1548	0.0524

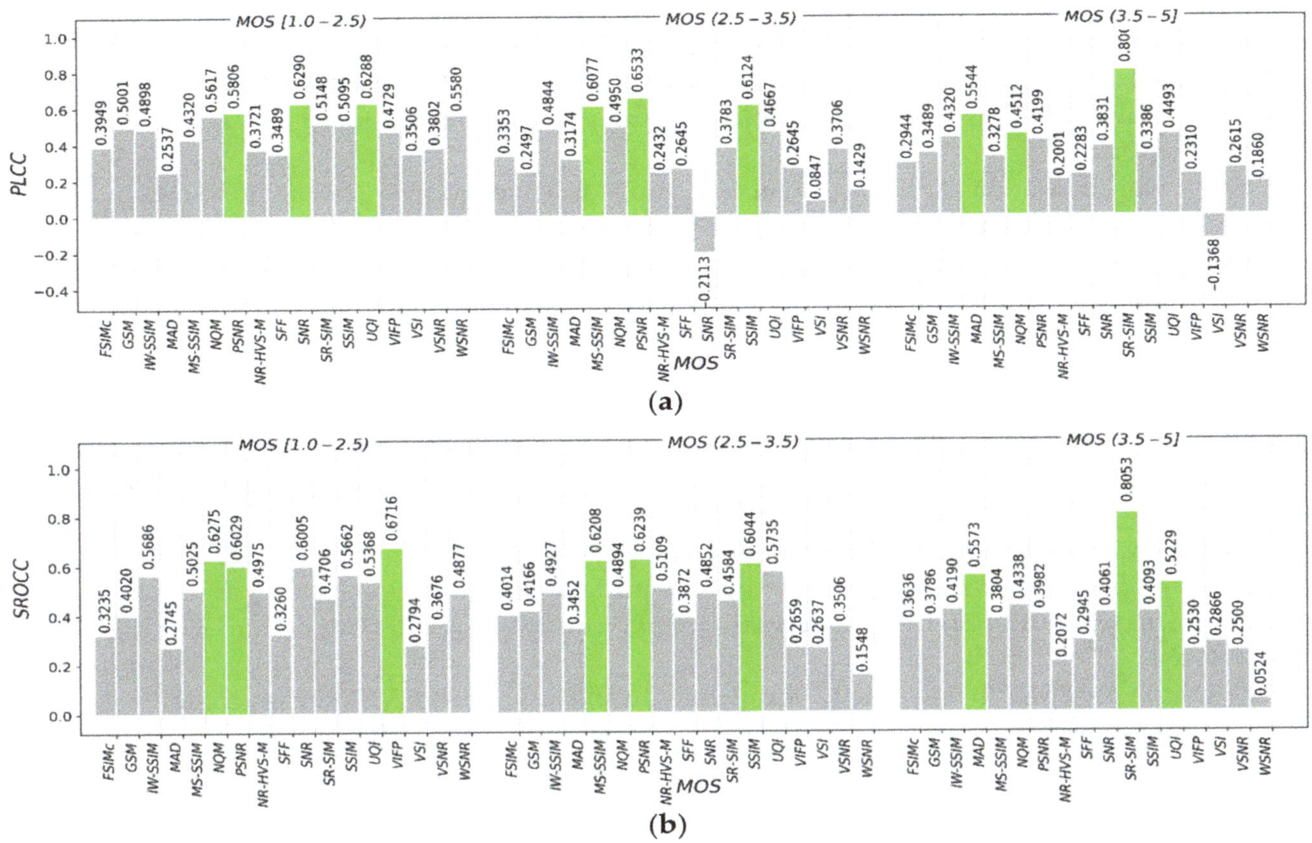

Figure 9. Comparison of the correlation values between different quality groups of *MOS* scores and FR-IQA measures (the top three best results are highlighted in light green for each quality group): (**a**) *PLCC*, and (**b**) *SROCC*.

Dividing results into three different quality groups allows for the more specialized comparison. The global correlation scores computed for the full dataset (Tables 7–9) do not allow identifying metrics that have a moderate correlation for all images (regardless of segmentation quality) from metrics that have a high correlation for images with above-average segmentation quality and low correlation for images with bad segmentation quality.

8. Discussion

The internal measures do not correlate well with perceived segmentation quality (the overall results in Table 7 and Figure 6a,d comparing to the external Table 8 and Figure 6b and FR-IQA measures (Table 9 and Figure 6c). Although the internal measures are used to evaluate results of clustering algorithms and help to select an optimal number of clusters, they were never intended to correlate with perceived image segmentation quality and relates more to how k-means is optimized. Thus, poor correlation is to be expected. In contrast, internal metrics combining cohesion and separation (such as *DBI*, *CHI*, and *SH*) seems to achieve slightly higher correlations, but as in Table 10 and Figure 7, only for segmentation results with high *MOS* scores. Note *DBI* shows a negative correlation as this measure returns lower scores for better clustering solutions. Overall, *SH* achieved moderate correlation for images in the *MOS* range (3.5–5].

The overall correlation scores for the external validation metrics are in a tight cluster (Table 8 and Figure 6b). As presented in Table 11 and Figure 8, lower *MOS* values correspond to the stronger correlations. This could be explained as the respondents agreeing more when determining the lower overall quality of the segmented image [7], while the opinions for better segmentation quality differs more. For the images in the low-quality group, *P*, $F_{1/2}$, and *JI* show a strong positive linear correlation,

$MOS > 0.7$. The $F_{1/2}$ and P metrics share a very similar correlation, as $F_{1/2}$ has a higher weight for P. Since there were no extended studies in this area, results in their entirety cannot be compared to previous studies. However, we observed *SROCC* values obtained by authors [7] for *JI* and F_1 metrics (0.848 and 0.848) used for single object segmentation are close to our overall *SROCC* values (0.811 and 0.802).

Best overall correlation for the IQA metric group was achieved by *PSNR*, *UQI*, *SSIM*, and *SR-SIM* ($MOS > 0.8$) (Table 9 and Figure 6c). All of them are also known to be very fast, according to the average calculation time for natural images from the TID2013 database [59]. These could be a reasonable choice for evaluating segmentation quality in terms of calculation speed and correlation with human perception.

It is widely accepted that the *PSNR* does not always agree with HVS assessing the quality of natural images distorted by various compression methods. From the overall results in Table 9 and Figure 6c, *PSNR* does not hold a significant advantage in the *MOS* score over most of other IQA metrics. Compared in the quality groups (Table 12 and Figure 9), we can state the similarity of *IW-SSIM* to *PSNR*. *IW-SSIM* is a relatively stable metric according to *PLCC*, and *PSNR* is stable only in the low-middle *MOS* (1.0–3.5) segmentation quality. The *UQI* metric is the most stable according to *SROCC*, while *PSNR* is only performing better for the low-middle segmentation quality and may be not an optimal choice depending on the segmentation situation. *SR-SIM* is a better choice for high-quality segmentations, *MOS* (3.5–5].

The reason for *PSNR* being able to compete with the advanced HVS metrics could be the nature of the segmented images, differing from the natural ones. The most distinct features of segmented images are clear contours and object shapes, uniform regions and colour, and absence of noise. HVS is sensitive to the structural information changes in the images. Therefore, metrics like *UQI* and *SSIM* are effective. HVS also largely relies on edge information for image interpretation [60]. In Reference [10], authors state that *PSNR* can be a good method to evaluate edge detection algorithms (for example, in BSR300 database).

The IQA metrics, in general, have a varying correlation to subjective assessment depending on the database, distortion type, image content, and image segmentation quality categories.

Authors of Reference [60] state the *PSNR* can sometimes have a strong correlation: 0.8756 (*SROCC*) and 0.8723 (*PLCC*) using natural images from the LIVE database. These results are very similar to our overall results: 0.8647 (*PLCC*) and 0.8608 (*SROCC*). PSNR does not outperform HVS-based metrics, which tend to be more stable across different databases of natural images and have even higher performance scores (*MOS* score).

Possible future directions for this research can include additional testing for RR, NR-IQA relations to subjective evaluation using larger-scale segmentation dataset(s) including multispectral satellite images. We suppose it is possible to select the prominent metrics for the combined measure, designed for improved correlation with subjective quality scores of the segmented images. To expand further, it can be interesting to evaluate the influence of algorithm selection, like using DBSCAN or another algorithm instead of k-means++, to the impact of correlation scores.

Depending on the goal or method, obtained segmentation results may have clusters assigned to a different colour value. For example, a segmented image consists of shades of a single colour versus very contrasting (or opposite) colour segments. Higher colour differences may impact some FR-IQA metric results and/or subjective evaluation itself. However, humans are able to distinguish more different levels of colour shades compared to gray shades [5]. Finally, determining a correlation among different groups of metrics might provide additional insight and a more diverse comparison.

9. Conclusions

This research aims to evaluate the correlation between the subjective and objective image quality metrics from the perspective of satellite image segmentation. Three broad classes of quality metrics were considered: internal and external cluster validation indices as well as FR-IQA measures.

From state-of-the-art technologies, we can conclude that there is no extensive research related to the assessment of the effectiveness of satellite image segmentation. For the segmentation quality test, we constructed our dataset of the satellite images with GT, based on "DeepGlobe Land Cover Classification Challenge" dataset.

From the experimental studies, several essential observations related to the assessment of the effectiveness of satellite image segmentation are made.

- When the segmentation results are diverse in perceived quality, then most external measures and FR-IQA metrics display very similar correlation with *MOS*.
- As perceived segmentation quality decreases, the correlation with *MOS* increases for the external quality measures.
- The *PSNR* metric achieved consistent results for low-middle quality segmentation, *MOS* range [1.0–3.5).
- The best metric for evaluating high-quality segmentation (*MOS* range (3.5–5]) was *SR-SIM*, achieving *SROCC*, and *PLCC* scores above 0.8, which also have low computational complexity.
- Since *PSNR* and *SR-SIM* complement each other covering full *MOS* range, they could be combined into a single measure.

The experimental studies show that dividing segmentation results into three different quality groups based on *MOS* allows the more specialized comparison of the objective quality metrics, according to perceived image quality.

Our study might provide insights to other research, where selecting the most suitable subjective metric from the HVS perspective is crucial. Herewith, our original results and obtained observations can be applied for improving current state-of-the-art segmentation methods.

Author Contributions: Conceptualization, G.K.-J. and E.J. Methodology, G.K.-J., E.J., R.B., T.L., and M.K. Software, E.J. Validation, G.K.-J., E.J., R.B., T.L., and M.K. Formal analysis, G.K.-J., E.J., R.B., T.L., and M.K. Investigation, G.K.-J. and E.J. Resources, E.J. Data curation, G.K.-J. and E.J. Writing—original draft preparation, E.J. Writing—review and editing, G.K.-J., E.J., R.B., T.L., and M.K. Supervision, G.K.-J., R.B., T.L., and M.K. Project administration, G.K.-J., R.B., T.L., and M.K. All authors have read and agreed to the published version of the manuscript.

References

1. Demir, I.; Koperski, K.; Lindenbaum, D.; Pang, G.; Huang, J.; Basu, S.; Hughes, F.; Tuia, D.; Raskar, R. Deep Globe 2018: A Challenge to Parse the Earth through Satellite Images. In Proceedings of the 2018 IEEE/CVF Conference on Computer Vision and Pattern Recognition Workshops (CVPRW), Salt Lake City, UT, USA, 18–22 June 2018; pp. 172–17209.
2. Helber, P.; Bischke, B.; Dengel, A.; Borth, D. EuroSAT: A Novel Dataset and Deep Learning Benchmark for Land Use and Land Cover Classification. *IEEE J. Sel. Top. Appl. Earth Obs. Remote Sens.* **2019**, *12*, 2217–2226. [CrossRef]
3. Chitade, A.; Katiyar, S. Color Based Image Segmentation Using K-Means Clustering. *Int. J. Eng. Sci. Technol.* **2010**, *2*, 5319–5325.
4. Chen, H.; Wang, S. Visible Color Difference-Based Quantitative Evaluation of Color Segmentation. *Vis. Image Signal Process. IEE Proc.* **2006**, *153*, 598–609. [CrossRef]
5. Gupta, P.; Srivastava, P.; Bhardwaj, S.; Bhateja, V.A. *Novel Full-Reference Image Quality Index for Color Images*; InConINDIA 2012, AISC 132; Satapathy, S.C., Avadhani, P.S., Abraham, A., Eds.; Springer: Berlin/Heidelberg, Germany, 2012; pp. 245–253.
6. Egiazarian, K.; Astola, J.; Lukin; Battisti, F.; Carli, M. *A New Full-Reference Quality Metrics Based on HVS*; Semantic Scholar: Seattle, WA, USA, 2006.
7. Shi, R.; Ngan, K.; Li, S.; Paramesran, R.; Li, H. Visual Quality Evaluation of Image Object Segmentation: Subjective Assessment and Objective Measure. *IEEE Trans. Image Process.* **2015**, *24*, 5033–5045. [CrossRef]

8. Csurka, G.; Larlus, D.; Perronnin, F. What is a Good Evaluation Measure for Semantic Segmentation? In Proceedings of the 24th British Machine Vision Conference, Bristol, UK, 9–13 September 2013.
9. Shi, R.; Ngan, K.N.; Li, S. Jaccard Index Compensation for Object Segmentation Evaluation. In Proceedings of the IEEE International Conference on Image Processing (ICIP), Paris, France, 27–30 October 2014; pp. 4457–4461.
10. Fardo, F.A.; Conforto, V.H.; Oliveira, F.C.; Rodrigues, P.S. A Formal Evaluation of PSNR as Quality Measurement Parameter for Image Segmentation Algorithms. *Comput. Vis. Pattern Recognit.* **2016**. ArXiv:1605.07116.
11. Marçal, A.R.; Rodrigues, A.; Cunha, M. Evaluation of Satellite Image Segmentation Using Synthetic Images. In Proceedings of the 2010 IEEE International Geoscience and Remote Sensing Symposium, Honolulu, HI, USA, 25–30 July 2010; pp. 2210–2213.
12. Wu, M.; Zhang, C.; Liu, J.; Zhou, L.; Li, X. Towards Accurate High Resolution Satellite Image Semantic Segmentation. *IEEE Access.* **2019**, *7*, 55609–55619. [CrossRef]
13. Singha, M.; Hemachandran, K. Color Image Segmentation for Satallite Images. *Int. J. Comput. Sci. Eng.* **2011**, *3*, 3756–3762.
14. Sirmaçek, B.; Unsalan, C. Road Detection from Aerial Images Using Color Features. In Proceedings of the 5th International Conference on Recent Advances in Space Technologies—RAST2011, Istanbul, Turkey, 9–11 June 2011.
15. Al-Ghrairi, A.; Abed, Z.H.; Fadhil, F.; Naser, F.K. Classification of Satellite Images Based on Color Features Using Remote Sensing. *Int. J. Comput. IJC* **2018**, *31*, 42–52.
16. Silva, R.D.; Minetto, R.; Schwartz, W.; Pedrini, H. Satellite Image Segmentation Using Wavelet Transforms Based on Color and Texture Features. *ISVC* **2008**, *Part II*, 113–122.
17. MacQueen, J.B. Some Methods for Classification and Analysis of Multivariate Observations. In *Proceedings of the 5th Berkeley Symposium on Mathematical Statistics and Probability*; University of California Press: Berkley, CA, USA, 1967; pp. 281–297.
18. Fränti, P.; Sieranoja, S. How much can k-means be improved by using better initialization and repeats? *Pattern Recognit.* **2019**, *93*, 95–112. [CrossRef]
19. Rousseeuw, P. Silhouettes: A Graphical Aid to the Interpretation and Validation of Cluster Analysis. *J. Comput. Appl. Math.* **1987**, *20*, 53–65. [CrossRef]
20. Arthur, D.; Vassilvitskii, S. K-Means++: The Advantages of Careful Seeding. In Proceedings of the 18th Annual ACM-SIAM Symposium on Discrete Algorithms (SODA) 2007, New Orleans, LA, USA, 7–9 January 2007.
21. Moghaddam, S.Z.; Monadjemi, A.; Nematbakhsh, N. Color Image Segmentation using Multi-thresholding Histogram and Morphology. *Int. J. Res. Rev. Comput. Sci.* **2012**, *3*, 1576–1579.
22. Martin, D.; Fowlkes, C.C.; Tal, D.; Malik, J. A Database of Human Segmented Natural Images and Its Application to Evaluating Segmentation Algorithms and Measuring Ecological Statistics. In Proceedings of the Eighth IEEE International Conference on Computer Vision, ICCV, Vancouver, BC, Canada, 7–14 July 2001; Volume 2, pp. 416–423.
23. The Assessment of the Segmentation Effectiveness of the Satellite Images. Available online: https://drive.google.com/drive/folders/10SqFSSCiUOA2gbJ_Y2l5gGhUcEoEJeej?usp=sharing (accessed on 29 October 2020).
24. Palacio-Niño, J.; Galiano, F. Evaluation Metrics for Unsupervised Learning Algorithms. *arXiv* **2019**. arxiv:905.05667.
25. Wang, Z.; Wang, E.; Zhu, Y. Image Segmentation Evaluation: A survey of Methods. *Artif. Intell. Rev.* **2020**, *53*, 5637–5674. [CrossRef]
26. Zhang, H.; Fritts, J.; Goldman, S.A. Image Segmentation Evaluation: A Survey of Unsupervised Methods. *Comput. Vis. Image Underst.* **2008**, *110*, 260–280. [CrossRef]
27. Desgraupes, B. *Clustering Indices*; University Paris Ouest. Lab Modal'X: Nanterre, France, 2016.
28. MATLAB Central File Exchange. Available online: https://www.mathworks.com/matlabcentral/fileexchange/46035-confusionmatstats-group-grouphat (accessed on 29 October 2020).
29. Tan, P.N.; Steinbach, M.; Kumar, V. *Introduction to Data Mining*; AddisonWesley: Boston, MA, USA, 2005.
30. Calinski, T.; Harabasz, J. A Dendrite Method for Cluster Analysis. *Commun. Stat. Theory Methods* **1974**, *3*, 1–27. [CrossRef]

31. Hartigan, J.A. *Clustering Algorihms. Probability and Mathematical Statistics*; John Wiley & Sons, Inc.: Hoboken, NJ, USA, 1975.
32. Xu, L. Bayesian Ying-Yang Machine, Clustering and Mumber of Clusters. *Pattern Recognit. Lett.* **1997**, *18*, 1167–1178. [CrossRef]
33. Davies, D.L.; Bouldin, D.W. A Cluster Separation Measure. *IEEE Trans. Pattern Anal. Mach. Intell.* **1979**, *PAMI-1*, 224–227. [CrossRef]
34. Gustriansyah, R.; Suhandi, N.; Antony, F. Clustering Optimization in RFM Analysis Based on K-Means. *Indones. J. Electr. Eng. Comput. Sci.* **2020**, *18*, 470–477. [CrossRef]
35. Zhang, L.; Li, H. SR-SIM: A Fast and High Performance IQA Index Based on Spectral Residual. In Proceedings of the 2012 19th IEEE International Conference on Image Processing, Lake Buena Vista, Orlando, FL, USA, 30 September–3 October 2012; pp. 1473–1476.
36. Wang, Z.; Bovik, A. A Universal Image Quality Index. *IEEE Signal Process. Lett.* **2002**, *9*, 81–84. [CrossRef]
37. Wang, Z.; Simoncelli, E.P.; Bovik, A. Multiscale Structural Similarity for Image Quality Assessment. In Proceedings of the Thrity-Seventh Asilomar Conference on Signals, Systems & Computers, Pacific Grove, CA, USA, 9–12 November 2003; Volume 2, pp. 1398–1402.
38. Wang, Z.; Li, Q. Information Content Weighting for Perceptual Image Quality Assessment. *IEEE Trans. Image Process.* **2011**, *20*, 1185–1198. [CrossRef] [PubMed]
39. Damera-Venkata, N.; Kite, T.; Geisler, W.; Evans, B.; Bovik, A. Image Quality Assessment Based on a Degradation Model. A publication of the IEEE Signal Processing Society. *IEEE Trans. Image Process.* **2000**, *9*, 636–650. [CrossRef] [PubMed]
40. Zhang, L.; Zhang, L.; Mou, X.; Zhang, D. FSIM: A Feature Similarity Index for Image Quality Assessment. *IEEE Trans. Image Process.* **2011**, *20*, 2378–2386. [CrossRef] [PubMed]
41. Wang, Z.; Bovik, A.; Sheikh, H.R.; Simoncelli, E.P. Image Quality Assessment: From Error Visibility to Structural Similarity. *IEEE Trans. Image Process.* **2004**, *13*, 600–612. [CrossRef] [PubMed]
42. Larson, E.C.; Chandler, D. Most Apparent Distortion: Full-Reference Image Quality Assessment and the Role of Strategy. *J. Electron. Imaging* **2010**, *19*, 011006.
43. Liu, A.; Lin, W.; Narwaria, M. Image Quality Assessment Based on Gradient Similarity. *IEEE Trans. Image Process.* **2012**, *21*, 1500–1512.
44. Mitsa, T.; Varkur, K.L. Evaluation of Contrast Sensitivity Functions for the Formulation of Quality Measures Incorporated in Halftoning Algorithms. In Proceedings of the IEEE International Conference on Acoustics, Speech, and Signal Processing, Minneapolis, MN, USA, 27–30 April 1993; Volume 5, pp. 301–304.
45. Chang, H.; Yang, H.; Gan, Y.; Wang, M. Sparse Feature Fidelity for Perceptual Image Quality Assessment. *IEEE Trans. Image Process.* **2013**, *22*, 4007–4018. [CrossRef]
46. Sheikh, H.; Bovik, A. Image Information and Visual Quality. *IEEE Trans. Image Process.* **2006**, *15*, 430–444. [CrossRef]
47. Chandler, D.; Hemami, S. VSNR: A Wavelet-Based Visual Signal-to-Noise Ratio for Natural Images. *IEEE Trans. Image Process.* **2007**, *16*, 2284–2298. [CrossRef]
48. Ponomarenko, N.; Silvestri, F.; Egiazarian, K.; Carli, M.; Astola, J.; Lukin, V. On Between-Coefficient Contrast Masking of DCT Basis Functions, CD-ROM. In Proceedings of the Third International Workshop on Video Processing and Quality Metrics for Consumer Electronics VPQM-07, Scottsdale, AZ, USA, 25–26 January 2007; p. 4.
49. Zhang, L.; Shen, Y.; Li, H. VSI: A Visual Saliency-Induced Index for Perceptual Image Quality Assessment. *IEEE Trans. Image Process.* **2014**, *23*, 4270–4281.
50. Min, X.; Gu, K.; Zhai, G.; Hu, M.; Yang, X. Saliency-Induced Reduced-Reference Quality Index for Natural Scene and Screen Content Images. *Signal Process.* **2018**, *145*, 127–136. [CrossRef]
51. MeTriX MuX Visual Quality Assessment Package. Available online: https://github.com/sattarab/image-quality-tools/tree/master/metrix_mux (accessed on 29 October 2020).
52. Ma, L.; Lin, W.; Deng, C.; Ngan, K.N. Image Retargeting Quality Assessment: A Study of Subjective Scores and Objective Metrics. *IEEE J. Sel. Top. Signal Process.* **2012**, *6*, 626–639. [CrossRef]
53. International Telecommunication Union (ITU). *Methodologies for the Subjective Assessment of the Quality of Television Images*; Document Rec. ITU-R BT.500-14, 10/2019; ITU: Geneva, Switzerland, 2020.

54. International Telecommunication Union (ITU). *Methods for the Subjective Assessment of Video Quality, Audio Quality and Audiovisual Quality of Internet Video and Distribution Quality Television in Any Environment*; Document Rec. ITU-T P.913, 03/2016; ITU: Geneva, Switzerland, 2016.
55. Huynh-Thu, Q.; Garcia, M.; Speranza, F.; Corriveau, P.J.; Raake, A. Study of Rating Scales for Subjective Quality Assessment of High-Definition Video. *IEEE Trans. Broadcast.* **2011**, *57*, 1–14. [CrossRef]
56. Wu, D.; Yuan, F.; Cheng, E. Underwater No-Reference Image Quality Assessment for Display Module of ROV. *Sci. Program.* **2020**, *2020*, 1–15.
57. Schober, P.; Boer, C.; Schwarte, L. Correlation Coefficients: Appropriate Use and Interpretation. *Anesth. Analg.* **2018**, *126*, 1763–1768. [PubMed]
58. Okarma, K. Quality Assessment of Images with Multiple Distortions using Combined Metrics. *Elektron. Elektrotechnika* **2014**, *20*, 128–131. [CrossRef]
59. Ieremeiev, O.; Lukin, V.; Okarma, K.; Egiazarian, K. Full-Reference Quality Metric Based on Neural Network to Assess the Visual Quality of Remote Sensing Images. *Remote. Sens.* **2020**, *12*, 2349. [CrossRef]
60. Zhai, G.; Min, X. Perceptual Image Quality Assessment: A survey. *Sci. China Inf. Sci.* **2020**, *63*, 211301. [CrossRef]

2

The Integration of Remote Sensing and Field Surveys to Detect Ecologically Damaged Areas for Restoration in South Korea

Kyungil Lee [1], Hyun Chan Sung [1], Joung-Young Seo [2], Youngjae Yoo [1], Yoonji Kim [1], Jung Hyun Kook [3] and Seong Woo Jeon [1,*]

1 Division of Environmental Science & Ecological Engineering, Korea University, 02841 145 Anam-ro, Seongbuk-gu, Seoul 02841, Korea; leedake@korea.ac.kr (K.L.); wona2015@naver.com (H.C.S.); lemonesty@korea.ac.kr (Y.Y.); yoonjik605@korea.ac.kr (Y.K.)
2 DNC Engineering, 69, Songdam 2-ro, Anjung-eup, Pyeongtaek-si 17941, Korea; dnceng@naver.com
3 Eco, Environment, Education, 583, Yangcheon-ro, Gangseo-gu, Seoul 07547, Korea; jhkook@ethree.co.kr
* Correspondence: eepps_korea@korea.ac.kr

Abstract: Ecological damage refers to the reduction in the value of the environment due to human activities such as development. The intensity of ecosystem damage is worsening worldwide. Although the importance of restoration projects to reduce ecosystem damage is increasing, they are difficult to carry out, owing to the absence of data and monitoring of damaged areas. In this study, ecologically damaged areas for restoration in South Korea were detected using remote sensing and field surveys. For the analysis, national standardized vector datasets and Google Earth images were used; field surveys were conducted from 2018 to early 2020. Our results showed that 62% of the ecological damage that occurred in South Korea existed in forest ecosystems; the damaged areas were mostly smaller than 50,000 m^2. Additionally, most of the causes and types of damage due to human activities such as development were soil erosion related. The results also suggest the importance of obtaining monitoring data on ecologically damaged areas and the importance of establishing an appropriate restoration plan using this data.

Keywords: ecological degradation; restoration; remote sensing; field survey

1. Introduction

Ecological damage is the reduction in the value of the environment to meet socio-economic needs, including land degradation, deforestation, and land pollution. Ecological damage can be caused by human anthropogenic activities or natural phenomena [1,2]. As per the Millennium Ecosystem Assessment, approximately 60% of the world's ecosystems have been damaged through human use; the damage caused by land degradation is estimated to cost US$4.3–20.2 trillion per year worldwide [3]. Furthermore, additional damage may occur owing to the negative impacts on biodiversity, human health, and food security in these damaged areas [4]. To prevent this damage, the goal of the ecological restoration of more than 15% of the world's damaged ecosystems has been set through the Convention on Biological Diversity; related research is also increasing [5–9].

As per existing research cases, the damage caused by urbanization and the social demand for ecosystem restoration are increasing, as is the importance of information about damaged areas and the policy role for restoration projects.

However, in the absence of restoration-related monitoring data, difficulties in policy decisions such as restoration site prioritization and restoration planning are presented [5,6,8,9]. Monitoring is recognized as a costly and time-consuming operation; thus, it is often not performed before and after a restoration project. However, if monitoring is not performed, the evaluation of damage and restoration recovery trajectories becomes impossible [10,11]. Additionally, for restoration site prioritization and planning—a limitation of current restoration projects—the location, size, and type of the damaged area must be identified in advance through monitoring: a restoration direction that considers that these aspects should be established. If the restoration project is executed sporadically owing to a lack of planning, an inefficient use of budget and manpower due to restoration project may occur, and the ecological connectivity of the restoration area may be insufficient [7].

The role of remote sensing in restoration planning and monitoring has increased, especially in the last decade, owing to improved sensor technology and data availability [12]. Its main advantages include its capabilities to provide complete information on land cover as well as analyze that information [13]. Remote sensing tools such as land-use and land-cover (LULC) data have been used to track land-cover change around protected areas [14–18]. Additionally, national standardized vector datasets have been developed for country-wide comparisons of land cover based on remotely sensed data, such as the National Land Cover Dataset by the US Geological Survey (USGS) [19]. However, most studies using remote sensing only have a limited study area such as protected area; there have been limitations that could not be overcome except by carrying out field surveys for the verification of the area where the change occurred and there is no basis for judging whether it was an illegal damage or a just natural change.

To establish a restoration direction and plan for the damaged area, it is necessary to collect information such as the cause, type, and extent of the damage and construct a dataset that can be used by the restoration-related ministries for each type [7,14]. In addition, the dataset for damaged areas is important for the environmental offsets policy, which is increasingly popular in conservation worldwide. Environmental offsets mean 'the measurable conservation outcomes resulting from actions designed to compensate for significant environmental impacts arising from development project despite appropriate prevention measures [20,21]. Restoration is one of the offset methods that allows for an easier measurement of loss versus gain compared to other methods and provides a unique opportunity to maintain spatial ecological connectivity. However, restoration can be complicated by logistical, social, and political constraints and requires early planning to ensure adequate baseline information to inform feasible restoration goals and practice [22]. Therefore, it is important to construct spatial information and a criterion of illegally damaged areas for legitimate restoration project plans and to advance field trials to provide aid in an assessment of feasibility. The technique of damaged area monitoring using both remote sensing and field survey would increase the data quality compared to a simple remote sensing monitoring, whilst reducing the cost compared to a simple filed survey for a specific region [23].

In this study, we aimed to introduce a two-step methodology to detect ecologically damaged areas and monitor them using both remote sensing and field surveys to build a damaged area database for legitimate restoration planning. We first used remote sensing data such as national standardized vector datasets and Google Earth data to analyze the illegally changed natural environment areas and then conducted field surveys to verify the derived results and determine the type of damage.

2. Materials and Methods

2.1. Study Area

The study area included the inland areas of South Korea: approximately 99,827 km^2 (latitude 34° N–38.5° N & longitude 126° E–129.5° E) divided into 16 regional local governments (Figure 1). The dominant land-cover categories of the study area are forest (67.7%), agricultural land (22.2%), urban areas (4.6%), and others, including grassland, water, barren land, and wetlands (5.5%) [24]. South Korea experiences damage to the natural environment in various forms and magnitudes owing to natural disasters such as flood due to rainfall, soil erosion due to wind and gravity, as well as human activities such as land development and road construction [7]. Additionally, socio-economic damage is also increasing owing to the deterioration of ecological health, disconnection with adjacent ecosystems, and inhibition of natural landscapes in the damaged area. However, since the locations of and information about the damaged areas are not known in advance, specific policies or prior plans are not established, and sporadic restoration projects are being carried out by government departments [7]. Figure 1 is the general land cover status of South Korea provided by the Ministry of Environment (MOE) [25].

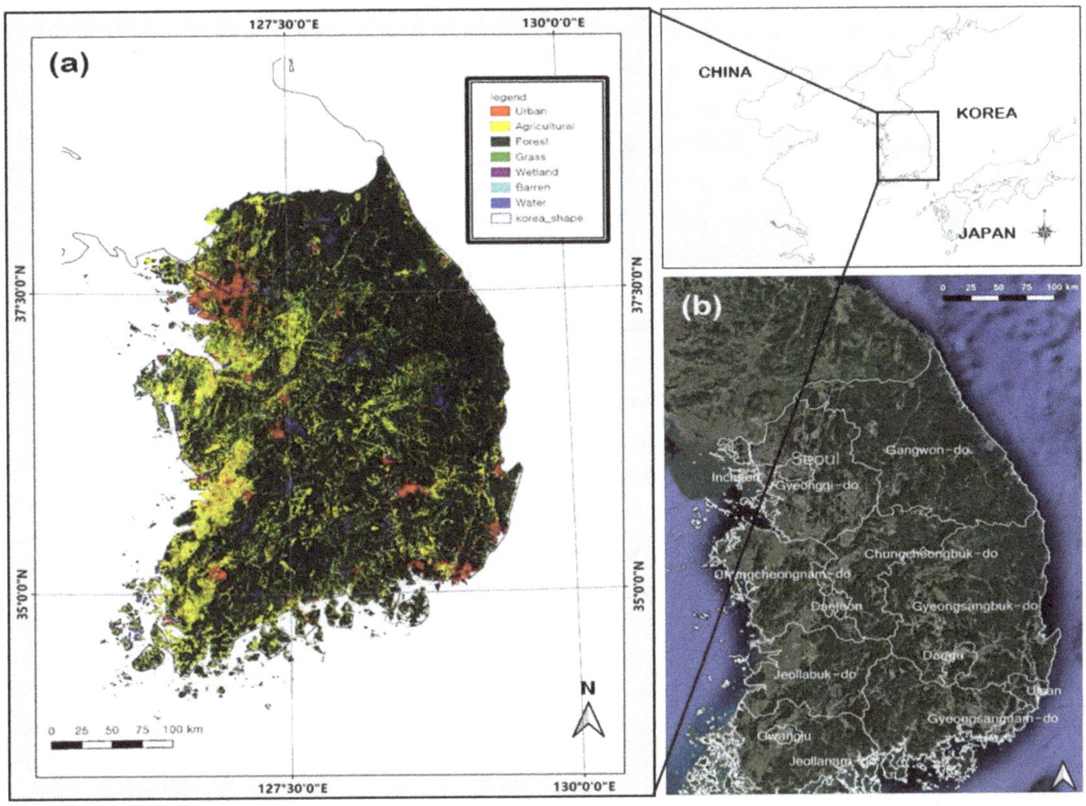

Figure 1. (**a**) The study area and the land cover status; (**b**) a satellite image of the study area showing the 16 regional local governments.

2.2. Materials

2.2.1. Land-Cover Data

To analyze the land-cover changes (LCC) in the study areas, the level 2 land-cover maps of 2004 and 2013 produced by the MOE of South Korea were used [25]. The level 2 land-cover maps were constructed using satellite images such as Indian Remote Sensing (IRS-1D), IKONOS imageries, and aviation images at a 5-m spatial resolution. First, land cover classification was performed twice using 5m resolution images and the constructed map was examined and supplemented through fieldwork; the criteria for classification accuracy should be at least 95% on average (a detailed explanation of the processing chain and validation methods used can be found at the Environmental Geographic Information Service in Korean) [25]. These remote sensing-based maps have been used by the scientific community for various purposes; however, it appears that limited research has been conducted to locate and quantify ecologically damaged areas using these maps.

The level 2 land-cover map classified the entire land into 22 land-cover types. To derive the ecologically damaged areas, the data from 22 categories of land cover were reclassified as "forest," "grass," "wetland," "agricultural area," and "barren area" according to this study's purpose (Table 1). Most related studies used land-cover maps constructed using 30-m resolution satellite images; however, here, we used 5-m resolution land-cover maps, so that more accurate and detailed LCC could be identified.

Table 1. Two-level hierarchy of land-cover categories in the 2004 and 2013 land-cover maps provided by the South Korean Ministry of Environment ("Original classification") and the reclassified categories used in this study ("Classification for this study").

Original Classification		Classification for This Study
Level 1	**Level 2**	
Urban Area	Residential Area	Developed Area
	Industrial Area	
	Commercial Area	
	Cultural, Sport, recreational Area	
	Traffic Area	
	Public Facilities	
Agricultural Land	Rice Paddy	Natural and Semi-Natural Environment Area
	Field	
	Cultivation under Structure	
	Orchard	
	Other Cultivation	
Forest	Broadleaf Forest	
	Coniferous Forest	
	Mixed Forest	
Grass	Natural Grass	
	Artificial Grass	
Wetland	Inland Wetland	
	Coastal Wetland	
Barren	Natural Barren	Potentially Damaged Area
	Artificial Barren	
Water	Inland Water	Water
	Seawater	

2.2.2. Land-Use Data

Change detection using LCC analysis allows scientists to look at changes in land use or land cover and disturbance using binary comparisons contrasting conditions between two discrete time periods [19,26]. However, to determine whether the land was illegally damaged, additional analysis is needed in an area where LCC occurred. The land-use zoning map was used for this analysis in this study.

Land-use zoning is one of the most effective measures to control various land-use activities. It first originated for the re-construction of disordered and undisciplined cities and later attracted substantial attention worldwide [27,28]. Some countries, like the United States, have employed municipal/country zoning ordinances to optimize residential, industrial, commercial, and ecological land use in rural and urban planning, but China introduced regional land-use zoning to resolve land-use conflicts between rural and urban development and keep agricultural land from the occupation of urban expansion [29–31]. In South Korea, the land-use zoning system was introduced for the same purpose as that in the United States; it is divided into four areas: urban areas, control areas, agricultural and forest areas, and natural environment conservation areas (Ministry of Land, Infrastructure and Transport [32] (Table 2).

Table 2. Division of land-use zoning in South Korea defined by South Korean Ministry of Land, Infrastructure and Transport.

Division	Description
Urban areas	Areas requiring systematic development, maintenance, management, preservation, etc., where the population and industries are concentrated, or such concentration is anticipated therein
Control areas	Areas to be systematically controlled corresponding to urban areas to accommodate the population and industries of urban areas or those requiring control corresponding to an agricultural and forest area or natural environment conservation area to promote the agricultural and forest industry and preserve the natural environment or forests
Agricultural and forest areas	Areas necessary to promote the agricultural and forest industry and preserve forests, such as agricultural promotion areas under the Farmland Act or conserved mountainous districts under the Mountainous Districts Management Act that do not belong to urban areas
Natural environment conservation areas	Areas necessary to preserve the natural environment, water resources, coastal areas, ecosystems, water supply resources, and cultural heritage assets and protect and foster fishery resources, etc.

Land-use zoning for sustainable development includes a set of sustainability objectives related to urban sprawl restriction and ecological environment restoration; each land-use zone type has been used to regulate land-use activities at the regional local government scale in South Korea, which is the major urban management scale [28,32]. Therefore, the land-use zoning map can be used as the basis for judging illegally damaged areas among the changed areas; this study utilized the land-use zoning map for 2018 provided by the National Spatial Data Infrastructure Portal (NSDI) [33].

2.2.3. The Area Subject to Environmental Impact Assessment Data (EIA Map)

The Environmental Impact Assessment (EIA) introduced by the United States in the late 1960s and early 1970s has been adopted widely in the rest of the world [34]. In South Korea, the EIA refers to an assessment conducted by surveying, forecasting, and assessing the appropriateness of a site location and the environmental impact of a development project to provide measures for environmental conservation (MOE, 2018). The EIA process has important purposes such as supporting decision making to prevent projects with undesirable environmental impacts and assisting in informing development decisions by instructing a consideration of alternatives (such as locations, scales, and conditions) [34]. The EIA and remote sensing have been used to address environmental problems, but few studies have linked the two fields [35–37]. In the case of studies that combined the two fields, most studies presented the potential of RS data such as satellite imagery to support the EIA; conversely, studies using EIA data to support decision making for RS were rare. The area subject to EIA is under review for development at the national level and can be the basis for judging the area where the land-cover change occurred legally.

In South Korea, information related to the EIA needs to be collected and disseminated through the Information Support System for Environmental Impact Assessment (EIASS) in point of law [38]. In this study, the spatial data of EIA projects conducted until 2018 were acquired through the EIASS and used in the exclusion analysis for the regions derived above [39]. All data used for damaged area detection were projected to the common Projected Coordinate System 'WGS 1984 UTM Zone (EPSG:2097), the same as the Google Earth data (Figure 2).

2.2.4. Google Earth Data

In this study, land-cover data from 2004 and 2013 were used, owing to the necessity of using the high-resolution land-cover map constructed for the entire country; at this time, the land-cover condition of the damaged area may differ from that of the present. To modify the differences due to time intervals, we used Google Earth data as the reference data. Google Earth is the most popular virtual globe software; it visualizes images as a global mosaic of the Earth using mid and high-resolution satellite and aerial imagery from multiple providers ranging from SPOT5 and Landsat up to aerial orthophotos [40]. The advantages of using Google Earth data are that they have a high resolution (65×65 cm), available for free, and describe the imagery date [41]. Google Earth's high-resolution data have been used as reference data in many studies; previous research proved the horizontal and vertical precision of these images, and images from 2018 were selected for this study [40–42].

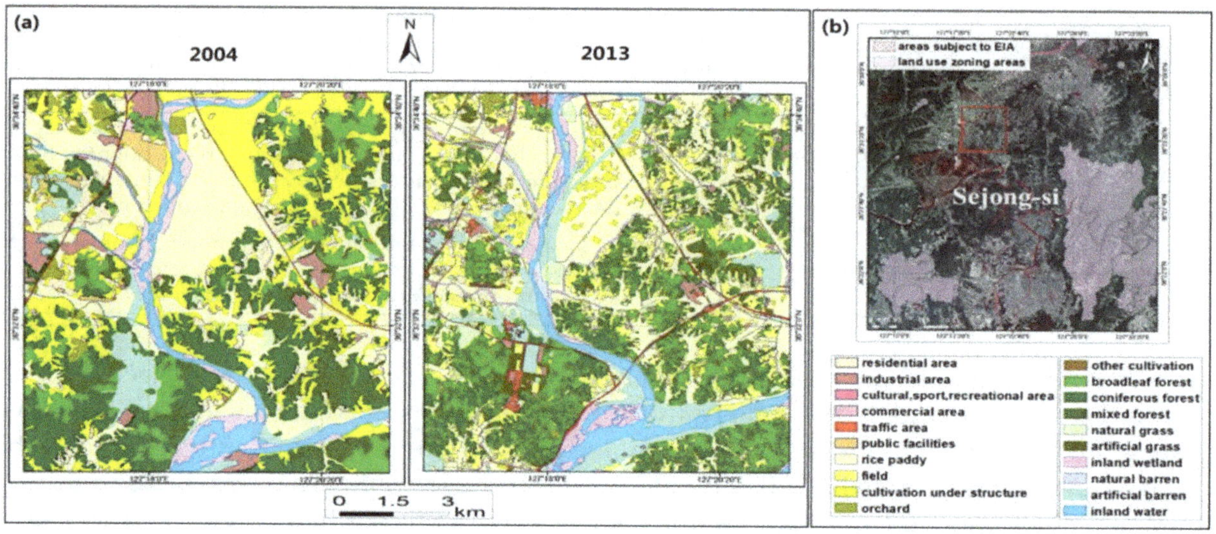

Figure 2. Examples of data used for damaged area detection in Sejong-si in South Korea. (**a**) Level 2 land cover in 2004 and 2013; (**b**) examples of areas subject to Environmental Impact Assessment (EIA) and land-use zoning areas (natural environment conservation areas) overlaid on Google Earth imagery.

2.3. Methods

2.3.1. Land-Cover Change Analysis

In South Korea, the legal definition of environmental damage includes conditions that inflict serious damage on intrinsic functions of the natural environment, for example, overhunting or over gathering wild animals or plants, destroying their habitats, disturbing the order of the ecosystem, impairing the natural scenery, washing away the topsoil, etc. [43]. In this study, the damaged area was defined as the area changed from a natural and seminatural environment area (agricultural land, forest, grassland, wetland) to a potentially damaged area (barren area) according to the legal definition. The areas categorized as natural environment area and potentially damaged area in the 2004 and 2013 Level 2 land-cover maps were used to calculate this.

Change detection has historically been used to look at changes in land cover using binary comparisons, contrasting conditions during two discrete time periods [19]. The over-lay procedure was used to obtain the changes in land cover during the specified time period. A two-way cross-matrix obtained by the procedure was used to describe the change in land-cover area in South Korea by each regional local government. Among the areas where changes occurred, areas that were changed from natural and semi-natural environment areas to potential damaged areas were extracted. Next, among the extracted areas, areas with less than 5000 m^2 were excluded because 5000 m^2 is the minimum area of the project subject to EIA in South Korea [44].

2.3.2. Analysis of Illegal Damaged Area

For areas derived through land cover change analysis, we conducted an overlap analysis using Land use zoning map and EIA map to classify that areas into the illegal and legal changes.

First, when the changed area belonged to urban areas and agricultural and forest areas through overlapping with the land use zoning map, the land cover was intentionally changed for development and agriculture purposes. Areas intentionally changed were not illegally damaged and excluded from the list of extracted areas. Contrastingly, if the land cover changed area is on control areas and natural environment conservation areas, there is a possibility that it has been illegally damaged because it is designated to protect the natural environment by the law [32]. The areas belonging to the two zones were not excluded and analyzed by overlapping with the EIA map to determine whether they were illegally damaged. If the area overlaps with the EIA map, development is in progress or can be developed in the future. Therefore, it was excluded from the list of damaged sites and modifications using google earth data were performed on the rest of the areas.

2.3.3. Modification Using Google Earth Data

The recent Google Earth data from 2018 were used to determine whether the extracted areas were damaged for real or not and to calculate the exact size of damaged area. Because of the necessity of using the high-resolution land cover map constructed for the entire country there may be a difference according to the time interval even if actual damage occurred at the area before.

First, we checked the current land cover of the polygon of the sites extracted as a damaged area and surrounding of environment. If the target site has already been developed or is difficult to see as a damaged area compared to the surrounding environment, it was excluded from the field survey target area (Figure 3a). Second, if it is judged to be a damaged area, but the polygon of damaged area does not match the with actual area, the polygon of the extracted area was modified to derive an accurate area (Figure 3b). Spatial overlay analysis to derive damaged areas and modification using Google Earth Data were performed using ArcGIS software 10.3 and a field survey was conducted on the list of damaged areas finally derived through modification.

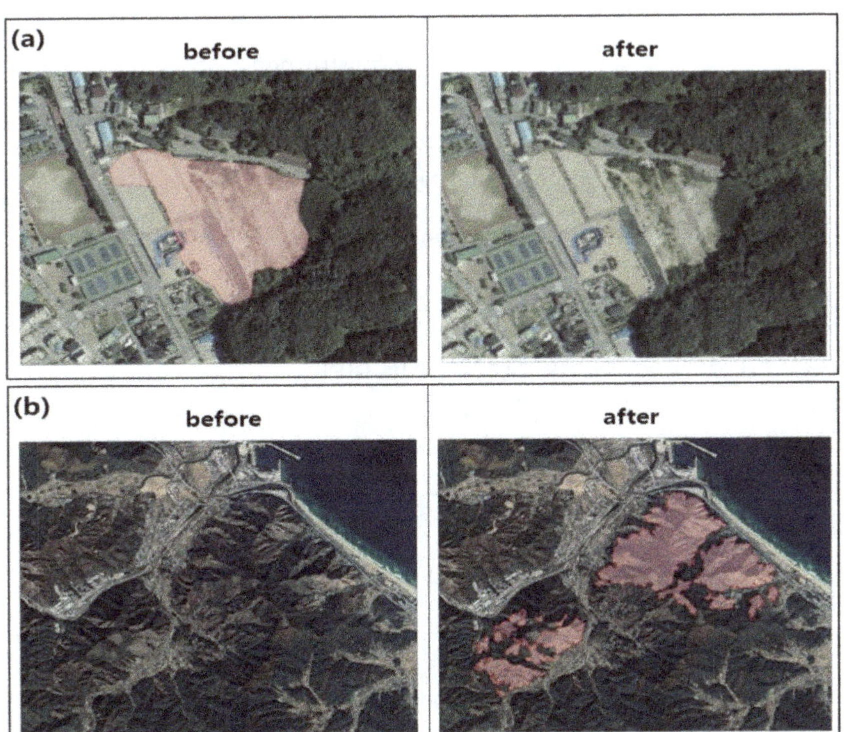

Figure 3. Examples of polygon modification: (**a**) excluding the developing areas; (**b**) modifying the area damaged by forest fire for accurate area calculation.

2.3.4. Field Survey

From 2018 to early 2020, field surveys were conducted for all the damaged areas detected through remote sensing to verify the identity of the damaged areas and the type of damage. To restore the ecosystem structure in the area where damage has occurred, the cause and status of the damage should be identified to establish the restoration direction and plan [7]. During field surveys, we collected the information about the areas that were damaged such as coordinates of the location, surrounding vegetation, damage type, range, cause, etc.

The causes of damage can be divided into natural damage and anthropogenic damage; the types of damage are classified according to the ecological structure of the target site [7]. In this study, the criteria for type of damage based on MOE [7] and Victoria Resources Online (VRO) were used in the field survey [45] (Table 3). The identification of damage occurrence, cause of damage for each area were determined based on investigations involving comparison with the surrounding environment, and the type of damage was determined through expert judgment based on the criteria. The steps from data construction to field survey for ecologically damaged areas detection are shown in Figure 4. Additionally, the land cover—accounting for more than 75% of the damaged area—was judged as the main ecosystem of the area.

Table 3. Field survey criteria for damage classification.

Division	Cause of the Damage	Type of Damage
Natural damage	1. Invasive alien species 2. Soil erosion by rainfall, wind, gravity 3. Water level fluctuation by heavy rain 4. Forest fire	1. Soil degradation - Soil salinity - Compaction - Acidification 2. Ecosystem decline
Anthropogenic damage	1. Road and railway construction 2. Development 3. Maintenance 4. Agricultural activities 5. Human activities 6. Military facilities 7. Secondary damage 8. Forest fire	- Catchment salinity - Vegetation decline 3. Soil erosion - Sheet and rill erosion - Wind erosion - Landslide - Stream bank erosion

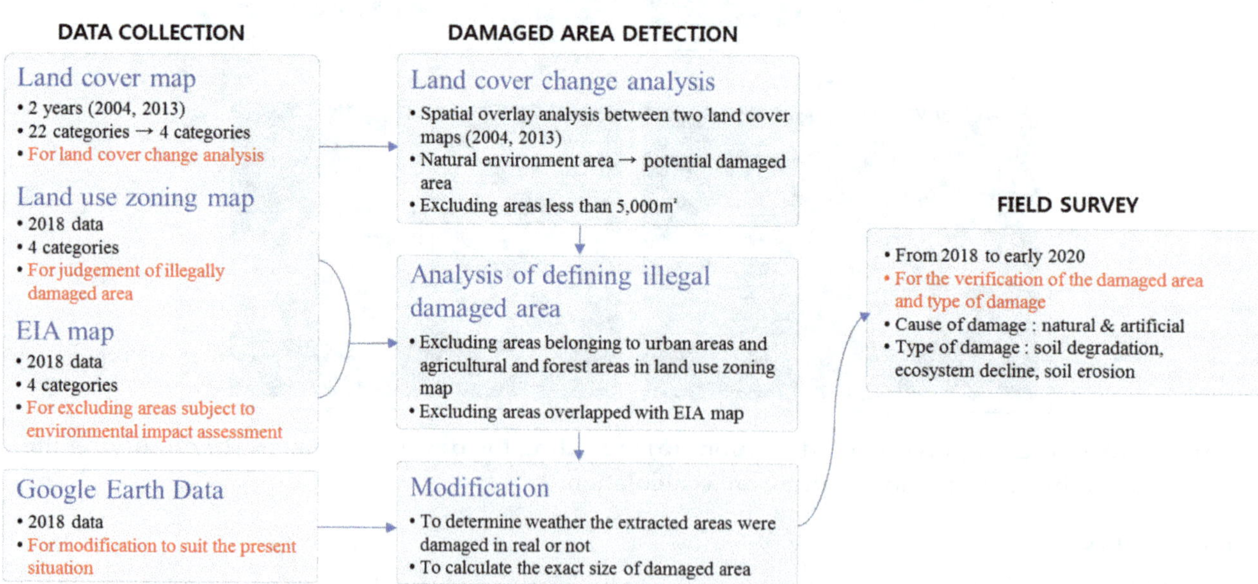

Figure 4. Utilization data and analysis flow chart for detecting ecological damaged areas.

3. Results

3.1. Damaged Area Detection

In total, 399 damaged areas over 5000 m^2 were detected using remote sensing in South Korea (Table 4) (Figure 5). Most areas where there was slight damage, such as Seoul, Busan, and Incheon, were promoted to metropolitan cities before and after the 1980s, and urbanization and development took place before other regions. Additionally, the number of development projects that have occurred in the last 10 years (2008–2017) is small compared to the other regions; thus, it is expected that there will be a lower possibility of anthropogenic damage [46] (see Table A1). There were more damaged areas in Gyeonggi-do (84 cases) and Gangwon-do (82 cases) than in the other regions. Gyeonggi-do is the region with the largest number and area of total development projects, including housing site development businesses, according to the metropolitan plan, and is considered to be the region with the highest possibility of anthropogenic damage. In the case of Gangwon-do, the ratio of forest area to total area is 81.5%, the highest in Korea; the average slope of the mountainous areas is also high. Thus, here, naturally occurring landslides and the anthropogenic deterioration of forests are expected to occur more frequently than in other regions [47].

Table 4. Distribution status of damaged areas in South Korea derived through remote sensing.

Division	5000–10,000 m2	10,000–50,000 m2	50,000–100,000 m2	100,000–150,000 m2	150,000–200,000 m2	200,000–250,000 m2	250,000–300,000 m2	300,000 m2~	Total (Row)
Seoul *	2	1	1	0	0	0	0	0	4 (2%)
Busan *	1	1	0	0	0	0	0	0	2 (1%)
Daegu *	0	0	0	0	0	0	0	0	0 (0%)
Incheon *	0	0	0	0	0	0	0	0	0 (0%)
Gwangju *	1	0	0	0	0	0	0	0	1 (1%)
Daejeon *	0	0	0	0	0	0	0	0	0 (0%)
Ulsan *	1	1	0	0	0	0	0	0	2 (1%)
Sejong *	0	0	0	0	0	0	0	0	0 (0%)
Gyeonggi-do	20	53	4	0	2	1	3	0	84 (21%)
Gangwon-do	25	44	5	4	0	1	0	3	82 (20%)
Chungcheong buk-do	20	29	3	0	0	0	0	0	52 (13%)
Chungcheong nam-do	10	14	1	1	0	0	0	0	26 (6%)
Jeolla buk-do	10	10	1	0	0	1	1	0	23 (5%)
Jeolla nam-do	17	20	6	1	0	0	0	0	44 (11%)
Gyeongsang buk-do	9	20	3	2	0	0	0	0	34 (8%)
Gyeongsang nam-do	17	21	2	3	0	0	0	2	45 (11%)
Total (column)	133 (33%)	215 (54%)	26 (6%)	11 (3%)	2 (1%)	3 (1%)	4 (1%)	5 (1%)	399 (100%)

*: metropolitan city in South Korea.

Considering the size of the damaged areas, approximately 54% of the damaged areas was identified with 10,000m^2–50,000m^2; this is more than half of the total damaged area. The number of damaged areas between 5000 m^2 and 10,000m^2 was 133, the second most abundant size category. Therefore, approximately 87% of the damage was analyzed to be relatively small; approximately 13% of the damaged areas was found to be damaged over 50,000 m^2.

3.2. Field Surveys

3.2.1. Status of Damaged Areas by Region

As per the field surveys conducted over 2 years, 288 areas were categorized as actually damaged areas (Table 5). Of the 399 damaged areas derived through remote sensing, approximately 72% of the areas were found to be actual damaged areas; it is judged that the methodology used in this study detected the damaged areas with good accuracy. As per the field surveys, approximately 90% of the damaged areas were found to have an area of 50,000 m^2 or less—almost the same as the ratio obtained through remote sensing. Contrastingly, in the case of distribution by region, in the analysis using remote sensing, the number of damaged areas in Gyeonggi-do and Gangwon-do was similar, but 35 areas of Gangwon-do were excluded from the field survey. In Gangwon-do, the forest area accounted for more than 80% of land cover; Rocky Mountains, quarries, and talus areas were widely distributed. In this study, natural environmental damage was defined as a change into a barren area; the damaged areas detected through remote sensing were often classified into the above three topographies. Rocky mountains and quarries were excluded as natural ecosystems that require management. The term talus has been used to define the landform of cliffs containing conical-pile form rock fragments and was excluded as a natural phenomenon [48] (Figure 6). Most of the other excluded areas are currently under development or have already been completed and post-processed. This was considered to be an error due to differences between the land-cover map construction time and the current land status.

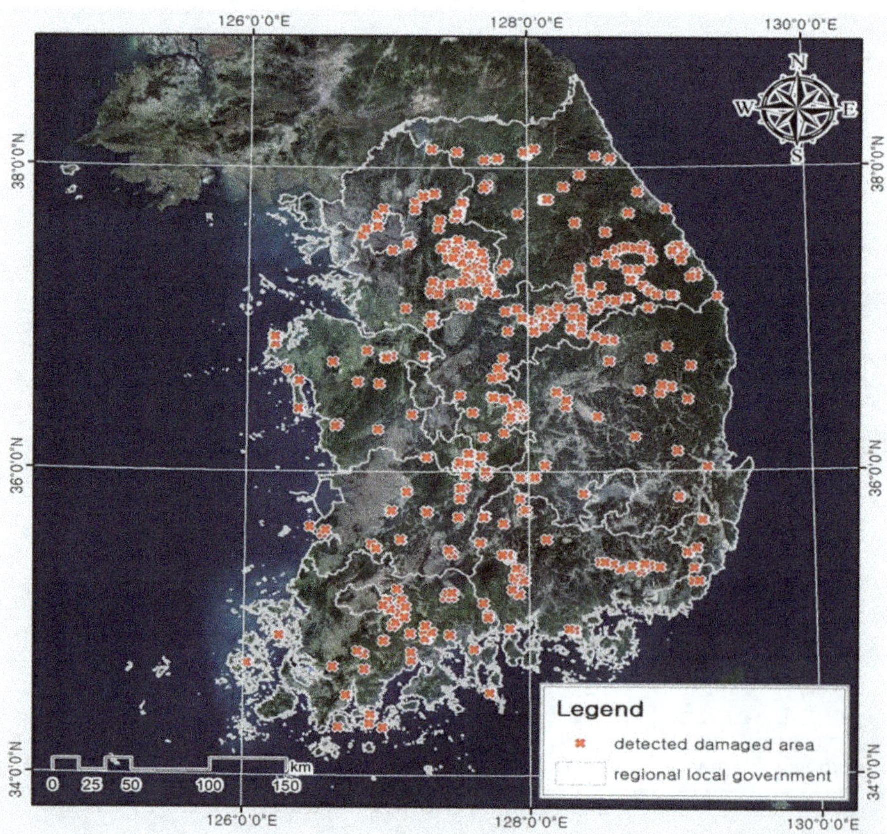

Figure 5. Distribution status of damaged areas in South Korea derived through remote sensing.

3.2.2. Status of Damaged Areas by Causes and Types

Through the field surveys, 62% of ecological damage occurred in the forest ecosystem, 14% was found in barren areas, and the rest were grassland and agricultural ecosystems, etc. It is considered that 63% of Korea's land cover is forests and considered to be due to a result of continuous development projects and illegal damage to forest lands. Contrastingly, the land-cover type with the least damage was urban areas. As with the results of the remote sensing analysis, it appears that there is slight damage in the areas where urbanization has already progressed (see Figure A1).

According to the results of the investigation into the cause of the damage, approximately 97% of the damage was caused by anthropogenic factors; there were almost eight cases of damage caused by natural causes (Figure 7b). In Gangwon-do, which is expected to have a large amount of forest damage, most areas that appeared to be natural talus areas, quarries, etc., rather than actually damaged areas, were excluded from the list of damaged areas.

In the case of damage type, approximately 66% of damage was soil erosion. Soil erosion is a damage type that is a problem worldwide and is mainly caused by forestry and agriculture (Figure 7d). The intensity of soil erosion occurrence is worsening owing to increases in population and human activities [49]. Through field surveys, most damage in Korea was identified as soil erosion due to deforestation and development targeting forest areas. There were 50 cases of soil degradation and ecosystem decline, which are other types of damage. This was therefore a relatively less important type of damage than soil erosion. These two types of damage are also caused by deforestation, salinization, habitat loss, and fragmentation caused by urbanization and human activities and can also partially occur in areas where soil erosion has occurred [50,51].

Table 5. Distribution status of damaged areas in South Korea derived through field surveys.

Division	5000–10,000 m2	10,000–50,000 m2	50,000–100,000 m2	100,000–150,000 m2	150,000–200,000 m2	200,000–250,000 m2	250,000–300,000 m2	300,000 m2~	Total (Row)
Seoul *	1	1	1	0	0	0	0	0	3 (1%)
Busan *	1	1	0	0	0	0	0	0	2 (1%)
Daegu *	0	0	0	0	0	0	0	0	0 (0%)
Incheon *	0	0	0	0	0	0	0	0	0 (0%)
Gwangju *	0	0	0	0	0	0	0	0	0 (0%)
Daejeon *	0	0	0	0	0	0	0	0	0 (0%)
Ulsan *	1	1	0	0	0	0	0	0	2 (1%)
Sejong *	0	0	0	0	0	0	0	0	0 (0%)
Gyeonggi-do	20	52	4	0	1	1	3	0	81 (28%)
Gangwon-do	18	27	1	0	0	1	0	0	47 (16%)
Chungcheong buk-do	16	18	0	0	0	0	0	0	34 (12%)
Chungcheong nam-do	5	10	1	0	0	0	0	0	16 (5%)
Jeolla buk-do	9	7	0	0	0	1	1	0	18 (6%)
Jeolla nam-do	13	16	5	0	0	0	0	0	34 (12%)
Gyeongsang buk-do	7	17	2	2	0	0	0	0	28 (10%)
Gyeongsang nam-do	7	12	0	2	0	0	0	2	23 (8%)
Total (column)	98 (34%)	162 (56%)	14 (4%)	4 (1%)	1 (1%)	3 (1%)	4 (1%)	2 (1%)	288 (100%)

*: metropolitan city in South Korea.

Figure 6. Examples of areas excluded from the damaged areas: (**a**) The area identified as talus in Gangwon-do; (**b**) the area where grassland was established in Gyeongsangnam-do; (**c**) the area where development is in progress in Gyeongsangnam-do.

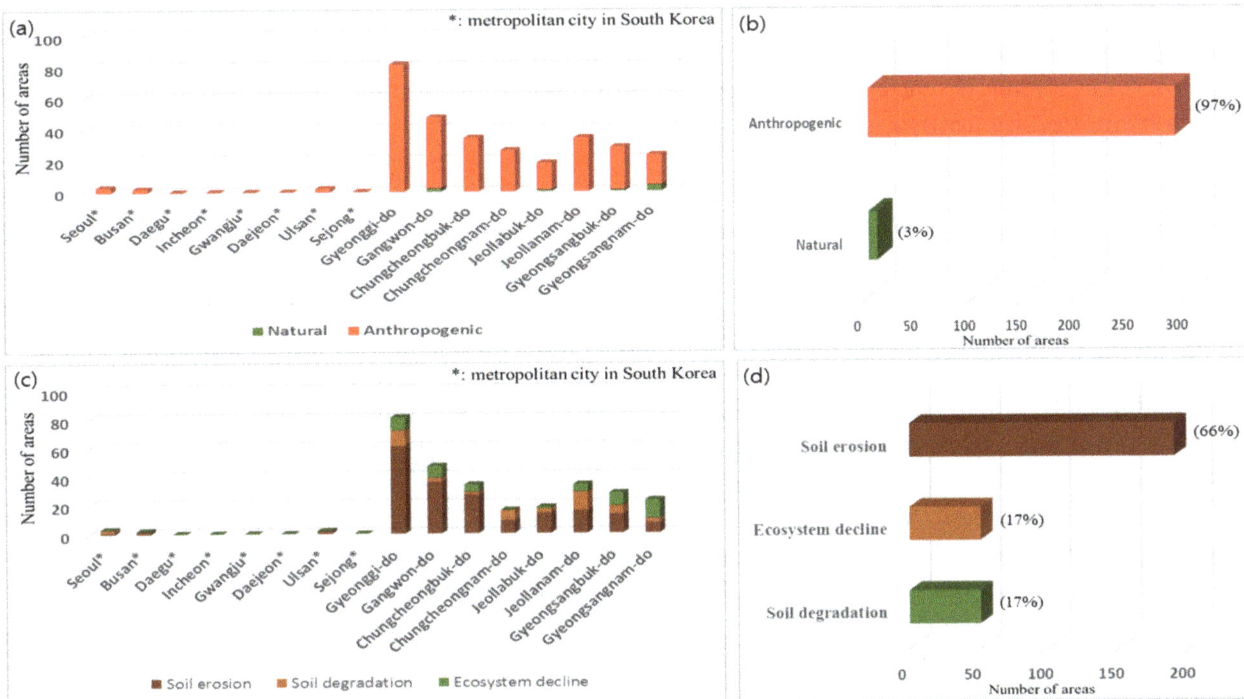

Figure 7. Distribution status of damaged areas in South Korea by cause and type: (**a**) Distribution of damaged areas by cause for each regional local government and; (**b**) total distribution of damaged areas by cause in South Korea; (**c**) Distribution of damaged areas by type for each regional local government and; (**d**) total distribution of damaged areas by type in South Korea.

In this study, the cause, type, and ecosystem of each damaged area were classified for the direction of restoration (Figure 8); however, when performing the actual restoration project, it is necessary to understand the complex type of damage to the damaged area and perform integrated and organic restoration considering the connection with surrounding ecosystems [7].

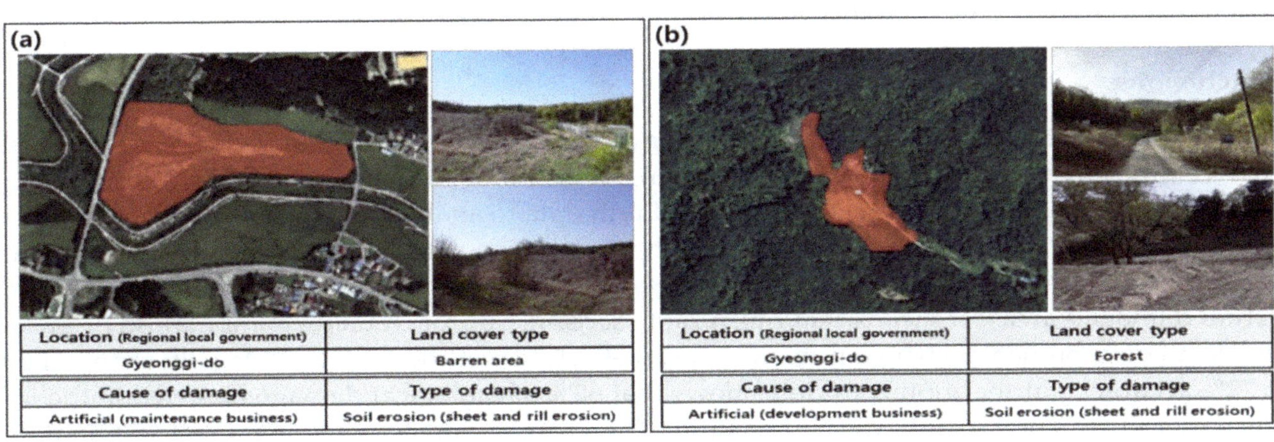

Figure 8. Examples of actual damage status derived through field surveys: (**a**) Damaged area where soil erosion occurred due to maintenance businesses in Gyeonggi-do; (**b**) Damaged area where soil erosion occurred due to development businesses in Gyeonggi-do.

4. Discussion

Damaged areas were derived through remote sensing using national standardized vector datasets and satellite images and verified through field surveys. Resultantly, more than 70% of the RS-derived damaged areas were found to be actual damage sites, suggesting the damage detection possibility of the methodology used here. Most of the damaged areas in South Korea were distributed in Gyeonggi-do and Gangwon-do; the areas were mostly less than 50,000 m^2. Moreover, most of the damage incurred was soil erosion due to anthropogenic causes. Soil erosion mainly occurs in forest ecosystems and agricultural ecosystems. It was confirmed that 62% of the actual damage occurred in forest ecosystems; most were damaged by anthropogenic causes.

Most of the previous studies related to remote sensing for restoration have been carried out to detect physical changes using satellite imagery targeting the area where the damage may have occurred [14–18]. However, with simple detection of physical changes, it is difficult to determine the cause, type, and legality of the damage, and there is a limit inhibiting policy use. In this study, by applying national standardized vector datasets institutionally used for change detection worldwide, such as LULC maps and EIA maps, we confirmed that it is possible to judge whether the detected damage is illegal. In addition, information necessary for the restoration project was collected through field surveys that were not conducted in previous studies. Therefore, here, a database of the current status of the illegally damaged area distribution and damage type for each area was established. It is necessary to establish an effective restoration plan by related ministries and regional local governments using this database. It can also help promote the potential to meet the environmental conservation and economic development of the environmental offsets policy by providing sites for developing/refining/informing restoration methods [20,22]. Additionally, based on these results, additional policy proposals are needed to prevent the ecological damage caused by development and illegal human activities targeting forests.

However, legal basis and criteria for ecologically damaged areas and the degree of dataset and thematic map construction may differ by country. Even if other countries' data is different from the data used here, it is judged that if the spatial data related to the zoning, protected area, and development activities established at the national level are used, it will be possible to derive damage in each country. The land-cover map used here has a high resolution, but the constructed time is different from the present. Thus, there are limitations in which the damaged areas are different from the current land cover; no additional damaged areas could be derived. To improve the accuracy of future research, it is necessary to utilize the latest high-resolution land-cover maps and additional legal thematic maps related to the damage of the ecological environment at the national level.

5. Conclusions

In this study, we applied two-step methodology to detect ecologically damaged areas and monitor them in South Korea for the period 2008–2017 using both remote sensing and field surveys. We used national standardized vector datasets and Google Earth data to analyze the illegally changed areas and then conducted field surveys for 2 years to verify the derived results and determine the type of damage. As a result, 288 damaged areas of the 399 damaged derived through remote sensing were found to be actual damaged areas. It means that our remote sensing approach using combination of imagery related with environmental change proved to be an effective method to detect illegally changed areas with 72% accuracy. Furthermore, our data shows that 62% of ecological damage occurred in the forest ecosystem and 66% of damage was soil erosion. Therefore, when establishing a restoration plan in South Korea in the future, it is necessary to expand the budget for ministries related to forest and to recruit soil erosion experts. Finally, we suggest that the proposed approach is a viable to tool to detect ecologically damaged areas but flexible application is required according to the situation of each damaged area, such as changing the combination of maps to be used.

Author Contributions: This paper represents a result of collaborative teamwork. Conceptualization, K.L. and S.W.J.; data curation, H.C.S., J.-Y.S. and J.H.K.; investigation, K.L. and J.-Y.S.; methodology, K.L. and Y.Y.; visualization, Y.Y., Y.K. and J.H.K.; writing-original draft, K.L.; filed survey, K.L., H.C.S., J.-Y.S., Y.Y., Y.K., J.H.K. and S.W.J.; project administration, supervision, and funding acquisition, S.W.J. All authors have read and agreed to the published version of the manuscript.

Appendix A

Table A1. Status of urban development projects in South Korea (2008–2017).

Division	Total		Ongoing		Finished	
	Number	Area (1000 m2)	Number	Area (1000 m2)	Number	Area (1000 m2)
Seoul *	10	9858	10	9858	-	-
Busan *	10	1557	2	1303	8	254
Daegu *	13	859	9	709	4	150
Incheon *	29	10,628	19	7305	10	3323
Gwangju *	13	2282	8	2085	5	197
Daejeon *	14	4412	8	949	6	3463
Ulsan *	11	4379	9	4043	2	336
Sejong *	1	232	1	232	-	-
Gyeonggi-do	138	47,278	104	38,460	34	8818
Gangwon-do	17	2283	10	1730	7	553
Chungcheongbuk-do	14	3296	12	2998	2	298
Chungcheongnam-do	54	13,634	25	6165	29	7469
Jeollabuk-do	10	8670	5	5038	5	3632
Jeollanam-do	20	7952	12	6456	8	1496
Gyeongsangbuk-do	40	10,668	24	8634	16	2034
Gyeongsangnam-do	51	16,424	26	11,042	25	5382
total (column)	445	144,412	284	107,007	161	37,405

*: metropolitan city in South Korea.

Appendix B

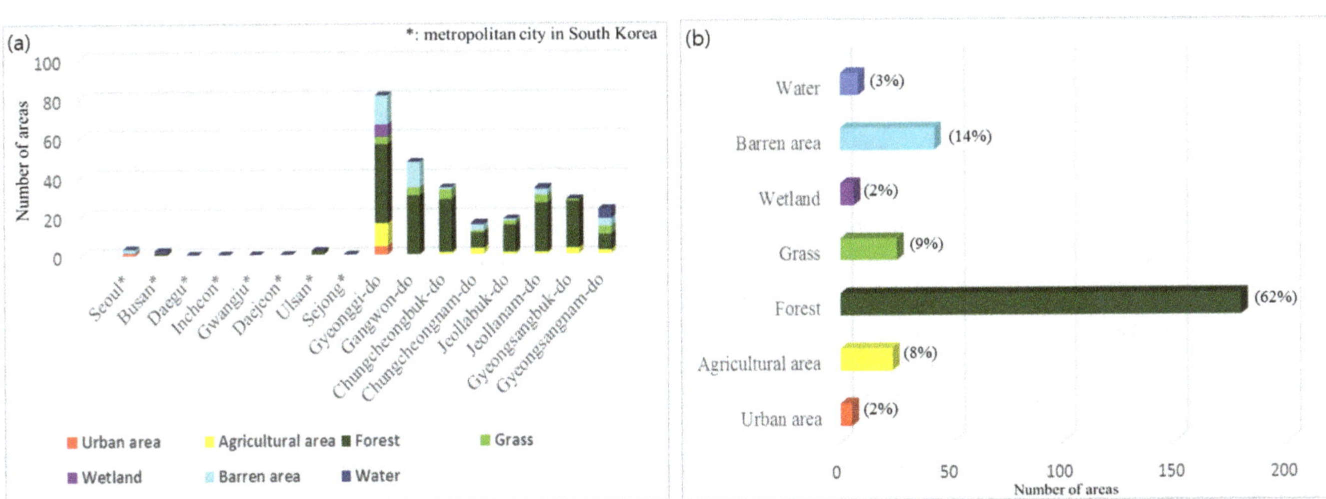

Figure A1. Distribution status of damaged areas in South Korea by ecosystem. (**a**) Distribution of damaged areas by ecosystem for each regional local government and (**b**) total distribution of damaged areas by ecosystem in South Korea.

References

1. Fereidoun, H.; Nourddin, M.S.; Rreza, N.A.; Mohsen, A.; Ahmad, R.; Pouria, H. The effect of long-term exposure to particulate pollution on the lung function of Teheranian and Zanjanian students. *Pak. J. Physiol.* **2007**, 3.
2. Reed, M.S.; Dougill, A.J.; Taylor, M.J. Integrating local and scientific knowledge for adaptation to land degradation: Kalahari rangeland management options. *Land Degrad. Dev.* **2007**, *18*, 249–268. [CrossRef]
3. Costanza, R.; De Groot, R.; Sutton, P.; Ploeg, S.; Anderson, S.; Kubiszewski, I.; Farber, S.; Turner, R.K. Changes in the global value of ecosystem services. *Glob. Environ. Chang.* **2014**, *26*, 152–158. [CrossRef]
4. Hobbs, R.J.; Cramer, V.A. Restoration ecology: Interventionist approaches for restoring and maintaining ecosystem function in the face of rapid environmental change. *Annu. Rev. Environ. Resour.* **2008**, *33*, 39–61. [CrossRef]
5. Aradóttir, Á.L.; Petursdottir, T.; Halldorsson, G.; Svavarsdottir, K.; Arnalds, O. Drivers of ecological restoration: Lessons from a century of restoration in Iceland. *Ecol. Soc.* **2013**, *18*, 14. [CrossRef]
6. Luuppala, L. *Ecological Restoration: Conceptual Analysis and Ethical Implications*; HELDA—University of Helsinki: Helsinki, Finland, 2015.
7. Ministry of Environment. *Research for the Systematic Restoration of the Damaged Natural Environment*; Republic of Korea, Ministry of Environment: Sejong City, Korea, 2011. (In Korean)
8. Suding, K.N. Toward an era of restoration in ecology: Successes, failures, and opportunities ahead. *Annu. Rev. Ecol. Evol. Syst.* **2011**, *42*, 465–487. [CrossRef]
9. Zhang, D.; Jia, Q.; Wang, P.; Zhang, J.; Hou, X.; Li, X.; Li, W. Analysis of spatial variability in factors contributing to vegetation restoration in Yan'an, China. *Ecol. Indic.* **2020**, *113*, 1062–1078. [CrossRef]
10. DeLuca, T.H.; Aplet, G.H.; Wilmer, B.; Burchfield, J. The unknown trajectory of forest restoration: A call for ecosystem monitoring. *J. For.* **2010**, *108*, 288–295.
11. Hooper, M.J.; Glomb, S.J.; Harper, D.D.; Hoelzle, T.B.; McIntosh, L.M.; Mulligan, D.R. Integrated risk and recovery monitoring of ecosystem restorations on contaminated sites. *Integr. Environ. Assess. Manag.* **2016**, *12*, 284–295. [CrossRef]
12. Nagendra, H.; Lucas, R.; Honrado, J.P.; Jongman, R.H.; Tarantino, C.; Adamo, M.; Mairota, P. Remote sensing for conservation monitoring: Assessing protected areas, habitat extent, habitat condition, species diversity, and threats. *Ecol. Indic.* **2013**, *33*, 45–59. [CrossRef]
13. Lu, D.; Mausel, P.; Brondizio, E.; Moran, E. Change detection techniques. *Int. J. Remote Sens.* **2004**, *25*, 2365–2401. [CrossRef]
14. Bellón, B.; Blanco, J.; Vos, A.D.; Roque, F.D.O.; Pays, O.; Renaud, P.C. Integrated Landscape Change Analysis of Protected Areas and their Surrounding Landscapes: Application in the Brazilian Cerrado. *Remote Sens.* **2020**, *12*, 1413. [CrossRef]
15. Kintz, D.B.; Young, K.R.; Crews-Meyer, K.A. Implications of land use/land cover change in the buffer zone of a national park in the tropical Andes. *Environ. Manag.* **2006**, *38*, 238–252. [CrossRef] [PubMed]
16. Reif, M.K.; Theel, H.J. Remote sensing for restoration ecology: Application for restoring degraded, damaged, transformed, or destroyed ecosystems. *Integr. Environ. Assess. Manag.* **2017**, *13*, 614–630. [CrossRef]
17. Scharsich, V.; Mtata, K.; Hauhs, M.; Lange, H.; Bogner, C. Analysing land cover and land use change in the Matobo National Park and surroundings in Zimbabwe. *Remote Sens. Environ.* **2017**, *194*, 278–286. [CrossRef]
18. Szantoi, Z.; Brink, A.; Buchanan, G.; Bastin, L.; Lupi, A.; Simonetti, D.; Mayaux, P.; Peedell, S.; Davy, J. A simple remote sensing-based information system for monitoring sites of conservation importance. *Remote Sens. Ecol. Conserv.* **2016**, *2*, 16–24. [CrossRef]
19. Willis, K.S. Remote sensing change detection for ecological monitoring in United States protected areas. *Biol. Conserv.* **2015**, *182*, 233–242. [CrossRef]
20. Bull, J.W.; Suttle, K.B.; Gordon, A.; Singh, N.J.; Milner-Gulland, E.J. Biodiversity offsets in theory and practice. *Oryx* **2013**, *47*, 369–380. [CrossRef]
21. Martin, N.; Evans, M.; Rice, J.; Lodhia, S.; Gibbons, P. Using offsets to mitigate environmental impacts of major projects: A stakeholder analysis. *J. Environ. Manag.* **2016**, *179*, 58–65. [CrossRef]
22. CSBI (Cross Sector Biodiversity Initiative). *A Cross-Sector Guide for Implementing the Mitigation Hierarchy*; Biodiversity Consultancy: Cambridge, UK, 2015.

23. Rhodes, C.J.; Henrys, P.; Siriwardena, G.M.; Whittingham, M.J.; Norton, L.R. The relative value of field survey and remote sensing for biodiversity assessment. *Methods Ecol. Evol.* **2015**, *6*, 772–781. [CrossRef]
24. Yoo, C.; Im, J.; Cho, D.; Yokoya, N.; Xia, J.; Bechtel, B. Estimation of All-Weather 1 km MODIS Land Surface Temperature for Humid Summer Days. *Remote Sens.* **2020**, *12*, 1398. [CrossRef]
25. Environmental Geographic Information System. Available online: http://egis.me.go.kr (accessed on 13 August 2018).
26. Cohen, W.B.; Goward, S.N. Landsat's role in ecological applications of remote sensing. *Bioscience* **2004**, *54*, 535–545. [CrossRef]
27. Natoli, S.J. Zoning and the development of urban land use patterns. *Econ. Geogr.* **1971**, *47*, 171–184. [CrossRef]
28. Xia, Y.; Liu, D.; Liu, Y.; He, J.; Hong, X. Alternative zoning scenarios for regional sustainable land use controls in China: A knowledge-based multi objective optimisation model. *Int. J. Environ. Res. Public Health* **2014**, *11*, 8839–8866. [CrossRef] [PubMed]
29. Qian, Z. Without zoning: Urban development and land use controls in Houston. *Cities* **2010**, *27*, 31–41. [CrossRef]
30. Rossi-Hansberg, E. Optimal urban land use and zoning. *Rev. Econ. Dyn.* **2004**, *7*, 69–106. [CrossRef]
31. Xu, X.; Hou, L.; Lin, H.; Liu, W. Zoning of sustainable agricultural development in China. *Agric. Syst.* **2006**, *87*, 38–62. [CrossRef]
32. Ministry of Land, Infrastructure and Transport. *National Land Planning and Utilization Act*; Republic of Korea, Ministry of Land, Infrastructure and Transport: Sejong City, Korea, 2017.
33. National Spatial Data Information Portal. Available online: http://www.nsdi.go.kr/ (accessed on 14 August 2018).
34. Li, J.C. *Environmental Impact Assessments in Developing Countries: An Opportunity for Greater Environmental Security*; USAID, FESS: Washington, DC, USA, 2008.
35. El-Gafy, M.A.; Abdelrazig, Y.A.; Abdelhamid, T.S. Environmental impact assessment for transportation projects: Case study using remote-sensing technology, geographic information systems, and spatial modeling. *J. Urban Plan. Dev.* **2011**, *137*, 153–158. [CrossRef]
36. Moufaddal, W.M. Use of satellite imagery as environmental impact assessment tool: A case study from the NW Egyptian Red Sea coastal zone. *Environ. Monit. Assess.* **2005**, *107*, 427–452. [CrossRef]
37. Saroglu, E.; Bektas, F.; Dogru, A.O.; Ormeci, C.; Musaoglu, N.; Kaya, S. Environmental impact analyses of quarries located on the Asian side of Istanbul using remotely sensed data. In Proceedings of the XXII International Cartographic Conference (ICC), Corunna, Spain, 3–8 July 2011.
38. Ministry of Environment. *Environmental Impact Assessment Act*; Republic of Korea, Ministry of Environment: Sejong City, Korea, 2018.
39. Environmental Impact Assessment Support System. Available online: https://www.eiass.go.kr (accessed on 18 August 2018).
40. Pulighe, G.; Baiocchi, V.; Lupia, F. Horizontal accuracy assessment of very high resolution Google Earth images in the city of Rome, Italy. *Int. J. Digit. Earth* **2016**, *9*, 342–362. [CrossRef]
41. Eskandari, S.; Reza Jaafari, M.; Oliva, P.; Ghorbanzadeh, O.; Blaschke, T. Mapping Land Cover and Tree Canopy Cover in Zagros Forests of Iran: Application of Sentinel-2, Google Earth, and Field Data. *Remote Sens.* **2020**, *12*, 1912. [CrossRef]
42. Benker, S.C.; Langford, R.P.; Pavlis, T.L. Positional accuracy of the Google Earth terrain model derived from stratigraphic unconformities in the Big Bend region, Texas, USA. *Geocarto Int.* **2011**, *26*, 291–303. [CrossRef]
43. Ministry of Environment. *Natural Environment Conservation Act*; Republic of Korea, Ministry of Environment: Sejong City, Korea, 2018.
44. Ministry of Environment. *Enforcement Decree of The Environmental Impact Assessment Act*; Republic of Korea, Ministry of Environment: Sejong City, Korea, 2019.
45. Victoria Resources Online (VRO). Available online: http://vro.agriculture.vic.gov.au (accessed on 28 September 2018).
46. MOLIT Statistics System. Available online: http://stat.molit.go.kr (accessed on 5 April 2020).
47. Lee, C.J.; Yoo, N.J. A Study on Devris Flow Landslide Disasters and Restoration at Inje of Kangwon Province, Korea. *J. Korean Soc. Hazard Mitig.* **2009**, *9*, 99–106.
48. Zhang, Q.; Zhang, T.; Liu, X. Index system to evaluate the quarries ecological restoration. *Sustainability* **2018**, *10*, 619. [CrossRef]

49. Pimentel, D.; Kounang, N. Ecology of soil erosion in ecosystems. *Ecosystems* **1998**, *1*, 416–426. [CrossRef]
50. Aulakh, M.S.; Sidhu, G.S. Soil degradation in India: Causes, major threats, and management options. In Proceedings of the MARCO symposium, Tsukuba, Japan, 26–28 August 2015; pp. 151–156.
51. Elmqvist, T.; Zipperer, W.G.; Guneralp, B. Urbanization, habitat loss, biodiversity decline: Solution pathways to break the cycle. In *Routledge Handbook of Urbanization and Global Environmental Change*; Seta, K., Solecki, W., Griffith, D., Corrie, A., Eds.; Routledge: London, UK, 2016; pp. 139–151.

3

Hypergraph Embedding for Spatial-Spectral Joint Feature Extraction in Hyperspectral Images

Yubao Sun, Sujuan Wang, Qingshan Liu *, Renlong Hang and Guangcan Liu

Jiangsu Key Laboratory of Big Data Analysis Technology, Jiangsu Collaborative Innovation Center on Atmospheric Environment and Equipment Technology, Nanjing University of Information Science and Technology, Nanjing 210044, China; sunyb@nuist.edu.cn (Y.S.); SuJuanWang_2018@163.com (S.W.); renlong_hang@163.com (R.H.); gcliu@nuist.edu.cn (G.L.)

* Correspondence: qsliu@nuist.edu.cn.

Academic Editors: Qi Wang, Nicolas H. Younan, Carlos López-Martínez, Xiaofeng Li and Prasad S. Thenkabail

Abstract: The fusion of spatial and spectral information in hyperspectral images (HSIs) is useful for improving the classification accuracy. However, this approach usually results in features of higher dimension and the curse of the dimensionality problem may arise resulting from the small ratio between the number of training samples and the dimensionality of features. To ease this problem, we propose a novel algorithm for spatial-spectral feature extraction based on hypergraph embedding. Firstly, each HSI pixel is regarded as a vertex and the joint of extended morphological profiles (EMP) and spectral features is adopted as the feature associated with the vertex. A hypergraph is then constructed by the K-Nearest-Neighbor method, in which each pixel and its most K relevant pixels are linked as one hyperedge to represent the complex relationships between HSI pixels. Secondly, the hypergraph embedding model is designed to learn a low dimensional feature with the reservation of geometric structure of HSI. An adaptive hyperedge weight estimation scheme is also introduced to preserve the prominent hyperedges by the regularization constraint on the weight. Finally, the learned low-dimensional features are fed to the support vector machine (SVM) for classification. The experimental results on three benchmark hyperspectral databases are presented. They highlight the importance of spatial–spectral joint features embedding for the accurate classification of HSI data. The weight estimation is better for further improving the classification accuracy. These experimental results verify the proposed method.

Keywords: feature extraction; hypergraph learning; morphological profiles; hyperedge weight estimation

---

1. Introduction

Hyperspectral imaging is an important mode of remote sensing imaging, which has been widely used in a diverse range of applications, including environment monitoring, urban planning, precision agriculture, geological exploration, etc. [1–3]. Most of these applications depend on the key problem of classifying the image pixels within hyperspectral imagery (HSI) into multiple categories, i.e., HSI classification, and extensive research efforts have been focused on this problem [4–9].

In HSI, each pixel contains hundreds of spectral bands from the visible to the infrared range of the electromagnetic spectrum. In general, the spectral signature of each pixel can be directly used as the feature for classification. However, due to the noise corruption and high correlation between spectral bands, the using of the spectral feature alone is often unable to obtain good classification results. It is well accepted that the HSI pixels within a small spatial neighborhood are often made up of the same materials. Thus, spatial contextual information is also useful for classification [10,11]. Landgrebe and Ketting proposed the well-known extraction and classification of homogeneous objects (ECHO) approach that partitioned the HSI pixels into homogeneous object and classified homogeneous object as different categories [12]. Later, Markov random field (MRF) modeling was widely adopted to capture the interpixel dependency through the neighbor system [13,14]. However, the optimization of MRF-based methods is very time-consuming. Due to the high dimensionality of HSI data, the computationally effective algorithm is desirable. In this sense, Pesaresi and Benediktsson [15] proposed the use of morphological transformations to build a morphological profile (MP) for extracting the structural information. Palmason et al. [16] extended the method proposed in [15] to the high-resolution hyperspectral data classification. They first extracted several principal components of the hyperspectral data. Then, the MP is constructed based on each selected principal component. At last, all MPs are jointed as extended MP (EMP), which is input into a neural network for classification. However, EMP was primarily designed for classification of urban structures and it did not fully utilize the spectral information in the data. Regrading this issue, Fauvel et al. [17] proposed fusing the morphological information and the original hyperspectral data, i.e., the two vectors of attributes are concatenated into one feature vector. The final classification is achieved by using a support vector machine classifier. Many other spectral and spatial joint features [18–22], such as 3D wavelet [18], spatial and spectral kernel [19], matrix-based discriminant subspace analysis [20], etc. are used for classification.

These joint features usually have a high dimension. In order to avoid the Hughes phenomenon, feature extraction and dimensionality reduction must be conducted before classification. Principal component analysis (PCA) and Fisher's linear discriminant analysis (LDA) [23] are two simple and effective approaches for dimension reduction. PCA aims at projecting the data along the directions of maximal variance. LDA is designed to generate the optimal linear projection matrix by maximizing the between-class distance while minimizing the within-class distance. Apart from these linear methods, many nonlinear versions have been developed, such as kernel PCA [24] and kernel LDA [25]. Some other feature extraction techniques have also been proposed, e.g., locality preserving projection (LPP) [26], independent component analysis (ICA) [27,28], and locally linear embedding (LLE) [29]. In particular, Yan et al. [30] proposed a general graph embedding (GE) model that seamlessly includes many existing feature extraction techniques. In this GE model, each data point is visualized as a vertex and a pairwise edge is used to represent the association relationship between two data points. They consider each feature extraction algorithm as an undirected weighted graph that describes geometric structures of data. GE algorithms have been widely explored for dimension reduction of HSI. Besides the geometric structures of data, sparsity is also explored to construct the graph embedding model. Luo et al. proposed constructing a graph with the sparse coefficients that reveals the sparse properties of data, and the transformation matrix is obtained for feature reduction [31]. In addition, by regarding different band sets as different views of land covers, multiview graph ensemble-based graph embedding is also utilized to promote the performance of graph embedding for hyperspectral image classification [32].

A hypergraph is a generalization of a pairwise graph. Different from pairwise graphs, each edge in a hypergraph is capable of connecting more than two vertices [33]. Thus, the complex relationships of the dataset can be captured by a hypergraph, and hypergraphs have been gaining more and more attention in recent years. Bu et al. [34] presented a hypergraph learning based music recommendation method with the use of hyperedges to exploit the complex social media information. A hypergraph semi-supervised learning model [35] was also proposed for image classification. Yuan et al. [36] utilized a hypergraph embedding model for HSI feature reduction, in which the spatial hypergraph models (SHs) are construed by selecting the K-nearest neighbors within the spatial region of the centroid pixel. Experimental results demonstrated that SH outperformed many existing feature extract methods for HSI classification, including raw spectral feature (RAW), PCA, LPP, LDA, nonparametric weighted feature extraction (NWFE) [37] and semi-supervised local discriminant analysis (SELD) [38]. However, SH is designed to learn the projection matrix for reducing the spectral feature. The spatial structure is not exploited for hypergraph embedding, which is not capable of simultaneously extracting the spectral-spatial features. Furthermore, the hyperedge weight is computed in advance and fixed in the hypergraph embedding procedure. As the discussion stated in [39,40], all of the hyperedges do not have the same effect on the learning procedure. Some hyperedges are not as informative as others. The hypergraph embedding should be enhanced by estimating the hyperedge weights adaptively.

In order to cope with these issues, we propose a novel algorithm for HSI spatial-spectral joint feature extraction. We combine the EMP and spectral features and adopt the KNN method to construct a hypergraph, where each sample and its K nearest neighbors are enclosed in one hyperedge. Similar to [36], a linear projection matrix $\mathbf{P}$ can be learnt by solving the hypergraph embedding model. However, in [36], the hyperedges' weights in the hypergraph embedded model are fixed. Inspired by [39,40], we introduce a scheme to update the weights adaptively to preserve the prominent hyperedge and further learn the low-dimensional structure. It helps improve the accuracy of the final HSI classification to a certain extent. Finally, the leaned low-dimensional features are fed to the SVM for classification. The flowchart of the proposed method is shown in Figure 1. Experiments conducted on three widely used types of HSI demonstrate that the proposed method achieves superior performance over many other feature extract methods for HSI classification.

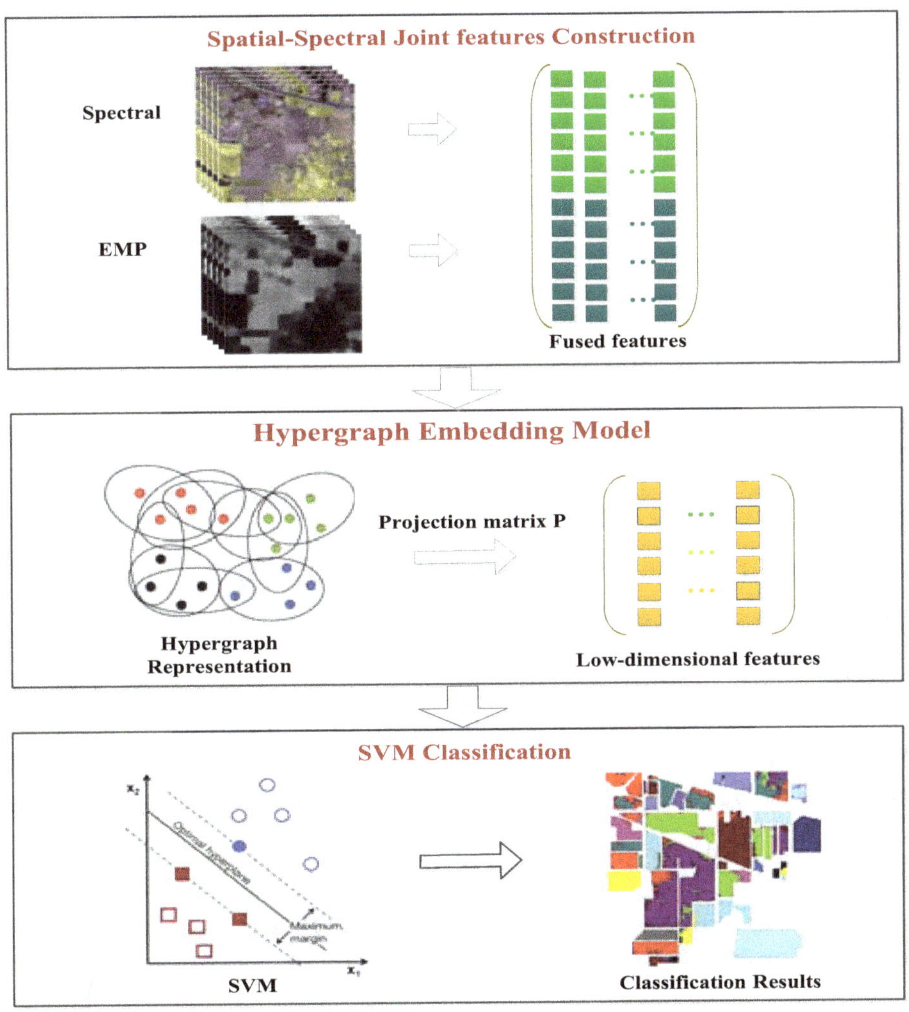

Figure 1. The flowchart of the proposed method.

2. Hypergraph Model

Denote a hypergraph as $\mathbf{G} = (\mathbf{V},\mathbf{E},\mathbf{W})$, which consists of a set of vertices $\mathbf{V}$, a family of hyperedge $\mathbf{E}$ and a weight matrix $\mathbf{W}$ of hyperedges. Different from pairwise graphs (For convenience, we call it a simple graph in the following), every hyperedge e_i can contain multiple vertices and is assigned a weight $w(e_i)$. As shown in Figure 2b, hyperedge e_1 is composed of vertices v_1, v_2 and v_3. e_2 is composed of vertices v_3 and v_4. e_3 is composed of vertices v_4, v_5, v_6 and v_7. $\mathbf{W}$ is a diagonal matrix of the hyperedge weights. The connection relationship of hypergraph **G** can be represented by an incidence matrix $\mathbf{H} \in \mathbf{R}^{|\mathbf{V}| \times |\mathbf{E}|}$, which can be defined as:

$$H_{ij} = \mathbf{H}(v_i, e_j) = \begin{cases} 1, & \text{if } v_i \in e_j, \\ 0, & \text{if } v_i \notin e_j. \end{cases} \tag{1}$$

The degree of vertex v and hyperedge e can be respectively represented as:

$$d(v_i) = \sum_{e_j \in \mathbf{E}} w(e)\mathbf{H}(v_i, e_j), \tag{2}$$

$$\delta(e_j) = \delta_j = \sum_{v_i \in \mathbf{V}} \mathbf{H}(v_i, e_j). \tag{3}$$

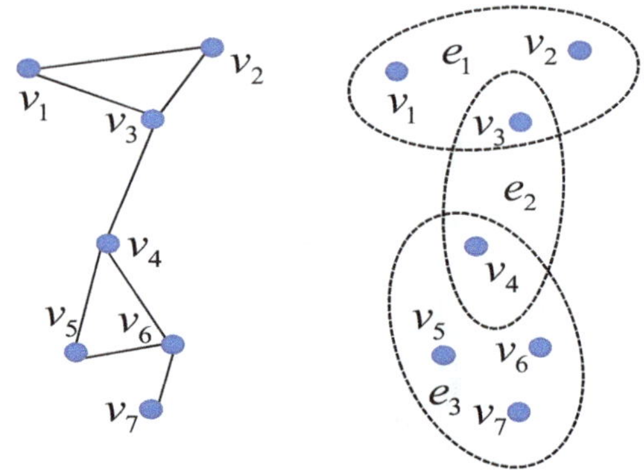

(a) Simple graph (b) Hypergraph **G**

	e_1	e_2	e_3
v_1	1	0	0
v_2	1	0	0
v_3	1	1	0
v_4	0	1	1
v_5	0	0	1
v_6	0	0	1
v_7	0	0	1

(c) Incidence matrix **H**

Figure 2. The example of graph and hypergraph (**a**) simple graph, each edge consists of only two data points; (**b**) hypergraph **G**, each hyperedge is marked by an ellipse and consists of at least two data points; (**c**) taking the seven vertices as example, **H** is the incidence matrix of **G**, whose values are usually binary.

According to the above definition, the main difference between hypergraphs and simple graphs is that every hyperedge can link more than two vertexes. Therefore, hypergraph is suitable to represent local group information and the high-order relationship of data. For example, considering seven vertices in Figure 2b, they are attributed to three groups and the corresponding incidence matrix is shown in Figure 2c. In terms of building a simple graph with these seven data points, the complex relations within the group are broken into multiple pairwise links. Some valuable information may be lost in this procedure; therefore, a simple graph can not describe the group structure well.

3. Hypergraph Embedding of Spatial-Spectral Joint Features

As shown in Figure 1, our algorithm mainly consists of three steps: spatial-spectral joint feature construction, hypergraph embedding and SVM classification.

3.1. Spatial-Spectral Joint Feature Construction

Following [16], we first extract several PCs from the original HSI $\mathbf{I}(x)$ and then build an MP from each of the PCs:

$$\mathbf{MP}(x) = \left\{ \mathbf{CP}_n(x), \ldots, \mathbf{I}(x), \ldots, \mathbf{OP}_n(x) \right\}, \tag{4}$$

where n is the number of the circular structural element (SE) with different radius sizes, $\mathbf{OP}_n(x)$ and $\mathbf{CP}_n(x)$ are the opening profile (OP) and the closing profile (CP) at the pixel x with an SE of a size n, respectively. Specifically, we have $\mathbf{CP}_0(x) = \mathbf{OP}_0(x) = \mathbf{I}(x)$. The MP of $\mathbf{I}$ contains the original image $\mathbf{I}$,

n opening profile and n closing profile. Therefore, each MP is a $(2n+1)$-dimensional vector. Finally, all MPs are stacked together in one as EMP:

$$\mathbf{EMP}(x) = \left\{ \mathbf{MP}_{\mathrm{PC}^1}(x), \mathbf{MP}_{\mathrm{PC}^2}(x), \ldots, \mathbf{MP}_{\mathrm{PC}^m}(x) \right\}, \tag{5}$$

where m represents the number of PCs. The EMP is defined as an $m(2n+1)$-dimensional vector.

After obtaining the EMP feature, we represent the spatial and spectral joint feature of the i-th HSI pixel as

$$v_i = \begin{bmatrix} x_i \\ \mathrm{EMP}(x_i) \end{bmatrix} \in R^{\mathrm{m(2n+1)}+d}, \tag{6}$$

where d is the number of the spectral bands. Denote the spectral features matrix of HSI as $\mathbf{X} = [\boldsymbol{x}_1, \boldsymbol{x}_2, \ldots, \boldsymbol{x}_N] \in \mathbf{R}^{d \times N}$, EMP matrix of HSI as $\mathbf{EMP} = [\mathrm{EMP}(x_1), \cdots, \mathrm{EMP}(x_N)]$, where x_i is the i-th pixel, and N is the number of HSI pixels. Then, the joint feature matrix of HSI can be represented as: $\mathbf{V} = \begin{bmatrix} \mathbf{X} \\ \mathbf{EMP} \end{bmatrix} \in R^{(\mathrm{m(2n+1)}+d) \times N}$.

3.2. Hypergraph Embedding

We take each pixel of HSI as a vertex and construct a hypergraph $\mathbf{G} = (\mathbf{V}, \mathbf{E}, \mathbf{W})$ to represent the correlation between HSI pixels. Each vertex v_i is associated with the spatial and spectral joint feature defined in Equation (6). The hypergraph G is constructed by the K-nearest neighbor method. In detail, each pixel $\boldsymbol{v}_i$ and its K nearest neighbors are enclosed as hyperedge e_i. Thus, hyperedge set $\mathbf{E} = \{e_1, e_2, \ldots, e_N\}$ contains N hyperedges. Meanwhile, the weight $w(e_i)$ of hyperedge e_i is defined as:

$$w(e_i) = \sum_{v_i, v_j \in e_i} \exp\left(-\frac{\left\| v_j - v_i \right\|_2^2}{2\sigma^2} \right), \tag{7}$$

where σ is the mean distance between all vertices and can be calculated by $\sigma = \frac{1}{N^2} \sum_i \sum_j d(\boldsymbol{v}_i, \boldsymbol{v}_j)$, $d(\boldsymbol{v}_i, \boldsymbol{v}_j)$ is the distance between vertex $\boldsymbol{v}_i$ and vertex $\boldsymbol{v}_j$. The degree of vertex $\boldsymbol{v}_i$ and the degree of hyperedge e_i can be computed by Equations (2) and (3), respectively. Based on this definition, the more "compact" hyperedge (local group) is assigned with a higher weight.

Denote $\mathbf{D}_v$ and $\mathbf{D}_e$ as two diagonal matrices of the vertex degrees and the hyperedge degrees, respectively, and $\mathbf{P} \in \mathbf{R}^{(m(2n+1)+d) \times u}$ (generally, $m(2n+1)+d >> u$) as the linear projection matrix. The objective of hypergraph embedding model is to learn the projection matrix $\mathbf{P}$ for reducing the feature dimension with the preservation of geometric property in the original space. The objective function is formulated as:

$$\begin{aligned} &\min_{\mathbf{P}^T \mathbf{V} \mathbf{D}_v \mathbf{V}^T \mathbf{P}=1} \frac{1}{2} \sum_{e \in \mathbf{E}} \sum_{v_i, v_j \in e} \frac{w(e) h(v_i, e) h(v_j, e)}{\delta(e)} \left\| \mathbf{P}^T v_i - \mathbf{P}^T v_j \right\|_2^2 \\ &= \frac{1}{2} \sum_{k=1}^{N} \sum_{i,j=1}^{N} \frac{w_k h_{ik} h_{jk}}{\delta_k} \left\| \mathbf{P}^T v_i - \mathbf{P}^T v_j \right\|_2^2 \\ &= \mathrm{trace}\left(\mathbf{P}^T \mathbf{V} \mathbf{L} \mathbf{V}^T \mathbf{P} \right), \end{aligned} \tag{8}$$

where $\mathbf{L} = \mathbf{D}_v - \mathbf{H}\mathbf{W}\mathbf{D}_e^{-1}\mathbf{H}^T$ is the hypergraph laplacian matrix. The constraint $\mathbf{P}^T \mathbf{V} \mathbf{D}_v \mathbf{V}^T \mathbf{P} = 1$ is used for scale normalization of the low-dimensional representations. This objective function induces the constraint that if v_i and v_j are similar and belong to the same hyperedge, they should also be adjacent in embedded space. In addition, an efficient hypergraph weight estimation scheme is proposed to preserve the prominent hyperedges. Assuming that $\boldsymbol{w} = (w_1, w_2, \ldots, w_N)^T$ is composed of the elements lying in the main diagonal of $\mathbf{W}$, we enforce $\mathbf{1}_N^T w = 1$ and add an l_2 norm regularizer on w. Then, our proposed embedding model is finally defined as:

$$\{\mathbf{P}^*, w^*\} = \arg \min_{\mathbf{P}^T\mathbf{V}\mathbf{D}_v\mathbf{V}^T\mathbf{P}=1} \left\{ \text{trace}\left(\mathbf{P}^T\mathbf{V}\mathbf{L}\mathbf{V}^T\mathbf{P}\right) + \lambda\|w\|^2 \right\} \quad \text{s.t. } \mathbf{1}_N^T w = 1. \tag{9}$$

3.3. Optimization Algorithm

The objective function Equation (9) is a multiple variables optimization problem, and it is non-convex with respect to w and $\mathbf{P}$ jointly. However, it is convex with either of them individually when the other is fixed. Thus, an alternative iteration strategy is adopted to get the solution of Equation (9). We first initialize w according to Equation (7). With w fixed, we optimize $\mathbf{P}$ according to Equation (8). The solution of Equation (8) is to find the eigenvectors corresponding to the first u largest eigenvalues of the matrix $\left(\mathbf{V}\mathbf{L}\mathbf{V}^T\right)^{-1}\left(\mathbf{V}\mathbf{D}_v\mathbf{V}^T\right)$.

Next, fix $\mathbf{P}$ and optimize w:

$$\arg\min_{w} \left\{ \text{trace}\left(\mathbf{P}^T\mathbf{V}\mathbf{L}\mathbf{V}^T\mathbf{P}\right) + \lambda\|w\|^2 \right\} \quad \text{s.t. } \mathbf{1}_N^T w = 1. \tag{10}$$

In this paper, we employ the Lagrangian algorithm to optimize the Equation (10). The Lagrangian function of the objective function (10) is defined as:

$$\begin{aligned} \psi(w, c) &= \text{trace}\left(\mathbf{P}^T\mathbf{V}\mathbf{L}\mathbf{V}^T\mathbf{P}\right) + \lambda w^T w + c\left(\mathbf{1}_N^T w - 1\right) \\ &= \frac{1}{2}\sum_{k=1}^{N}\sum_{i,j=1}^{N} \frac{w_k h_{ik} h_{jk}}{\delta_k} \left\|\mathbf{P}^T v_i - \mathbf{P}^T v_j\right\|_2^2 + \lambda w^T w + c\left(\mathbf{1}_N^T w - 1\right). \end{aligned} \tag{11}$$

The partial derivatives of ψ w.r.t. $w_i, i = 1, 2, \cdots, M$ are given by:

$$\frac{\partial\psi(w, c)}{\partial w_k} = \frac{1}{2}\sum_{i,j=1}^{N} \frac{h_{ik} h_{jk}}{\delta_k} \left\|\mathbf{P}^T v_i - \mathbf{P}^T v_j\right\|_2^2 + 2\lambda w_k + c = 0. \tag{12}$$

By simplifying Equation (12), w_k can be calculated as:

$$w_k = -\frac{\frac{1}{2}\sum_{i,j=1}^{N} \frac{h_{ik} h_{jk}}{\delta_k} \left\|\mathbf{P}^T v_i - \mathbf{P}^T v_j\right\|_2^2 + c}{2\lambda}. \tag{13}$$

According to the constraint $1_N^T w = 1$, the Lagrange multiplier can be calculated as:

$$c = -\frac{1}{N}\left[\frac{1}{2}\sum_{k=1}^{N}\sum_{i,j=1}^{N} \frac{h_{ik} h_{jk}}{\delta_k} \left\|\mathbf{P}^T v_i - \mathbf{P}^T v_j\right\|_2^2 + 2\lambda\right]. \tag{14}$$

By substituting Equation (14) into Equation (13), we can obtain w finally.

Following this iteration process, w and $\mathbf{P}$ are alternately optimized until the maximal iteration number is reached or the relative difference of objective function value of Equation (9) is smaller than a given tolerance const ε, i.e.,

$$\frac{|f(t+1) - f(t)|}{|f(t)|} \leqslant \varepsilon, \tag{15}$$

where $f(t+1)$ and $f(t)$ is the function value of Equation (9) at iteration $t+1$ and t, respectively. In addition, we can obtain the final projection matrix $\mathbf{P}^*$. At last, the joint feature set $\mathbf{V}$ is reduced as a low-dimensional feature set $\mathbf{Y} = \left[(\mathbf{P}^*)^T v_1, \ldots, (\mathbf{P}^*)^T v_N\right]$, which is then transmitted into an SVM classifier. Based on the above analysis, the proposed method can be summarized in Algorithm 1.

Algorithm 1: The proposed method (denoted as SSHG*) for HSI classification.

Input: Data matrix $\mathbf{X}$, the reduced dimensionality u, the nearest neighbors number K and regularization parameter λ.

Output: The class-label vector f.

1 Normalize all the features to [0,1].

2 Build the MP from each of the PCs: $\mathbf{MP}(x) = \{\mathbf{CP}_n(x), \ldots, \mathbf{I}(x), \ldots, \mathbf{OP}_n(x)\}$.

3 Obtain the EMP by stacking all MPs: $\mathbf{EMP}(x) = \{\mathbf{MP}_{\mathrm{PC}^1}(x), \mathbf{MP}_{\mathrm{PC}^2}(x), \ldots, \mathbf{MP}_{\mathrm{PC}^m}(x)\}$.

4 Represent the new stacked joint feature set as:
$\mathbf{V} = [\mathbf{X}; \mathbf{EMP}] = [v_1, v_2, \ldots, v_N] \in \mathbf{R}^{(m(2n+1)+d)\times N}$.

5 Compute the incidence matrix $\mathbf{H} \in \mathbf{R}^{|\mathbf{V}|\times|\mathbf{E}|}$by KNN, set $\mathbf{H}(v,e) = 1$, if $v \in e$, otherwise, $\mathbf{H}(v,e) = 0$.

6 Construct the hypergraph G and Calculate the weight of hyperedge e_i:
$w(e_i) = \sum_{v_j \in e_i} \exp\left(-\frac{\|v_j - v_i\|_2^2}{2\sigma^2}\right)$, the vertex degree: $d(v_j) = \sum_{e_i \in \mathbf{E}} w(e_i)\mathbf{H}(v_j, e_i)$, and the hyperedge degree: $\delta(e_i) = \sum_{v_j \in \mathbf{V}} \mathbf{H}(v_j, e_i)$.

7 Obtain the projection matrix $\mathbf{P}$ by optimizing Equation (8)

8 Solve Equation (10) and obtain the hyperedge weights computed as Equation (13).

9 With the new hyperedge weights, update $\mathbf{D}_v$, $\mathbf{L}$ and $\mathbf{W}$.

10 Repeat the steps 7–9 until the convergence criterion 15 is met or the maximal iteration number is reached.

11 Find the final projection matrix $\mathbf{P}^*$.

12 Project the joint feature set into a low-dimensional feature set: $\mathbf{Y} = \left[(\mathbf{P}^*)^T v_1, \ldots, (\mathbf{P}^*)^T v_N\right]$.

13 Feed the learned low-dimensional feature set $\mathbf{Y}$ into the SVM for classification.

14 **return** the class-label vector f.

4. Experiments and Discussion

4.1. Data Sets

In order to verify the performance of our proposed method, we conduct the experiments on the following three benchmark datasets.

(1) Indian Pines data set—the first data set was acquired by the AVIRIS sensor over the Indian Pines test site in Northwestern Indiana, USA. The size of the image is 145 pixels $\times$ 145 pixels with a spatial resolution of 20 m per pixel. Twenty water absorption bands (104–108, 150–163, 220) were removed, and the 200-band image is used for experiments. Sixteen classes of interest are considered.

(2) Pavia University data set—the second data set was acquired by the ROSIS sensor during a flight campaign over Pavia, northern Italy. The size of the image is 610 pixels $\times$ 340 pixels with a spatial resolution of 1.3 m per pixel. Twelve channels were removed due to noise. The remaining 103 spectral bands are processed. Nine classes of interest are considered.

(3) Botswana data set—the third data set was acquired by the NASA EO-1 satellite over the Okavango Delta, Botswana, in 2001. The size of the image is 1476 pixels $\times$ 256 pixels with a spatial resolution of 30 m per pixel. Uncalibrated and noisy bands that cover water absorption features were removed, and the remaining 145 bands are used for experiment. Fourteen classes of interest are considered.

4.2. Experimental Setting

In order to demonstrate the effectiveness of adaptive weight estimation, we implement our algorithm as two versions. One is SSHG, which only utilizes the KNN hypergraph model for dimension reduction of the stacked feature set without adaptive weight estimation. The other is SSHG* shown in Algorithm 1. They are compared with the following feature extraction methods: (1) the method by using PCA to extract spectral features (denoted as PCA); (2) the method by using EMP features without dimension reduction (denoted as EMP); (3) the method [17] stacking the EMP and the spectral features as feature without dimension reduction (denoted as EMPSpe); and (4) the spatial hypergraph embedding method proposed in [36] (denoted as SH). In order to facilitate comparisons with these competing feature extraction methods, we adopt the overall accuracy (OA), the average accuracy (AA), the per-class accuracy and Kappa coefficient (κ) to evaluate the classification performance. Furthermore, the SVM classifier with Gaussian kernel is adopted to classify all of the aforementioned feature data of these feature extraction methods. The grid search tool is used to select the parameters of the optimal penalty term and Gaussian kernel variance in SVM within the given sets $\{2^{-10}, ..., 2^{10}\}$ and $\{2^{-10}, ..., 2^{10}\}$, respectively. The one-against-all strategy is adopted for multi-class classification. Regarding the three data sets, we select 15 samples from each class randomly to form a training set and the remaining samples are used as the test set. The training sample selection and the classification process are repeated ten times to reduce the bias induced by random sampling. We retain the average results. The parameters setting of SH is the same as the original paper [36]. With respect to our algorithm, the tolerance const ε is set as 1×10^{-3} and the regularization parameter λ is set as 100. The number of nearest neighbors K is selected as 10, 15, 5 for Indian Pines, Pavia University and Botswana data sets, respectively.

4.3. Experimental Results

The classification results of various methods upon three types of HSI are reported in Tables 1–3, respectively. The best results are highlighted with bold fonts. The number in brackets corresponds to the optimal dimensionality of reduced features. Classification maps of these different approaches are shown in Figures 3–5, respectively. According to the experimental results, our proposed method achieves the highest OA, AA, and κ among all of the competing methods, which shows the effectiveness of our feature extraction algorithm. The effectiveness of our SSHG method owes much to the hypergraph embedding of spatial and spectral joint features.

Table 1. Classification accuracy of various algorithms on the Indian Pines image.

Class	PCA (25)	EMP (27)	EMPSpe (227)	SH (22)	SSHG (44)	SSHG* (44)
1	91.61	98.71	**99.03**	94.87	98.06	98.06
2	47.36	61.46	64.28	**82.59**	72.53	73.96
3	48.60	78.75	77.14	73.50	84.06	**84.85**
4	68.29	95.90	91.76	91.32	96.76	**97.21**
5	75.75	87.78	88.85	**92.12**	89.83	90.32
6	85.37	91.48	92.36	**98.22**	93.93	94.04
7	91.54	99.23	99.23	**100**	**100**	**100**
8	79.52	98.47	98.92	98.31	99.57	**99.63**
9	96.00	**100**	**100**	**100**	**100**	**100**
10	56.22	74.23	71.61	**87.51**	76.81	77.68
11	49.62	69.51	71.02	64.41	**75.65**	75.57
12	45.43	75.67	77.40	84.31	84.33	**84.79**
13	93.47	98.68	99.00	**99.49**	99.37	99.37
14	69.55	93.25	94.83	94.84	97.57	**97.58**
15	46.42	95.96	95.85	75.07	97.74	**97.76**
16	89.62	97.56	98.46	98.75	99.74	**99.87**
OA	58.90	79.14	79.88	82.33	84.36	**84.75**
AA	70.90	88.54	88.73	89.71	91.62	**91.92**
kappa	53.88	76.42	77.24	80.06	82.27	**82.73**

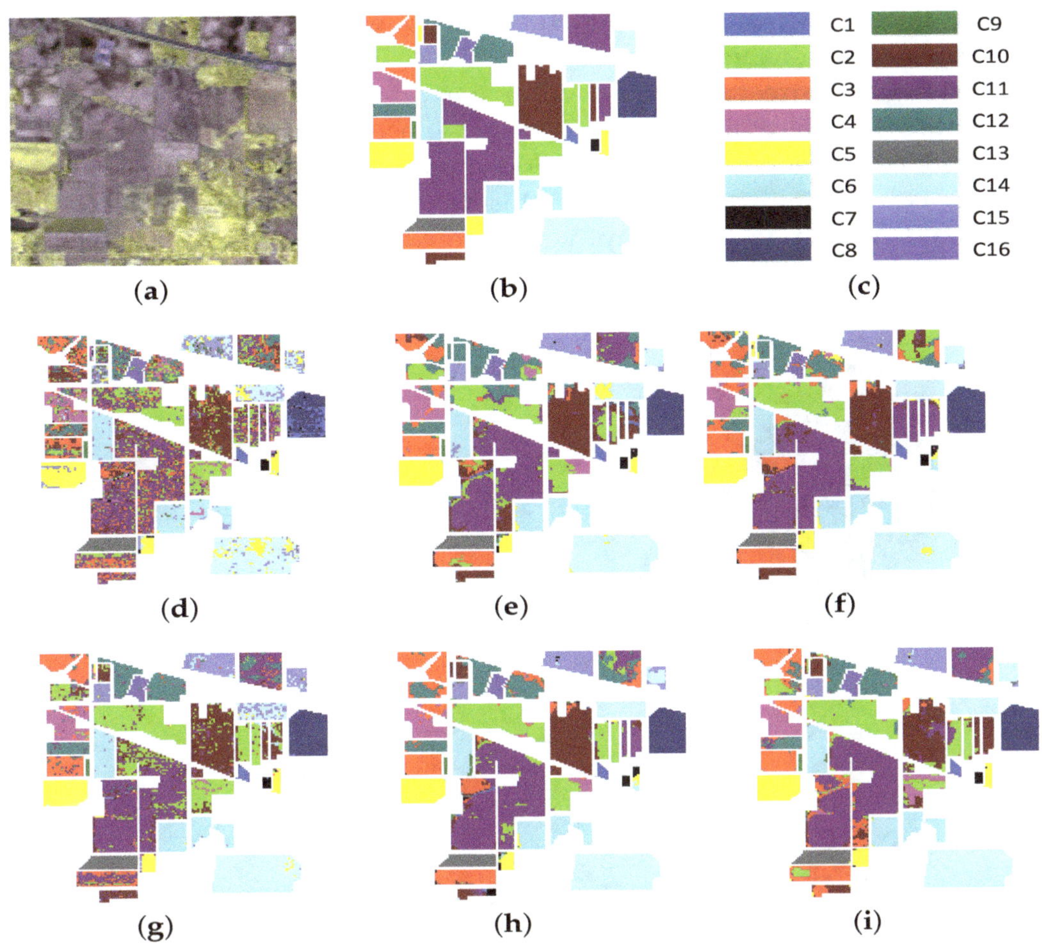

Figure 3. Indian Pines. (**a**) three-channel color composite image with bands 65, 52, 36; (**b**,**c**) ground-truth map and class labels; (**d**–**i**) classification maps of PCA, EMP, EMPSpe, SH, SSHG, SSHG*, respectively.

Table 2. Classification accuracy of various algorithms on the Pavia university image.

Class	PCA (10)	EMP (27)	EMPSpe (130)	SH (30)	SSHG (46)	SSHG* (46)
1	66.21	82.40	81.57	70.33	81.67	**82.70**
2	65.14	83.44	84.09	82.13	**92.02**	91.44
3	70.00	77.04	77.79	72.37	**80.47**	80.08
4	85.26	97.42	**97.44**	89.58	93.93	94.90
5	99.37	99.76	99.75	99.61	99.79	**99.80**
6	69.16	78.91	80.16	**91.76**	86.50	89.63
7	90.45	94.07	93.28	92.68	94.16	**94.44**
8	71.34	**86.12**	85.30	72.16	83.07	84.06
9	**99.72**	96.04	97.44	99.51	98.26	98.15
OA	70.59	84.77	85.05	81.88	89.01	**89.43**
AA	79.63	88.35	88.53	85.57	89.99	**90.58**
kappa	63.20	80.38	80.78	76.80	85.64	**86.24**

Comparing the EMP and EMPSpe method, we can find that EMPSpe method is always slightly better than EMP due to the fusion of EMP and spectral features for classification. As mentioned in [17], the stacked EMP and spectral features are transformed to low dimensional features by the decision boundary feature extraction (DBFE) and NWFE methods before classification. However, the DBFE and NWFE did not bring about the effective improvement of algorithm performance. SH utilized the hypergraph embedding model for feature reduction. Compared with PCA, the SH method has much better classification performance, which verifies the capacity of the hypergraph to capture the intrinsic complex relationships between HSI pixels. However, SH utilized only the spectral similarity for finding the nearest neighbors within a given spatial region. The superiority of SSHG over SH demonstrates that the embedding of EMP and spectral features is better for HSI classification. Specifically, our SSHG method can extract the rich spatial structures in the Pavia University data and achieve the maximum improvement upon this data. SSHG* obtains better classification results than SSHG, which demonstrates that adaptive hypergraph weight estimation is also beneficial for improving the classification accuracy.

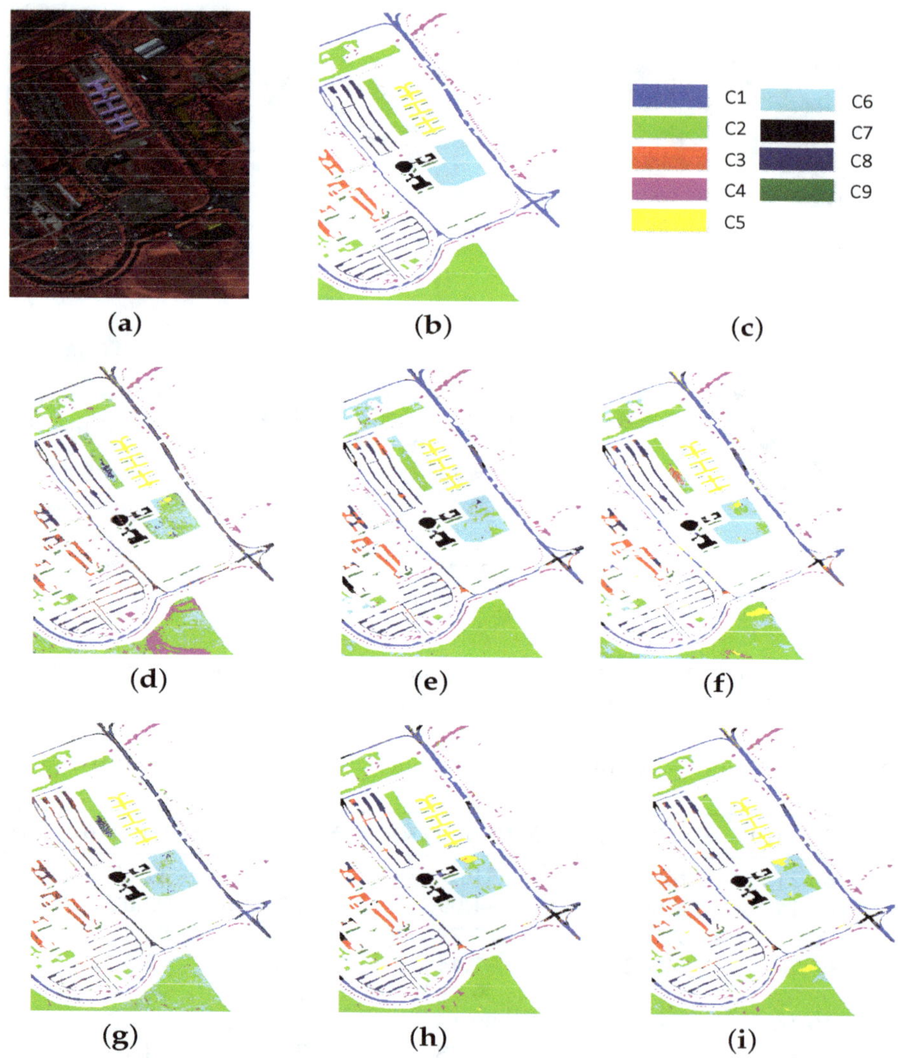

Figure 4. Pavia university. (**a**) three-channel color composite image with bands 102, 56, 31; (**b**,**c**) ground-truth map and class labels; (**d**–**i**) classification maps of PCA, EMP, EMPSpe, SH, SSHG, SSHG*, respectively.

There are two parameters, i.e., K and u, in our proposed method. The parameter K is the number of nearest neighbors, which determines how many pixels are included in the hyperedge. u is the dimensionality of the embedded low-dimensional feature. To evaluate their effects on the classification performance, we conduct the experiments on the above three datasets. We firstly fix the reduced dimensionality as $u = 40$ and evaluate the influence of different K on the OA. As seen in Figure 6, when K is set as 10, 15, 5 for Indian Pines, Pavia University and Botswana data sets, respectively, the OA achieves the highest value. Taken as a whole, $[5, 15]$ is usually a good range for the selection of parameter K. We then fix the K as 10, 15, 5 for the three datasets, respectively, and evaluate the influence of different us on the OA. Figure 7 shows the changes of OA with the reduced dimensions on three types of HSI. We can see that the inflection point of classification results is around the dimensionality 25 for these three HSIs, and there was no significant improvement on the classification results if the dimension continues to grow up.

Table 3. Classification accuracy of various algorithms on the Botswana image.

Class	PCA (22)	EMP (27)	EMPSpe (172)	SH (25)	SSHG (34)	SSHG* (34)
1	**100**	99.92	99.89	**100**	**100**	**100**
2	96.51	**100**	97.99	**100**	99.68	98.05
3	96.19	94.79	95.85	99.15	96.76	**100**
4	99.00	95.85	98.83	**99.50**	98.41	93.27
5	81.10	79.76	82.32	82.86	91.79	**96.38**
6	69.29	81.73	88.34	81.89	96.37	**99.22**
7	96.31	97.70	99.20	98.77	99.72	**99.95**
8	98.40	99.63	99.48	99.47	**100**	97.42
9	79.93	92.34	94.47	96.32	98.86	**99.79**
10	95.28	98.33	97.98	99.57	**99.92**	97.97
11	83.45	97.24	95.19	97.59	94.97	**99.88**
12	93.98	99.94	99.88	88.55	**100**	99.49
13	89.33	99.60	98.37	94.47	**99.92**	99.75
14	98.75	99.25	98.35	**100**	91.36	99.63
OA	89.83	94.69	95.65	95.10	97.79	**98.38**
AA	91.25	95.43	96.15	95.58	97.70	**98.63**
kappa	88.98	94.24	95.36	94.68	97.60	**98.24**

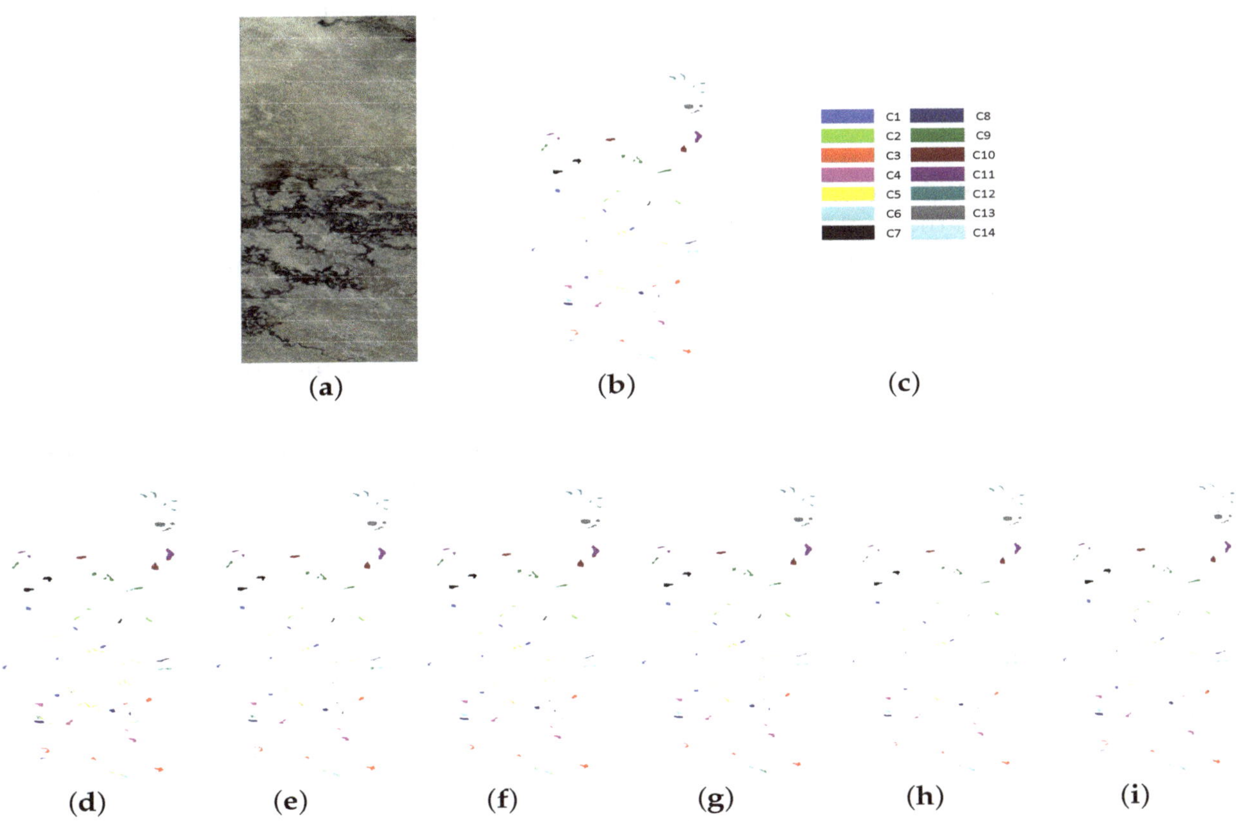

Figure 5. Botswana. (**a**) three-channel color composite image with bands 65, 52, 36; (**b**,**c**) ground-truth map and class labels; (**d**–**i**) classification maps of PCA, EMP, EMPSpe, SH, SSHG, SSHG*, respectively.

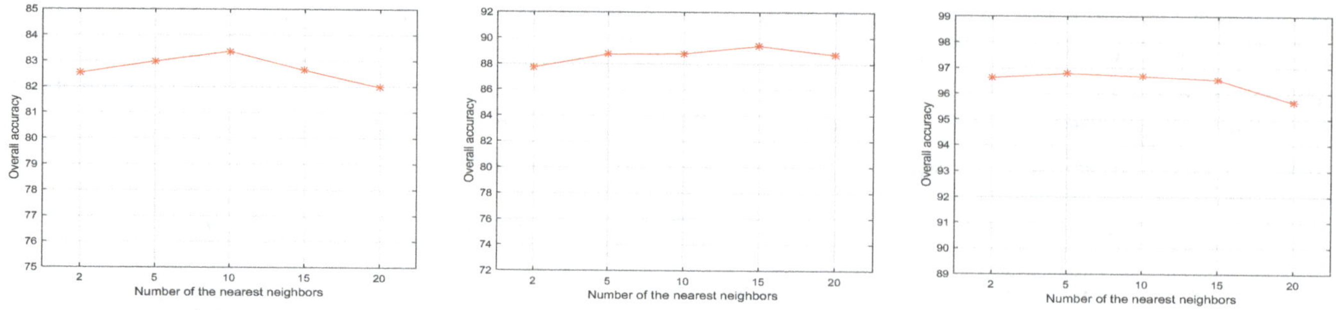

(a) (b) (c)

Figure 6. Effects of the number K of nearest neighbors on OA. (**a**) Indian Pines; (**b**) Pavia University; (**c**) Botswana.

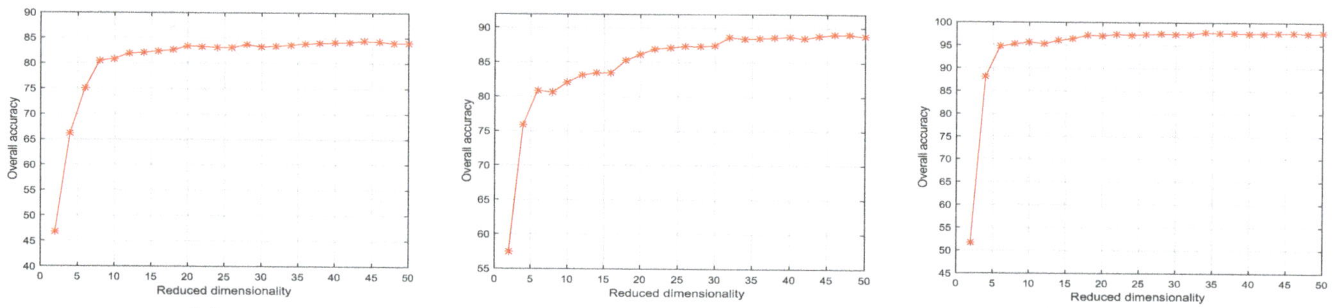

(a) (b) (c)

Figure 7. Effects on the reduced dimensions. (**a**) Indian Pines; (**b**) Pavia University; (**c**) Botswana.

5. Conclusions

In this paper, we propose a novel algorithm for spatial-spectral feature extraction based on hypergraph learning. A hypergraph is constructed by the KNN method and the embedding operation is conducted to transform the joint EMP and spectral features into the low-dimensional representation. Meanwhile, an efficient hypergraph weight estimation scheme is adopted to preserve the prominent hyperedges. Classification is performed with SVM using the embedded features. The experimental results on three benchmark hyperspectral datasets verify that our embedded representation can enhance the classification accuracy effectively. The hypergraph weight estimation can further improve the accuracy of HSI classification.

Acknowledgments: This work was supported in part by the Natural Science Foundation of China under Grant Numbers: 61672292, 61532009, 61622305, 61502238, 61300162 and, in part, by the Six Talent Peaks Project of Jiangsu Province, China, under Grant DZXX-037.

Author Contributions: Yubao Sun and Sujuan Wang contributed equally to this work. They proposed the algorithm and performed the experiments. Qingshan Liu supervised the study, analyzed the results and gave insightful suggestions for the manuscript. Sujuan Wang and Yubao Sun drafted the manuscript. Guangcan Liu and Renlong Hang contributed to the revision of the manuscript. All authors read and approved the submitted manuscript.

References

1. Clement, A. Advances in remote sensing of agriculture: context description, existing operational monitoring systems and major information needs. *Remote Sens.* **2013**, *5*, 949–981, doi:10.3390/rs5020949.
2. Shafri, H. Z. M.; Taherzadeh, E.; Mansor, S.; Ashurov, R. Hyperspectral remote sensing of urban areas: an overview of techniques and applications. *Res. J. Appl. Sci. Eng. Technol.* **2012**, *4*, 1557–1565.

3. Abbate, G.; Fiumi, L.; De Lorenzo, C.; Vintila, R. Avaluation of remote sensing data for urban planning. Applicative examples by means of multispectral and hyperspectral data. In Proceedings of the GRSS/ISPRS Joint Workshop on Remote Sensing and Data Fusion over Urban Areas, Berlin, Germany, 22–23 May 2003; pp. 201–205.
4. Wu, Z.; Wang, Q.; Plaza, A.; Li, J.; Sun, L. Parallel spatial-spectral hyperspectral image classification with sparse representation and markov random fields on GPUs. *IEEE J. Sel. Top. Appl. Earth Obs. Remote Sens.* **2015**, *8*, 2926–2938, doi:10.1109/JSTARS.2015.2413931.
5. Yuan, Y.; Lin, J.; Wang, Q. Hyperspectral image classification via multitask joint sparse representation and stepwise MRF optimization. *IEEE Trans. Cybern.* **2016**, *46*, 2966–2977, doi:10.1109/TCYB.2015.2484324.
6. Wang, Q.; Lin, J.; Yuan, Y. Salient band selection for hyperspectral image classification via manifold ranking. *IEEE IEEE Trans. Neural Netw. Learn. Syst.* **2016**, *27*, 1279–1289, doi:10.1109/TNNLS.2015.2477537.
7. Hang, R.; Liu, Q.; Sun, Y.; Yuan, X.; Pei, H.; Plaza, J.; Plaza, A. Robust matrix discriminative analysis for feature extraction from hyperspectral images. *IEEE J. Sel. Top. Appl. Earth Obs. Remote Sens.* **2017**, *10*, 2002–2011, doi:10.1109/JSTARS.2017.2658948.
8. Wu, Z.; Li, Y.; Plaza, A.; Li, J.; Xiao, F.; Wei, Z. Parallel and distributed dimensionality reduction of hyperspectral data on cloud computing architectures. *IEEE J. Sel. Top. Appl. Earth Obs. Remote Sens.* **2016**, *9*, 2270–2278, doi:10.1109/JSTARS.2016.2542193.
9. Sun, Y.; Hang, R.; Liu, Q.; Zhu, F.; Pei, H. Graph-Regularized low rank representation for aerosol optical depth retrieval. *Int. J. Remote Sens.* **2016**, *37*, 5749–5762, doi:10.1080/01431161.2016.1249302.
10. Fauvel, M.; Tarabalka, Y.; Benediktsson, J.A.; Chanussot, J.; Tilton, J.C. Advances in spectral-spatial classification of hyperspectral images. *Proc. IEEE* **2013**, *101*, 652–675, doi:10.1109/JPROC.2012.2197589.
11. Yuan, Y.; Lin, J.; Wang, Q. Dual-Clustering-Based hyperspectral band selection by contextual analysis. *IEEE Trans. Geosci. Remote Sens.* **2016**, *54*, 1431–1445, doi:10.1109/TGRS.2015.2480866.
12. Kettig, R.L.; Landgrebe, D.A. Classification of multispectral image data by extraction and classification of homogeneous objects. *IEEE Trans. Geosci. Electron.* **1976**, *14*, 19–26, doi:10.1109/TGE.1976.294460.
13. Descombes, X.; Sigelle, M.; Preteu, F. GMRF parameter estimation in a non-stationary framework by a renormalization technique: application to remote sensing imaging. *IEEE Trans. Image Process.* **1999**, *8*, 490–503, doi:10.1109/83.753737.
14. Jackson, Q.; Landgrebe, D.A. Adaptive bayesian contextual classification based on markov random fields. *IEEE Trans. Geosci. Remote Sens.* **2002**, *40*, 2454–2463, doi:10.1109/TGRS.2002.805087.
15. Pesaresi, M.; Benediktsson, J.A. A new approach for the morphological segmentation of high-resolution satellite imagery. *IEEE Trans. Geosci. Remote Sens.* **2001**, *39*, 309–320, doi:10.1109/36.905239.
16. Benediktsson, J.A.; Palmason, J.A.; Sveinsson, J.R. Classification of hyperspectral data from urban areas based on extended morphological profiles. *IEEE Trans. Geosci. Remote Sens.* **2005**, *43*, 480–491, doi:10.1109/TGRS.2004.842478.
17. Fauvel, M.; Benediktsson, J.A.; Chanussot, J.; Sveinsson, J.R. Spectral and spatial classification of hyperspectral data using svms and morphological profiles. *IEEE Trans. Geosci. Remote Sens.* **2008**, *46*, 3804–3814, doi:10.1109/TGRS.2008.922034.
18. Guo, X.; Huang, X.; Zhang, L. Three-Dimensional wavelet texture feature extraction and classification for multi/hyperspectral imagery. *IEEE Geosci. Remote Sens. Lett.* **2014**, *11*, 2183–2187, doi:10.1109/LGRS.2014.2323963.
19. Li, L.; Marpu, P.R.; Plaza, A.; Bioucas-Dias, J.M. Generalized composite kernel framework for hyperspectral image classification. *IEEE Trans. Geosci. Remote Sens.* **2013**, *51*, 4816–4829, doi:10.1109/TGRS.2012.2230268.
20. Hang, R.; Liu, Q.; Song, H.; Sun, Y. Matrix-based discriminant subspace ensemble for hyperspectral image spatial–spectral feature fusion. *IEEE Trans. Geosci. Remote Sens.* **2016**, *54*, 783–794, doi:10.1109/TGRS.2015.2465899.
21. Fang, L.; Li, S.; Kang, X.; Benediktsson, J.A. Spectral-spatial hyperspectral image classification via multiscale adaptive sparse representation. *IEEE Trans. Geosci. Remote Sens.* **2014**, *52*, 7738–7749, doi:10.1109/TGRS.2014.2318058.
22. Chen, Y.; Lin, Z.; Zhao, X.; Wang, G.; Gu, Y. Deep learning-based classification of hyperspectral data. *IEEE J. Sel. Top. Appl. Earth Obs.* **2014**, *7*, 2094–2107, doi:10.1109/JSTARS.2014.2329330.
23. Du, Q. Modified fisher's linear discriminant analysis for hyperspectral imagery. *IEEE Geosci. Remote Sens. Lett.* **2007**, *4*, 503–507, doi:10.1109/LGRS.2007.900751.

24. Fauvel, M.; Chanussot, J.; Benediktsson, J.; Atli, N. Kernel principal component analysis for the classification of hyperspectral remote sensing data over urban areas. *EURASIP J. Adv. Signal Process.* **2009**, *2009*, 1–14, doi:10.1155/2009/783194.
25. Li, W.; Prasad, S.; Fowler, J.E. Decision fusion in kernel-induced spaces for hyperspectral image classification. *IEEE Trans. Geosci. Remote Sens.* **2014**, *52*, 3399–3411, doi:10.1109/TGRS.2013.2272760.
26. He, X.; Niyogi, P. Locality preserving projections. In Proceedings of the Advances in Neural Information Processing Systems (NIPS), Vancouver, BC, Canada, 8–13 December 2003; pp. 186–197.
27. Villa, A.; Benediktsson, J.A.; Chanussot, J.; Jutten, C. Hyperspectral image classification with independent component discriminant analysis. *IEEE Trans. Geosci. Remote Sens.* **2011**, *49*, 4865–4876, doi:10.1109/TGRS.2011.2153861.
28. Mura, M.D.; Villa, A.; Benediktsson, J.A.; Chanussot, J. Classification of hyperspectral images by using extended morphological attribute profiles and independent component analysis. *IEEE Geosci. Remote Sens. Lett.* **2011**, *8*, 542–546, doi:10.1109/LGRS.2010.2091253.
29. Roweis, S.T.; Saul, L.K. Nonlinear dimensionality reduction by locally linear embedding. *Science* **2000**, *290*, 2323–2326, doi:10.1126/science.290.5500.2323.
30. Yan, S.; Xu, D.; Zhang, B.; Zhang, H.J. Graph embedding and extensions: A general framework for dimensionality reduction. *IEEE Trans. Pattern Anal. Mach. Intell.* **2007**, *29*, 40–51, doi:10.1109/TPAMI.2007.250598.
31. Luo, F.; Huang, H.; Liu, J.; Ma, Z. Fusion of graph embedding and sparse representation for feature extraction and classification of hyperspectral imagery. *Photogramm. Eng. Remote Sens.* **2017**, *83*, 37–46, doi:10.14358/PERS.83.1.37.
32. Chen, P.; Jiao, L.; Liu, F.; Zhao, J.; Zhao, Z. Dimensionality reduction for hyperspectral image classification based on multiview graphs ensemble. *J. Appl. Remote Sens.* **2016**, *10*, 030501, doi:10.1117/1.JRS.10.030501.
33. Zhou, D.; Huang, J.; Schölkopf, B. Learning with hypergraphs: clustering, classification, and embedding. In Proceedings of the Advances in Neural Information Processing Systems, Vancouver, BC, Canada, 3–6 December 2007; pp. 1601–1608.
34. Bu, J.; Tan, S.; Chen, C.; Wang, C.; Wu, H.; Zhang, L.; He, X. Music recommendation by unified hypergraph: combining social media information and music content. In Proceedings of the 18th ACM international conference on Multimedia, Firenze, Italy, 25–29 October 2010; pp. 391–400.
35. Liu, Q.; Sun, Y; Wang, C.; Liu, T.; Tao, D. Elastic net hypergraph learning for image clustering and semi-supervised classification. *IEEE Trans. Image Process.* **2017**, *26*, 452–463, doi:10.1109/TIP.2016.2621671.
36. Yuan, H.; Tang, Y.Y. Learning with hypergraph for hyperspectral image feature extraction. *IEEE Trans. Geosci. Remote Sens. Lett.* **2015**, *12*, 1695–1699, doi:10.1109/LGRS.2015.2419713.
37. Kuo, B.C.; Landgrebe, D.A. Nonparametric weighted feature extraction for classification. *IEEE Trans. Geosci. Remote Sens.* **2004**, *42*, 1096–1105, doi:10.1109/TGRS.2004.825578.
38. Liao, W.; Pizurica, A.; Scheunders, P.; Philips, W.; Pi, Y. Semisupervised local discriminant analysis for feature extraction in hyperspectral images. *IEEE Trans. Geosci. Remote Sens.* **2013**, *51*, 184–198, doi:10.1109/TGRS.2012.2200106.
39. Pliakos, K.; Kotropoulos, C. Weight estimation in hypergraph learning. In Proceedings of the IEEE International Conference on Acoustics, Speech and Signal Processing, South Brisbane, Australia, 19–24 April 2015; pp. 1161–1165.
40. Gao, Y.; Wang, W.; Zha, Z.J.; Shen, J.; Li, X.; Wu, X. Visual-textual joint relevance learning for tag-based social image search. *IEEE Trans. Image Process.* **2013**, *22*, 363–376, doi:10.1109/TIP.2012.2202676.

4

Topic Modelling for Object-Based Unsupervised Classification of VHR Panchromatic Satellite Images Based on Multiscale Image Segmentation

Li Shen [1,2,*], Linmei Wu [1,2], Yanshuai Dai [1,2], Wenfan Qiao [1,2] and Ying Wang [1,2]

1 State-Province Joint Engineering Laboratory of Spatial Information Technology for High-Speed Railway Safety, Southwest Jiaotong University, Chengdu 611756, China; linmay23@yeah.net (L.W.); 15708455802@163.com (Y.D.); swjtu_qiaowenfan@163.com (W.Q.); elaine.wy@foxmail.com (Y.W.)

2 Faculty of Geosciences and Environmental Engineering, Southwest Jiaotong University, Chengdu 611756, China

* Correspondence: lishen@swjtu.edu.cn.

Academic Editor: Qi Wang

Abstract: Image segmentation is a key prerequisite for object-based classification. However, it is often difficult, or even impossible, to determine a unique optimal segmentation scale due to the fact that various geo-objects, and even an identical geo-object, present at multiple scales in very high resolution (VHR) satellite images. To address this problem, this paper presents a novel unsupervised object-based classification for VHR panchromatic satellite images using multiple segmentations via the latent Dirichlet allocation (LDA) model. Firstly, multiple segmentation maps of the original satellite image are produced by means of a common multiscale segmentation technique. Then, the LDA model is utilized to learn the grayscale histogram distribution for each geo-object and the mixture distribution of geo-objects within each segment. Thirdly, the histogram distribution of each segment is compared with that of each geo-object using the Kullback-Leibler (KL) divergence measure, which is weighted with a constraint specified by the mixture distribution of geo-objects. Each segment is allocated a geo-object category label with the minimum KL divergence. Finally, the final classification map is achieved by integrating the multiple classification results at different scales. Extensive experimental evaluations are designed to compare the performance of our method with those of some state-of-the-art methods for three different types of images. The experimental results over three different types of VHR panchromatic satellite images demonstrate the proposed method is able to achieve scale-adaptive classification results, and improve the ability to differentiate the geo-objects with spectral overlap, such as water and grass, and water and shadow, in terms of both spatial consistency and semantic consistency.

Keywords: very high resolution (VHR) satellite image; topic modelling; object-based image analysis; image segmentation; unsupervised classification; multiscale representation

1. Introduction

Recent advances in remote sensing technology, particularly those relating to spatial resolution, are helping to make detailed observations of the Earth's surface possible. However, the resulting vast amounts of very high resolution (VHR) satellite images pose a challenge for automatic classification, due to the large amount of information and with-class variance characteristics of this kind of images [1].

It is widely acknowledged that, compared to their pixel-based counterparts, object-based classification methods, which can take advantage of both spectral and spatial information, are probably more appropriate for VHR satellite images [2–4]. In a typical object-based classification framework, image segmentation is usually an initial and vital step. The goal of segmentation is to partition an original image into a set of non-overlapping homogeneous segments, which are regarded as higher-level units that are more meaningful and efficient for subsequent processing. Thus, the classification accuracy of object-based classification is dependent, to a large extent, on the quality of image segmentation [5]. In order to characterize image structures at different scales, multiscale segmentation (MS) is often used to conduct a series of segmentation maps at multiple scales from fine to coarse ones, sequentially, by varying the scale parameter (SP) [6]. As pointed out by Dragut et al. [6], the SP controls the average segment size, i.e., a smaller value of the SP produces segmentations with small regions and detailed structures, and a larger value allows for more merges, thus preserving only large segments and coarse features. Therefore, the SP needs to be appropriately determined in order to create segments that can match the actual boundaries of landscape features of various sizes as much as possible [7]. However, there exist several problems in practical applications: (1) the determination of the optimal SP, in many cases, still relies on a trial-and-error optimization, which is time-consuming when it applies to complex image scenes [8]; (2) it is often impossible to determine a unique optimal SP for a whole scene or each geo-object, due to the fact that different geo-objects and even an identical geo-object may appear at different scales in the same image. Any specific SP is likely to cause over-segmentation of some parts of the image, but under-segmentation of other parts.

To overcome these shortcomings, many approaches that attempt to directly model multiple segmentations have been proposed instead of seeking and using an optimal one from multiple candidate segmentation maps [9–12]. In this direction, Russel et al. [10] used multiple segmentations to discover objects in natural image collections. The authors assumed while none of the SP settings can achieve the optimal image segmentation, some segments in some of the segmentations appear to be correct. Therefore, instead of selecting any particular optimal SP, they utilized multiple segmentation results to identify objects of various sizes by the learning machine. In a similar framework, Akcay et al. [13] proposed automatic detection of geospatial objects using multiple hierarchical segmentations. Afterwards, Santos et al. [14] presented a kind of boost-classifier adapted to MS for supervised classification, in which a sequential strategy of training for weak classifiers was adopted based on a hierarchy of image segmentations from fine to coarse. However, both [13] and [14] are built on hierarchical image segmentation [15], which cannot be directly obtained in many cases due to the reason that most image segmentation methods do not consider the hierarchical structures. These observations motive us to develop a novel object-based unsupervised classification based on MS, in which multiple segmentation maps are jointly utilized by means of topic modelling. In the proposed framework, the original image does not require to be hierarchically partitioned to form multiple segmentation maps.

Attracted by the success of topic models, e.g., latent Dirichlet allocation (LDA) [16] and its relatives [17,18], in text analysis community, there has been an increasing interest in applying such models for semantically-driven understanding of satellite images, such as image annotation [19–21], object detection [13,22], scene classification [23–25], and image classification or clustering [26–29]. Among various advantages provided by topic models is their ability to deal with polysemy [16]. For example, the word "bachelor" could refer to a kind of degree or an unmarried man. Based on co-occurrence of words in the context, topic models are able to capture the polysemous use of words. This characteristic offers a potential solution to cope with the perplexing, but so common, phenomenon in VHR satellite image classification, i.e., different geo-objects with nearly identical spectra. For image classification or clustering, the square image blocks with fixed-size or segments with arbitrary shape are viewed as the documents [27,29], and the pixels or local patches extracted from the images are regarded as the words in topic modeling. Using the probabilistic latent semantic analysis (pLSA) model, Yi et al. [26] presented a novel semantic clustering algorithm for VHR satellite images. The experiment results confirmed the advantage of topic modelling. However, in the proposed framework, a pre-processing step was required to partition the original satellite image into image documents, and an accompanying post-processing step was required to combine the allocated multiple labels of a pixel into a unique one using a certain voting rule. To address this problem, Tang et al. [27] developed a msLDA model, which built an automatic framework that combined a latent Dirichlet allocation with a multiscale image representation of a panchromatic satellite image. The msLDA archived an adaptive smoothing on clustering results. Nevertheless, its application usually introduces a computational bottleneck, due to the manner in which image documents are generated, where each pixel and its surrounding pixels within the square neighborhood constitute a document. Shu et al. [30] presented a nonparametric Bayesian hierarchical model to conduct unsupervised classification of VHR panchromatic satellite images by considering over-segments as documents. However, it also suffers from the same problems as the traditional object-based methods mentioned previously, i.e., it is modelled based on single-scale segmentation.

In this paper, a novel object-based unsupervised classification for VHR panchromatic satellite images using multiple image segmentation maps via the LDA model is presented. The proposed approach consists of four components: (1) a multiscale image segmentation component that allows characterizing of image structures at different scales; (2) a topic model component that learns the grayscale histogram distribution for each geo-object and the mixture distribution of geo-objects in each segment in an unsupervised manner; (3) a category label allocation component that classifies each segment by the ranking of probability-based similarities; and (4) an automatic application framework component that integrates multiple classification results at different scales into a unique one. It should be noted that while the proposed method still needs to determine the range of scales for creating a multiscale segmentation representation, the work has been greatly simplified compared to searching for the optimal image segmentation. The main contribution of the proposed method is an automatic framework of combining a topic model with a multiscale image segmentation representation to model both the co-occurrence of various geo-objects and multiscale structures.

This paper extends and improves on a preliminary work [31], which presents our initial ideas and results. In this paper: (1) a novel strategy of integrating multiple classification results at different scales into a unique one is added, which can ensure an adaptive smoothing classification result can be achieved; (2) a constraint specified by the mixture distribution of geo-objects, which can characterize the co-occurrence relationships of various geo-objects, is incorporated to correct the KL similarity between the histogram distribution of each segment and that of each geo-object; and (3) a more thorough presentation of introduction, methodology, experimental analysis, and discussion is conducted.

The remainder of this paper is organized as follows. In Section 2, the proposed methodology is presented in detail. Experimental results and related discussion are given in Section 3. Finally, the conclusion is drawn in Section 4.

2. Methodology

In this section, we present our approach for performing the object-based unsupervised classification of panchromatic satellite images, using the LDA model.

For the proposed method, a key prerequisite is to create a MS representation of the original satellite image with varied scales. Since the goal of MS is to produce enough segmentation maps of the image so as to have a high probability of acquiring better segments that can correspond to potential geo-objects, we do not need image segmentation at each scale to be exactly correct. Any method that can create a reasonable MS of a satellite image may meet the requirement.

Given the MS representation, the proposed method is composed of the following three steps: Firstly, the LDA model is utilized to learn the grayscale histogram distribution for each geo-object and the mixture distribution of geo-objects within each segment in an unsupervised manner. Then, the histogram distribution of each segment is compared with that of each geo-object using the Kullback-Leibler (KL) divergence measure [10], which is further weighted with a constraint specified by the mixture distribution of geo-objects. Each segment is allocated a geo-object category label with the minimum KL divergence. Finally, the scale-adaptive unsupervised classification map is achieved by integrating the multiple classification results at different scales. The general scheme is shown in Figure 1.

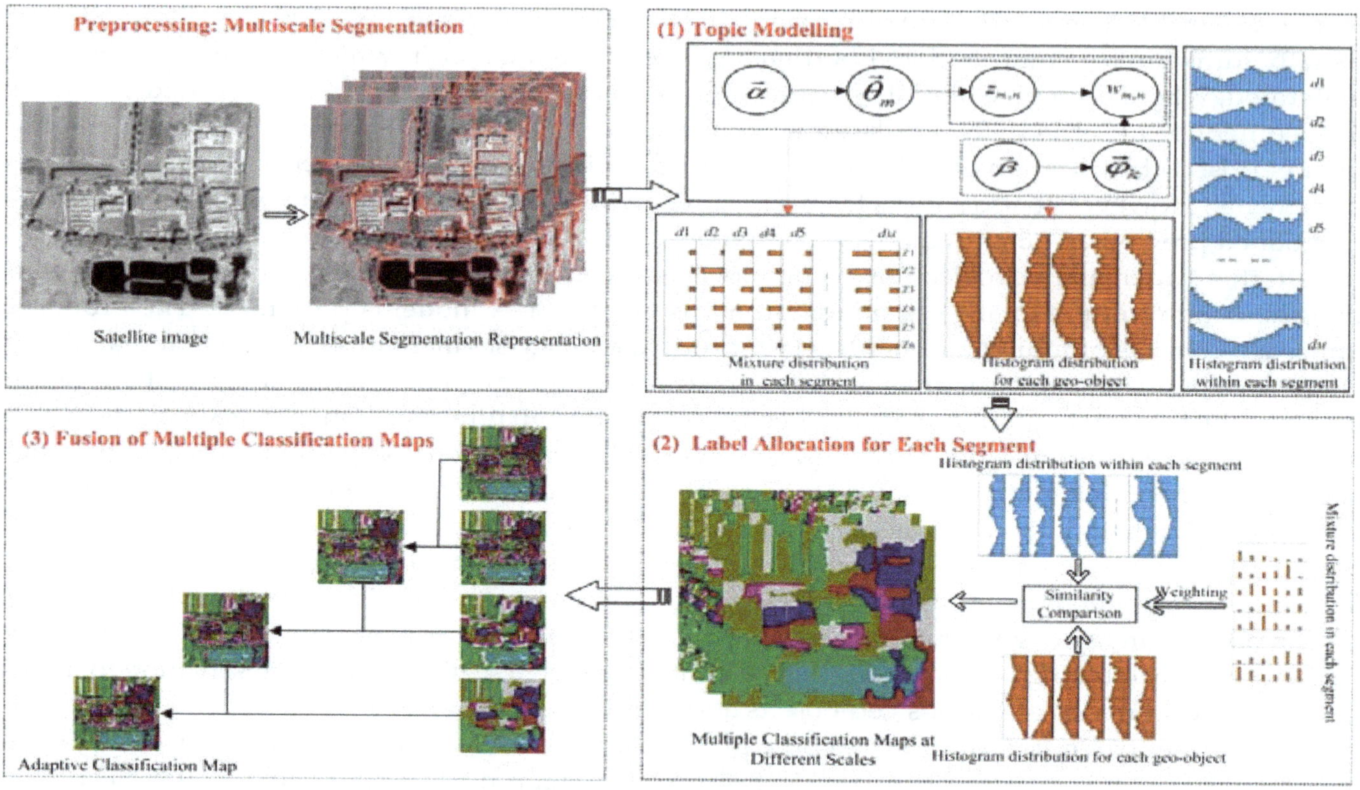

Figure 1. Flowchart of the proposed method.

2.1. Topic Modelling

2.1.1. Latent Dirichlet Allocation

The LDA is a generative hierarchical Bayesian probabilistic model, which is originally proposed to model collections of discrete data, such as text documents and natural images [32]. In this model, each document is viewed as a finite mixture of various latent topics. Each topic, in turn, is a probability distribution over words. For a corpus of M documents, the LDA assumes the following generative process for the m-th document of length N:

- For the k-th element of K topics, sample the topic-specific term distribution $\vec{\phi}_k$ according to the Dirichlet distribution, i.e., $\vec{\phi}_k \sim \text{Dirchlet}(\vec{\beta})$, where $\vec{\beta}$ is the hyperparameter.
- Sample the topic mixture $\vec{\theta}_m$ according to the Dirichlet distribution, i.e., $\vec{\theta}_m \sim \text{Dirchlet}(\vec{\alpha})$, where $\vec{\alpha}$ is the hyperparameter.
- For each word $w_{mn} \in \{w_{m1}, w_{m2}, \ldots, w_{mN}\}$, sample a topic z_{mn} according to the multinomial distribution, i.e., $z_{mn} \sim \text{Multinomial}(\vec{\theta}_m)$, and sample a word w_{mn} according to the multinomial distribution , i.e., $w_{mn} \sim \text{Multinomial}(\vec{\phi}_{z_{mn}})$.

Both variational inference and Gibbs sampling have been used to infer and estimate the parameters of the LDA. To solve the inferential problem, we need to calculate the posterior probability of the hidden variables given a corpus, i.e., $P(\vec{z}|\vec{w}, \vec{\alpha}, \vec{\beta})$. The general formulation of a Gibbs sampler for such latent-variable models becomes:

$$P(z_i|\vec{z}_{-i}, \vec{w}, \vec{\alpha}, \vec{\beta}) = \frac{P(\vec{z}, \vec{w}|\vec{\alpha}, \vec{\beta})}{\int\limits_Z P(\vec{z}, \vec{w}|\vec{\alpha}, \vec{\beta})dz_i}, \quad (1)$$

which can be approximated by interactively sampling each of the K topics using the chain rule and noting that $\vec{w} = \left\{\vec{w}_{-i}, w_i = v\right\}$ and $\vec{z} = \left\{\vec{z}_{-i}, z_i = k\right\}$:

$$\begin{aligned} P(z_i|\vec{z}_{-i}, \vec{w}, \vec{\alpha}, \vec{\beta}) &= \frac{P(\vec{z},\vec{w}|\vec{\alpha},\vec{\beta})}{P(\vec{z}_{-i},\vec{w}|\vec{\alpha},\vec{\beta})} \\ &\propto \frac{n_{k,-i}^{(v)}+\beta_t}{\sum\limits_{v'=1}^{V}(n_k^{(v')}+\beta_{v'})-1} * \frac{n_{m,-i}^{(k)}+\alpha_k}{\sum\limits_{k'=1}^{K}(n_m^{(k')}+\alpha_{k'})}, \end{aligned} \quad (2)$$

where the counts $n_{\cdot,-i}^{(\cdot)}$ represents the number of the words or topics with exception of index i.

For a new topic-word pair $(\widetilde{z} = k, \widetilde{w} = v)$ that is observed in a document $d(\widetilde{w}) = d(\widetilde{z}) = m$, given the state $(\widetilde{z}, \widetilde{w})$, the multinomial parameters can be determined by:

$$\phi_{k,v} = \frac{n_k^{(v)} + \beta_v}{\sum\limits_{v'=1}^{V}(n_k^{(v')} + \beta_{v'})}, \quad (3)$$

$$\theta_{m,k} = \frac{n_m^{(k)} + \alpha_k}{\sum\limits_{k'=1}^{K}(n_m^{(k')} + \alpha_{k'})}, \quad (4)$$

where V is the number of words in the vocabulary, α_k and β_v stand for the k-th element and v-th element of hyperparameter $\vec{\alpha}$ and $\vec{\beta}$, respectively. For details regarding the derivation, please refer to [33].

2.1.2. Build an Analogue of Text-Related Terms in the Image Domain

To borrow techniques used in the text domain to satellite images, the first issue that needs to be addressed is how to build an analogue of text terms in the image domain. For the proposed method, we follow the definition in [27,29]

- Word: a unique grayscale value of a pixel is defined as a word;
- Vocabulary: the unique grayscale values of the satellite image form the vocabulary;
- Document: each segment is regarded as a document, thus, all segments of multiple segmentation maps at different scales constitute the corpus;
- Topic: each topic corresponds to a specific geo-object category.

Given multiple segmentation maps at different scales, we can use the LDA model to learn K grayscale histogram distributions for K geo-objects, i.e., $\{\vec{\phi}_k\}_{k=1}^{K}$, and M mixture distributions of geo-objects within M segments, i.e., $\{\vec{\theta}_m\}_{m=1}^{M}$. Furthermore, M grayscale histogram distributions within M segments, i.e., $\left\{\vec{\pi}_m\right\}_{m=1}^{M}$, can also be easily obtained just by counting the frequencies of different grayscale values.

2.2. Label Allocation for Each Segment

Although there may be some differences in size due to the scale effect, the segments that correspond to the identical geo-object should have similar grayscale histogram distributions. As a consequence, the category label allocation for each segment is determined according to the following rule: the histogram distribution of each segment is compared with those of K geo-objects, respectively. The similarity between two types of distributions is measured using the KL divergence, which has been proved to be effective in measuring the probability-based similarity. The category label of the geo-object, which has the minimum KL divergence with the segment, is allocated as the label of the segment. Given the segment d_m, its category label c_{d_m} is mathematically given by:

$$c_{d_m} = \arg\min_{1 \le k \le K} KL(\vec{\pi}_m, \vec{\phi}_k), \tag{5}$$

where $KL(\vec{\pi}_m, \vec{\phi}_k)$ denotes the symmetrical KL divergence between two discrete distributions, i.e., $\vec{\pi}_m$ and $\vec{\phi}_k$. To be specific, the symmetrical KL divergence in the discrete form can be represented as the following:

$$KL(\vec{\pi}_m, \vec{\phi}_k) = \frac{\sum_{v=1}^{V} \{\pi_m^v In(\frac{\pi_m^v}{\phi_k^v}) + \phi_k^v In(\frac{\phi_k^v}{\pi_m^v})\}}{2}, \tag{6}$$

where π_m^v refers to the v-th element of $\vec{\pi}_m$, which describes the histogram distribution within the segment m. Likewise, ϕ_k^v refers to the v-th element of $\vec{\phi}_k$, which describes the histogram distribution for the geo-object k.

Furthermore, it is easy to understand: with the increase of the mixture proportion of a certain geo-object within the segment, the probability of classifying the segment as the corresponding geo-object should be improved accordingly. The mixture distribution of geo-objects within each segment, i.e., $\{\vec{\theta}_m\}_{m=1}^{M}$, learned by topic modelling, exactly provides the mixture proportion features. Thus, Equation (5) can be further weighted with a constraint specified by the mixture distribution of geo-objects:

$$c_{d_m} = \arg\min_{1 \le k \le K} KL(\vec{\pi}_m, \vec{\phi}_k) \cdot u_m^k, \tag{7}$$

where $\vec{u}_m$ is empirically set to $-In(\vec{\theta}_m)$, and u_m^k denotes the weight with the geo-object k.

Overall, by means of Equation (7), the category label of each segment is determined jointly by both the grayscale distribution and mixture distribution of geo-objects, which can characterize the co-occurrence relationships of various geo-objects.

2.3. Fusion of Multiple Classification Maps

After finishing step 2.2, multiple unsupervised classification maps at different scales are achieved. However, due to the existence of multiscale effects in VHR satellite images, any classification map based the single-scale segmentation cannot take into account the various granularities of the geo-objects, e.g., the narrow roads may be suitable to be extracted at a fine scale, while the large-area field should be classified at a coarse scale. Therefore, the final classification map is achieved by integrating the multiple classification maps at different scales, in order to achieve a scale-adaptive unsupervised classification.

Given the scale range $\{1, 2, \ldots, S\}$, S classification maps at multiple scales can be obtained by the means of topic modelling presented above. In other words, each pixel i in the original satellite image will be allocated S category labels. Let $d_{i,s} (s \in \{1, 2, \ldots S\})$ denote the segment covering the pixel i at the scale s, and $k(d_{i,s})$ denote the category label of the segment $d_{i,s}$, the category label of the pixel i is given by:

$$\begin{cases} s^* = \arg\min_{1 \le s \le S} KL(\vec{\pi}_{d_{i,s}}, \vec{\phi}_{k(d_{i,s})}), \\ c_i = k(d_{i,s^*}). \end{cases} , \tag{8}$$

where $\vec{\pi}_{d_{i,s}}$ denotes the grayscale histogram distribution within the segment $d_{i,s}$, $\vec{\phi}_{k(d_{i,s})}$ denotes the grayscale histogram distribution for the geo-object $k(d_{i,s})$, and s^* denotes the optimal scale for the pixel i.

3. Results and Discussion

In this section, we firstly describe the experimental images. Then, we introduce the quantitative evaluation methods for the experimental results and the state-of-art methods for comparison, and the parameter settings are also given in detail. Thirdly, we compare the performance of different approaches for three typical of geographical scenes in terms of both qualitative and quantitative aspects. The computational efficiency for different approaches is also discussed. Finally, we analyze the effects of scale setting in the proposed method on the classification results.

3.1. Experiment Data

In order to assess the effectiveness of the proposed approach, three panchromatic satellite images with different scenes and spatial resolutions are used. The first data is a panchromatic Mapping Satellite-1 image with 1600 × 1600 pixels and 2 m spatial resolution, which was acquired on 13 August 2012 and covers an area of Miyun District, Beijing, China. As shown in Figure 2a, five major types of geo-objects, i.e., building, road, water, grass, and ground, occur in this image. The second data is a panchromatic QuickBird image with 900 × 900 pixels and 0.6 m spatial resolution, as shown in Figure 2c. It was acquired on 22 April 2006 and located in Tong Zhou district of Beijing, China. There are six major types of geo-objects distributed in the image, including building, road, water, shadow, tree, and field. The third data is a panchromatic ZiYuan-3 (ZY-3) image that was acquired over an area of Tanggu District, Tianjin, China, on 15 August 2015. The image size is 3500 × 3500 pixels, and the spatial resolution is 2.1 m. As shown in Figure 2e, the image is made up of five major types of geo-objects, including building, road, water, grass, and field. We manually annotated all the original images at the pixel level as ground truth label data through visual interpretation. The corresponding ground truth maps for three satellite images are shown in Figure 2b,d,f, respectively.

3.2. Experiment Setup

3.2.1. Methods for Comparison with the Proposed Approach

To evaluate the effectiveness on three aspects of classification accuracy, spatial smoothness, and semantic consistency, the performance of the proposed approach is compared with that of four state-of-the-art unsupervised classification methods based on image segmentation: (1) the spectral-spatial ISODATA, where the pixel-based ISODATA classification is followed by a majority voting within the adaptive neighborhoods defined by the over-segmentation (termed as O_ISODATA) [34]; (2) the spectral-spatial LDA, similar to O_ISODATA, where the same over-segmentation is applied to the classification result of the LDA model using just the single-scale image segmentation map as corpus [30] (termed as O_LDA); (3) the msLDA proposed in [27]; and (4) the HDP_IBP proposed in [30].

For convenience, the proposed approach is referred to as the mSegLDA.

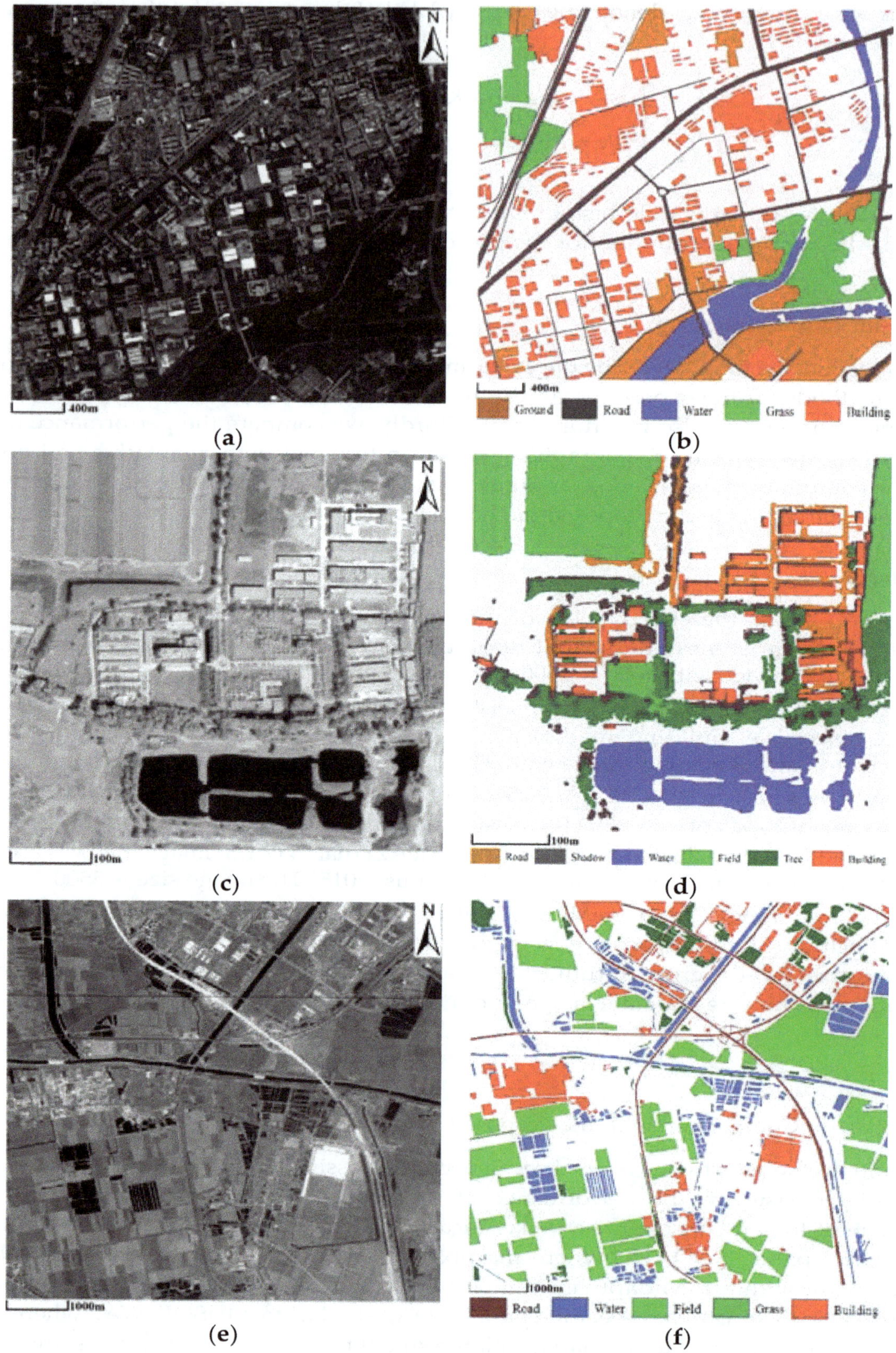

Figure 2. Experimental datasets. (**a**) Mapping Satellite-1 panchromatic image; (**b**) ground truth map of the Mapping Satellite-1 image; (**c**) QuickBird panchromatic image; (**d**) ground truth map of the QuickBird image; (**e**) ZY-3 panchromatic image; and (**f**) ground truth map of the ZY-3 panchromatic image.

3.2.2. Evaluation Criteria

In our experiments, two quantitative criteria, as well as visual inspection, are utilized to evaluate the unsupervised classification results, i.e., overall accuracy (OA) and overall entropy (OE).

- Overall accuracy (OA): OA, which serves as a quantitative measurement of the agreement between the classification result and the ground truth map, is one of the most widely used statistics for evaluating the classification accuracy. OA can be calculated by dividing the total correctly-classified pixels by the total number of pixels checked by the ground truth map, and is given as $OA = N_{correct}/N_{total}$, where $N_{correct}$ is the total number of correct pixels, and N_{total} is total number of pixels.
- Overall entropy (OE): entropy is an information theoretical criterion that is able to measure the homogeneity of the classification results. OE is defined as a linear combination of the class entropy, which describes how the pixels of the same geo-object are presented by the various clusters created, and the cluster entropy, which reflects the quality of the individual clusters in terms of the homogeneity of the pixels in a cluster. Generally speaking, a smaller overall entropy value corresponds to the classification map with a higher homogeneity. For details regarding, please refer to [27,35].

3.2.3. Parameter Setting

In order to produce multiple segmentation maps at different scales, this paper uses the entropy rate superpixel segmentation (ERSS) algorithm [36], which has been proven to be both effective and efficient. It should be noted that any method that can create a reasonable MS of satellite image may meet the requirement of the proposed approach. The ERSS algorithm utilizes the number of segments to control the scale size of image segmentation, i.e., a large number of segments may result in fine-scale segmentation, and conversely, a small number of segments will generate coarse-scale segmentation. For the Mapping Satellite-1 image, the QuickBird image and the ZY-3 image, the number of scales S is set to 6, 9, and 11, respectively, and the corresponding numbers of segments include {100, 200, 500, 800, 1000, 1500}, {100, 200, 500, 800, 1000, 1500, 2000, 2500, 3000}, and {2000, 3500, 5000, 6500, 8000, 9500, 11,000, 12,500, 14,000, 15,500, 17,000}.The range of scales, reflected by the numbers of segments, should be able to characterize multiscale structures in the images as much as possible. For both O_ISODATA and O_LDA, the over-segmentation map with 1500 segments, 2500 segments, and 17,000 segments are used in the three images.

Furthermore, topic model based methods initializes the Dirichlet priors as symmetric priors empirically, i.e., $\alpha = 0.1$, $\beta = 0.01$ for the Mapping Satellite-1 image, $\alpha = 0.01$, $\beta = 0.8$ for the QuickBird image, and $\alpha = 50/K$, $\beta = 0.01$ for the ZY-3 image. The number of geo-objects K is set to 6, 7, and 6 according to the distributions of geo-object classes, respectively.

3.3. Comparison of Classification Results

The classification results of various methods for three satellite images are shown in Figures 3–5, where each geo-object is represented by a different color.

3.3.1. Mapping Satellite-1 and QuickBird Images

From visual inspection, all the unsupervised classification results seem to be compact. The obvious speckle noise or the isolated pixel patches which are often found in the results of pixel-based classification approaches are greatly eliminated. Thus, the advantages of enforcing spatial consistency over the classification by means of performing image segmentation are confirmed. On the other hand, obvious misclassification between water and grass in Figure 3b,g, and water and shadow in Figure 4b,g can be observed in the classification results of the ISODATA. Two types of geo-objects, i.e., grass and shadow, are entirely incorrectly identified as water. This phenomenon can be explained by the fact that there exist obvious spectral overlaps between water and grass in Figure 2a, and water and shadow in Figure 2c. For this reason, the ISODATA which groups image pixels merely according to their grayscale values and, thus, are not able to differentiate various geo-objects with similar spectra well. While the O_ISODATA that conducts pixel-based ISODATA classification, followed by spatial regularization using the segmentation map can ensure spatial continuity within segments, it does not change the essential mechanism of the ISODATA.

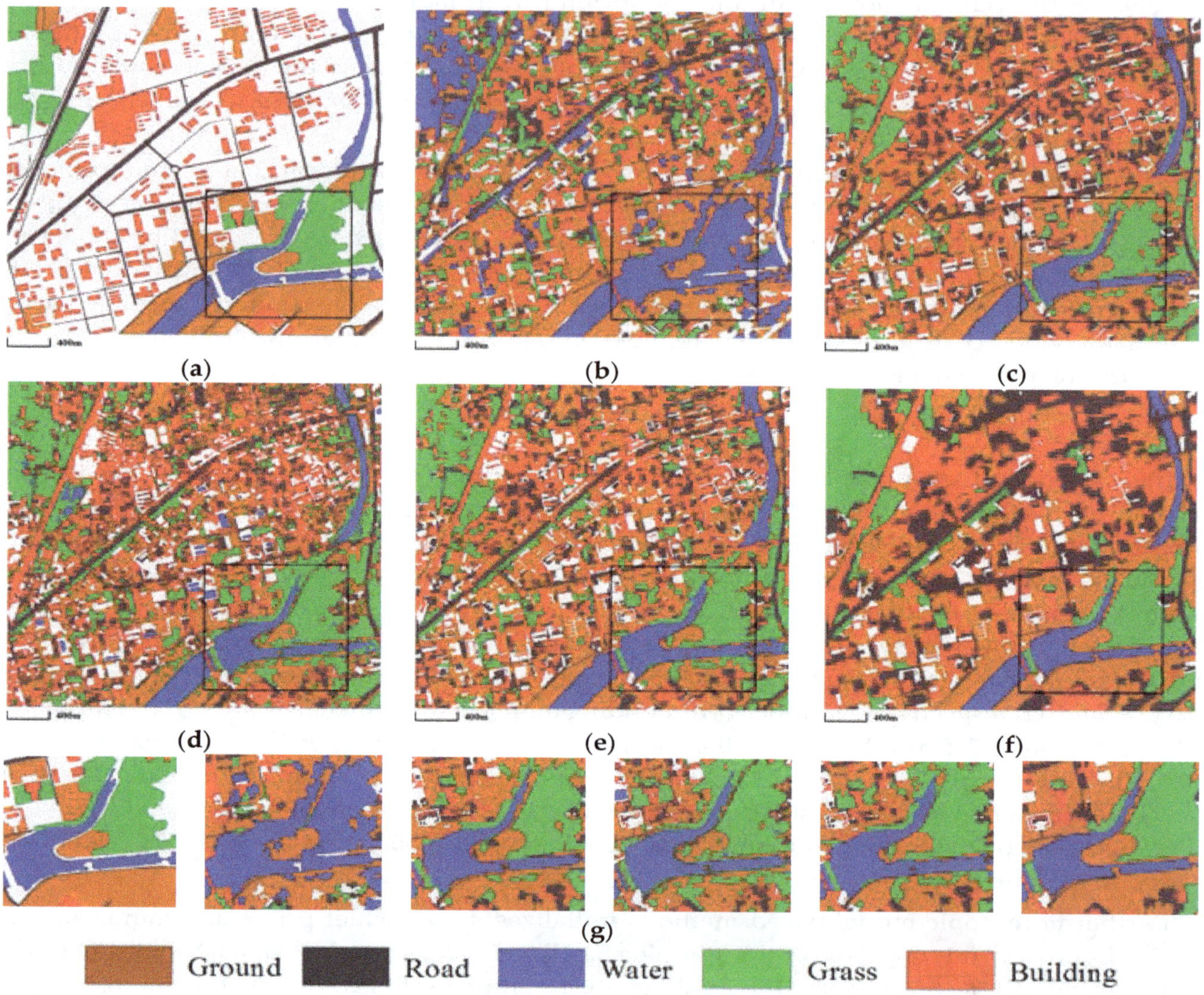

Figure 3. Classification results of various methods for the Mapping Satellite-1 image. (**a**) Ground truth map; (**b**) O_ISODATA; (**c**) O_LDA; (**d**) msLDA; (**e**) HDP_IBP; (**f**) mSegLDA; and (**g**) Details of (**a**–**f**).

However, benefitting from topic modelling, for the topic model-based approaches, the co-occurrence information, characterized by the mixture distribution of various geo-objects within each segment, can be utilized to correctly recognize different objects with similar spectra. As illustrated in Figures 3c–g and 4c–g, two types of geo-objects, i.e., grass and shadow, are well separated from water in the classification results of the topic model based approaches.

Furthermore, because the O_LDA is built on a single-scale over-segmentation map, it lacks a mechanism to model multiscale features of various-objects and, thus, cannot realize an adaptive smoothing on classification results. As shown in Figure 4c, the smoothing effect on the fine-scale road is proper, but is not sufficient for the large-scale field. Additionally, the majority voting scheme adopted by the O_LDA within the adaptive neighborhoods defined by the over-segmentation also results in the relatively fragmented classification map. In order to encode the scale-adaptive classification ability, the msLDA combines the topic model with a multiscale image representation derived by convoluting a given image with a variable-scale Gaussian into an automatic framework by embedding both image block and scale selections, and the HDP_IBP introduces the hierarchical spatial information, particularly the high-level scene cues, into the classification. As shown in Figures 3d–e and 4d–e, both the HDP_IBP and msLDA improve the adaptive smoothing effect on the classification results to a certain extent compared with the O_LDA. However, the improvement is still limited. As a contrast, as shown in Figure 3f,g and Figure 4f,g the proposed mSegLDA could realized a more significant self-adaptive spatial regularization on classification results according to various geo-object types at different scales, i.e., the large-scale geo-object (e.g., field) is heavily smoothed, resulting in a more homogenous classification, and the fine-scale geo-object (e.g., road) accepts a slight smoothing, thus preserving detailed structures and edge patterns.

From quantitative evaluation, both OA and OE in different classification results for two experiment images are calculated, as shown in Tables 1 and 2, respectively. The mSegLDA approach yields the best classification accuracies and the lowest values of OE, compared to other methods, indicating the proposed method can achieve a better classification performance on the whole.

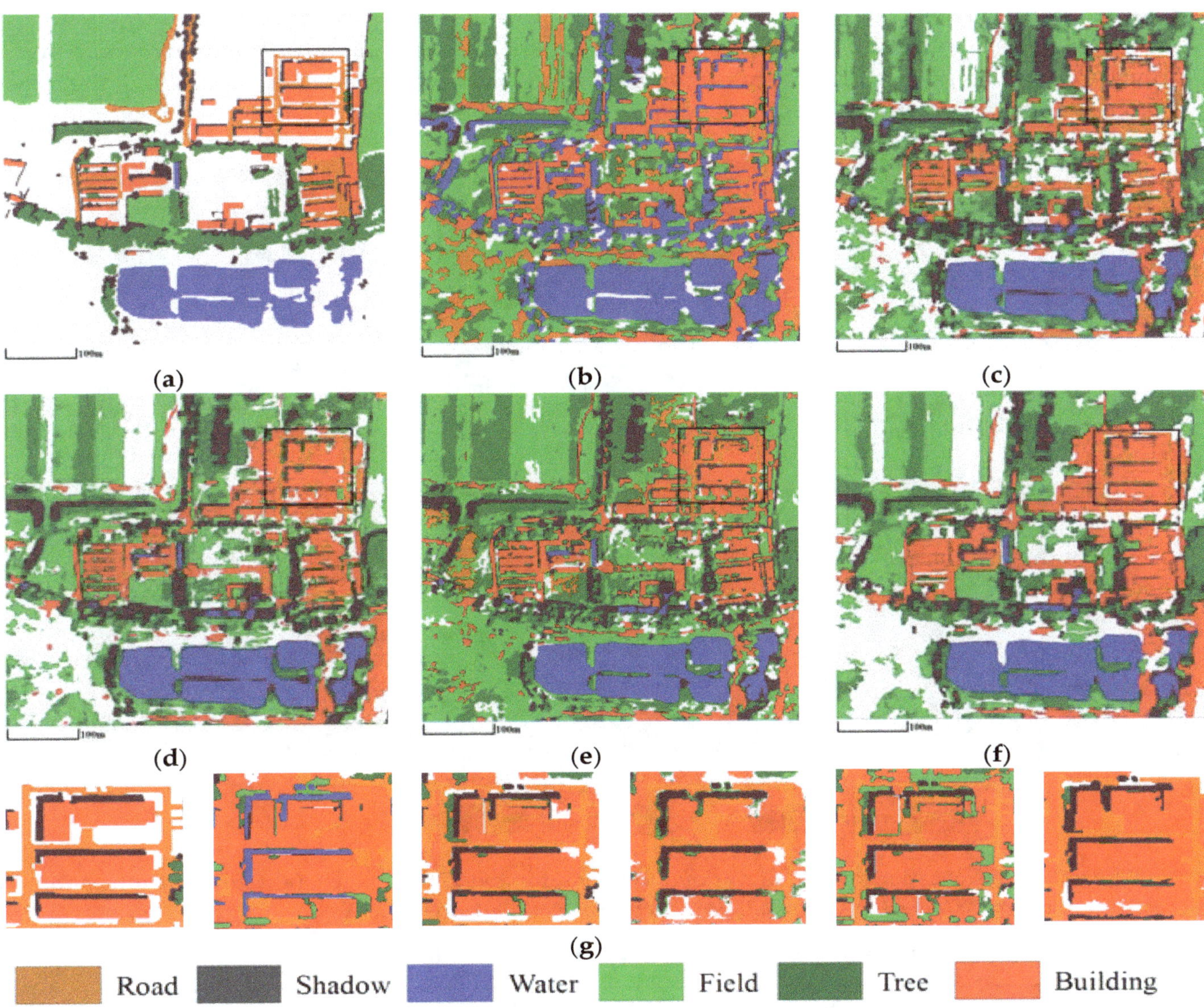

Figure 4. Classification results of various methods for the QuickBird image. (**a**) Ground truth map; (**b**) O_ISODATA; (**c**) O_LDA; (**d**) msLDA; (**e**) HDP_IBP; (**f**) mSegLDA; and (**g**) Details of (**a**–**f**).

Table 1. OA and OE of various methods for the Mapping Satellite-1 image.

	O_ISODATA	O_LDA	msLDA	HDP_IBP	mSegLDA
OA	48.6	69.3	68.4	70.1	72.0
OE	0.89	0.79	0.84	0.76	0.72

Table 2. OA and OE of various methods for the QuickBird image.

	O_ISODATA	O_LDA	msLDA	HDP_IBP	mSegLDA
OA	49.8	65.1	68.2	65.5	74.3
OE	0.96	0.84	0.80	0.84	0.71

3.3.2. ZY-3 Image

In this subsection, a ZY-3 satellite image of a large scene covering over 50 km^2 is used to evaluate the effectiveness of the proposed mSegLDA. As shown in Figure 5, the complexity of the scene significantly increases. For example, various geo-objects, and even an identical geo-object (e.g., building or field), are present at different sizes in the image. However, the proposed mSegLDA is still able to realize a more significant self-adaptive spatial regularization on classification maps according to various geo-object types at different scales, compared to other methods. As shown in Table 3, the largest value of OA and the lowest value OE are also obtained by the mSegLDA.

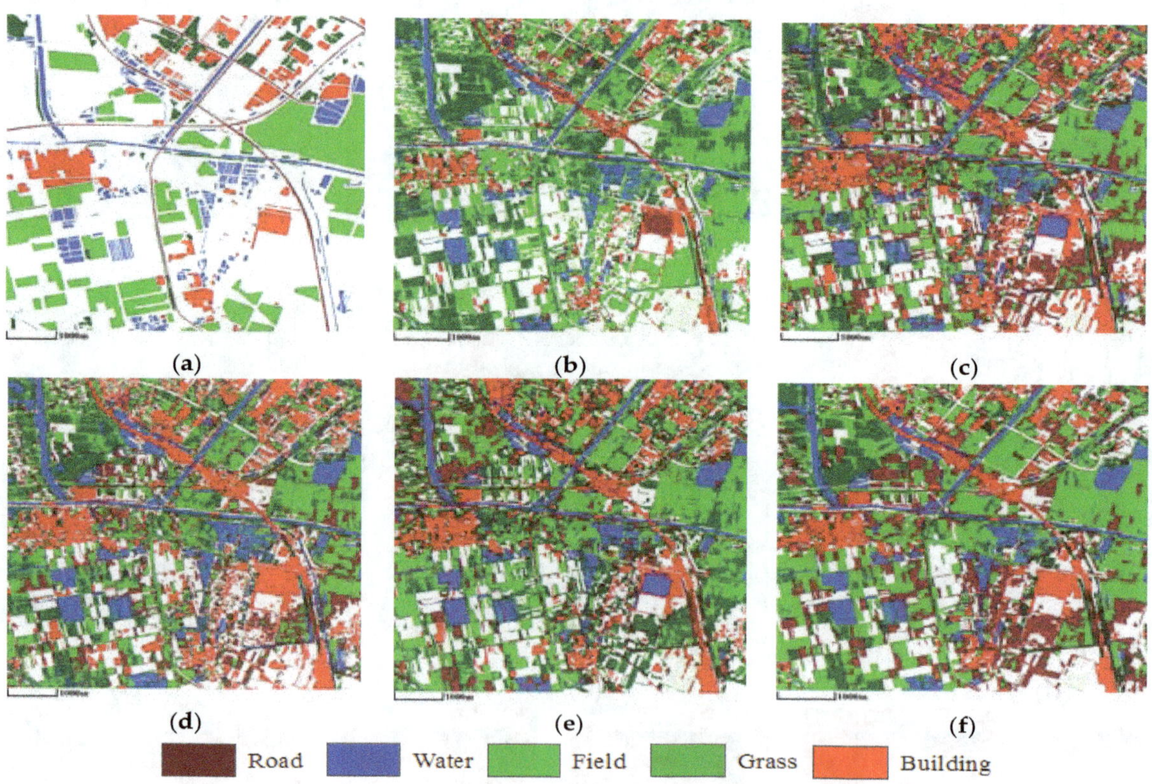

Figure 5. Classification results of various methods for the ZY-3 image. (**a**) Ground truth map; (**b**) O_ISODATA; (**c**) O_LDA; (**d**) msLDA; (**e**) HDP_IBP; and (**f**) mSegLDA.

Table 3. OA and OE of various methods for the ZY-3 image.

	O_ISODATA	O_LDA	msLDA	HDP_IBP	mSegLDA
OA	63.0	58.9	61.6	59.2	65.7
OE	0.78	0.93	0.85	0.87	0.74

3.4. Analysis of Computational Efficiency

The computational efficiency of the mSegLDA is also compared with that of other three topic model-based methods using the QuickBird image as an example, i.e., the O_LDA, the msLDA and the HDP_IBP. All the methods are coded using MATLAB R2013b (The MathWorks, Inc., Natick, MA, USA), and have been performed on a PC with an Intel (R) Core (TM) i7-4710MQ 2.50 GHz CPU and 8.00 GB RAM. As can be seen in Table 4, the O_LDA and the HDP_IBP spend a relatively less amount of running time compared to the msLDA and the mSegLDA due to their modelling mechanism, i.e., utilizing only a single-scale segmentation map as a corpus. Instead, since the mSegLDA needs to use multiple segmentation maps for the topic model inference and calculate a large number of KL divergences between segments, it is less efficient than the O_LDA and the HDP_IBP. However, the computation efficiency of the mSegLDA is better than that of the msLDA, because each pixel and its surrounding pixels within the square neighborhood represent a document in the msLDA, resulting in the significant increase in the number of documents and accompanying extensive computation.

Table 4. Running time of various methods for the QuickBird image.

Methods	Running Time (in Seconds)
O_LDA	797
msLDA	5056
HDP_IBP	766
mSegLDA	2413

In order to speed up the execution of the proposed mSegLDA, the Gibbs sampling component of the mSegLDA, which is computationally expensive, has been written using C++ MEX code. The running time of the mSegLDA for the QuickBird image is approximately 65 s.

3.5. Analysis of Scale Setting

3.5.1. Influence of Different Settings on the Range of Scales

As the number of scales S may influence the classification results, we, therefore, analyze how the performance of the proposed mSegLDA behaves with different settings on the range of scales using the Mapping Satellite-1 and QuickBird images. Since the ERSS algorithm utilizes the number of segments to control the scale size of image segmentation, the number of segments could be equivalent to the size of scales. Given the candidate scale set {100, 200, 500, 800, 1000, 1500, 2000, 2500, 3000, 3500, 4000}, for both the Mapping Satellite-1 and QuickBird images, a set of experiments for two images with different settings on the range of scales by adding one scale every time, i.e., {100, 200}, {100, 200, 500}, {100, 200, 500, 800} . . . , {100, 200, 500, 800, 1000, 1500, 2000, 2500, 3000, 3500, 4000}, are modelled using the mSegLDA.

Figures 6 and 7 show the values of OA and OE against the different settings on the range of scales. As can be seen, as the number of scales S increases, there is an approximately monotonic increase in OA and decrease in OE, and the classification performance remains relatively stable when S is larger than 6 for the Mapping Satellite-1 and 9 for the QuickBird image. This is due to the reason that, the mSegLDA approach needs to create a series of segmentation maps at multiple scales from fine to coarse ones for modelling, and the ideal multiscale image segmentation representation is expected to be able to represent all structural patterns at different scales as much as possible. Hence, a too small S, e.g., {100, 200}, means that only large-scale structure information can be characterized. As the increase of S, e.g., {100, 200, 500, 800, 1000, 1500} and {100, 200, 500, 800, 1000, 1500, 2000, 2500, 3000}, the fine to coarse range of scales makes it possible to characterize multiscale features of various geo-objects. On the other hand, although it can also ensure all structural patterns at different scales are presented, a larger value of S increases the computational efficiency.

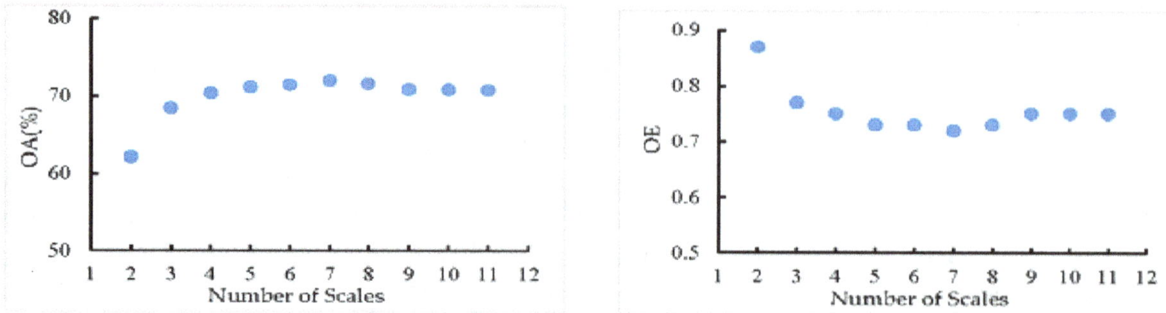

(a) (b)

Figure 6. OA and OE versus the number of scales for the Mapping Satellite-1 image. (**a**) Influence on OA; and (**b**) influence on OE.

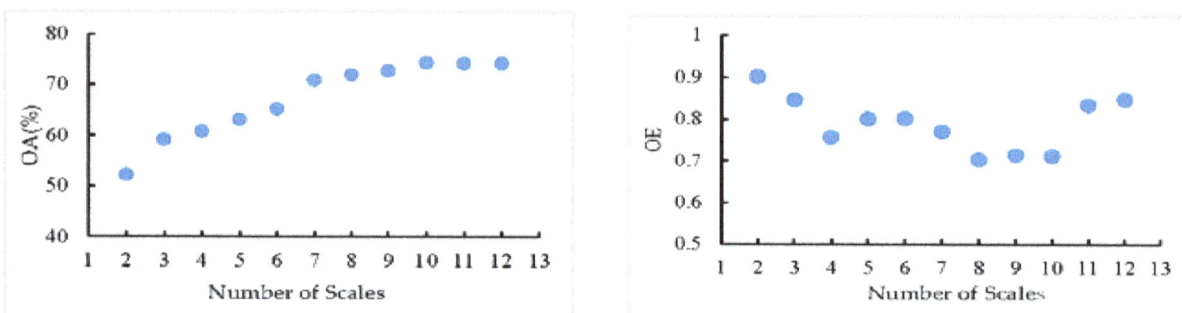

(a) (b)

Figure 7. OA and OE versus the number of scales for the QuickBird image. (**a**) Influence on OA; and (**b**) influence on OE.

Following the above rules, the range of scales for three experiment images are set to {100, 200, 500, 800, 1000, 1500}, {100, 200, 500, 800, 1000, 1500, 2000, 2500, 3000}, and {2000, 3500, 5000, 6500, 8000, 9500, 11,000, 12,500, 14,000, 15,500, 17,000}, respectively.

3.5.2. Special Cases of the mSegLDA

In this subsection, we analyze several special cases of the mSegLDA for the Mapping Satellite-1 and QuickBird images qualitatively and quantitatively to evaluate the effectiveness of modelling multiple segmentations by setting the number of scales to 1, i.e.,

- Case #1: the mSegLDA based on a single-segmentation map with 100 segments;
- Case #2: the mSegLDA based on a single-segmentation map with 200 segments;
- Case #3: the mSegLDA based on a single-segmentation map with 500 segments;
- Case #4: the mSegLDA based on a single-segmentation map with 800 segments;
- Case #5: the mSegLDA based on a single-segmentation map with 1000 segments;
- Case #6: the mSegLDA based on a single-segmentation map with 1500 segments;
- Case #7: the mSegLDA based on a single-segmentation map with 2000 segments;
- Case #8: the mSegLDA based on a single-segmentation map with 2500 segments; and
- Case #9: the mSegLDA based on a single-segmentation map with 3000 segments.

As shown in Figures 8 and 9, a smaller number of segments results in a more heavily smoothed classification result. However, it also filters the detailed patterns of certain geo-objects. On the contrary, a larger number of segments produces a relatively fragmented result. The advantage of integrating multiscale segmentation maps for modelling is further confirmed, as shown Tables 5 and 6.

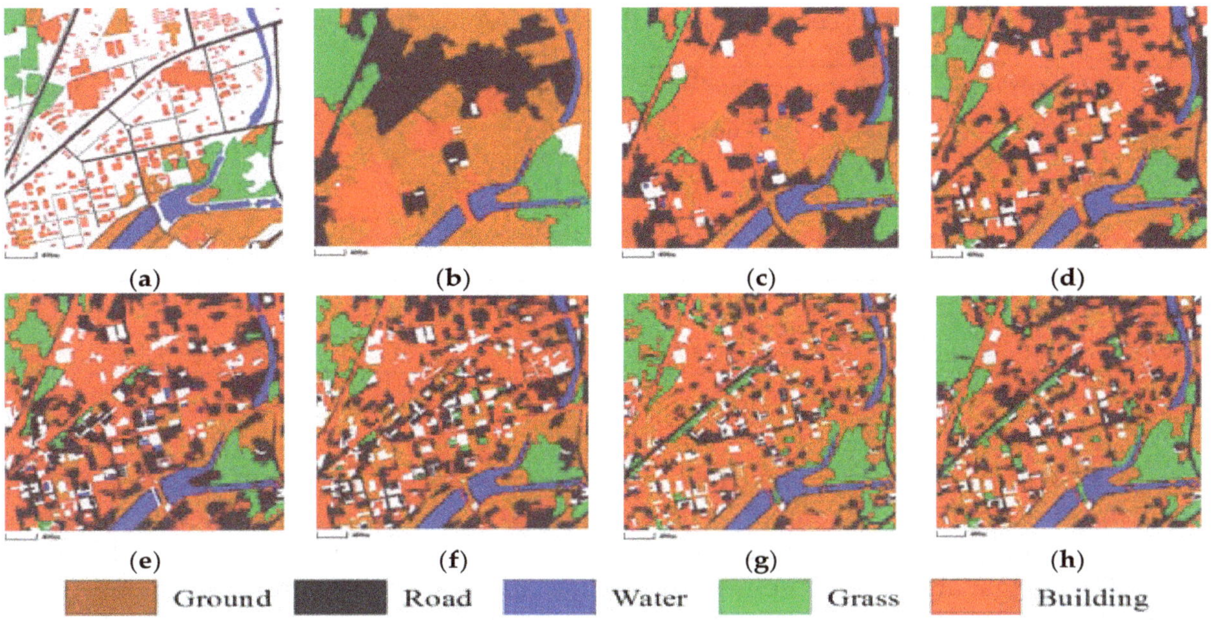

Figure 8. Classification results of six special cases of the mSegLDA for the Mapping Satellite-1 image. (**a**) Ground truth map; (**b**) Case #1; (**c**) Case #2; (**d**) Case #3; (**e**) Case #4; (**f**) Case #5; (**g**) Case #6; and (**h**) mSegLDA.

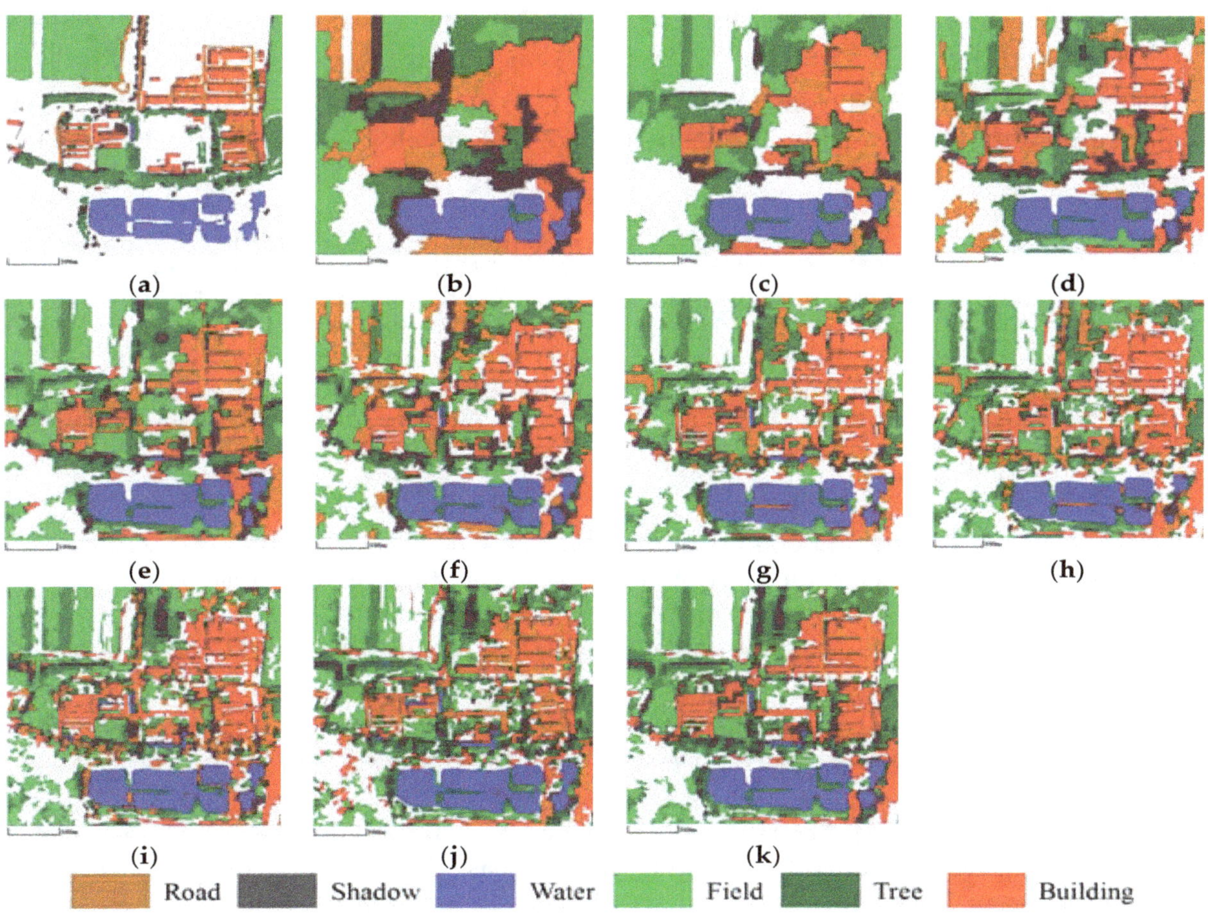

Figure 9. Classification results of nine special cases of the mSegLDA for the QuickBird image. (**a**) Ground truth map; (**b**) Case #1; (**c**) Case #2; (**d**) Case #3; (**e**) Case #4; (**f**) Case #5; (**g**) Case #6; (**h**) Case #7; (**i**) Case #8; (**j**) Case #9; and (**k**) mSegLDA.

Table 5. OA and OE of six special cases of the mSegLDA for the Mapping Satellite-1 image.

	Case #1	Case #2	Case #3	Case #4	Case #5	Case #6	mSegLDA
OA	64.1	62.6	66.2	66.1	68.1	66.0	72.0
OE	0.87	0.89	0.83	0.83	0.81	0.81	0.72

Table 6. OA and OE of nine special cases of the mSegLDA for the QuickBird image.

	Case #1	Case #2	Case #3	Case #4	Case #5	Case #6	Case #7	Case #8	Case #9	mSegLDA
OA	57.8	67.7	59.7	60.1	65.4	67.1	63.7	66.5	0.71	74.3
OE	0.87	0.87	0.76	0.75	0.85	0.74	0.74	0.79	0.78	0.72

4. Conclusions

This paper has presented a novel unsupervised object-based approach named mSegLDA for the classification of VHR panchromatic satellite images. The approach addresses the issues that: (1) various structural patterns at different scales are usually presented simultaneously in the same scene of a VHR image; and (2) different geo-objects may have nearly identical spectra. Our major contribution is to propose an automatic framework that combines the latent Dirichlet allocation with a multiscale image segmentation representation to model both the co-occurrence of various geo-objects and multiscale structures. Experimental results using VHR panchromatic satellite images with different scenes and spatial resolutions indicate that the proposed approach can achieve scale-adaptive classification results, and improve the ability to differentiate the geo-objects with spectral overlap, such as water and grass, and water and shadow.

Furthermore, the proposed framework still needs to determine the range of scales for creating multiscale segmentation empirically. In the future, we will develop an automatic strategy to obtain the appropriate range.

Acknowledgments: This work is supported in part by the National Key Research and Development Program of China under grant 2016YFB0501403, in part by the National Natural Science Foundation of China under grant 41401374, in part by the National Basic Research Program (973 program) of China under grant 2012CB719901, and in part by the Fundamental Research Funds for the Central Universities under grant 2682016CX079.

Author Contributions: Li Shen and Linmei Wu conceived and designed the experiments; Linmei and Yanshuai Dai performed the experiments; Yanshuai Dai, Wenfan Qiao and Ying Wang analyzed the data, and Li Shen wrote the manuscript.

References

1. Benediktsson, J.A.; Chanussot, J.; Moon, W.M. Advances in very-high-resolution remote sensing. *Proc. IEEE* **2013**, *101*, 566–569. [CrossRef]
2. Blaschke, T. Object based image analysis for remote sensing. *ISPRS J. Photogramm. Remote Sens.* **2010**, *65*, 2–16. [CrossRef]
3. Myint, S.W.; Gober, P.; Brazel, A.; Grossman-Clarke, S.; Weng, Q. Per-pixel vs. Object-based classification of urban land cover extraction using high spatial resolution imagery. *Remote Sens. Environ.* **2011**, *115*, 1145–1161. [CrossRef]
4. Blaschke, T.; Hay, G.J.; Kelly, M.; Lang, S.; Hofmann, P.; Addink, E.; Feitosa, R.Q.; van der Meer, F.; van der Werff, H.; van Coillie, F.; et al. Geographic object-based image analysis-towards a new paradigm. *ISPRS J. Photogramm. Remote Sens.* **2014**, *87*, 180–191. [CrossRef] [PubMed]
5. Judah, A.; Hu, B.; Wang, J. An algorithm for boundary adjustment toward multi-scale adaptive segmentation of remotely sensed imagery. *Remote Sens.* **2014**, *6*, 3583–3610. [CrossRef]
6. Dragut, L.; Csillik, O.; Eisank, C.; Tiede, D. Automated parametrisation for multi-scale image segmentation on multiple layers. *ISPRS J. Photogramm. Remote Sens.* **2014**, *88*, 119–127. [CrossRef] [PubMed]

7. Yang, J.; He, Y.; Weng, Q. An automated method to parameterize segmentation scale by enhancing intrasegment homogeneity and intersegment heterogeneity. *IEEE Geosci. Remote Sens. Lett.* **2015**, *12*, 1282–1286. [CrossRef]
8. Zhang, L.; Jia, K.; Li, X.; Yuan, Q.; Zhao, X. Multi-scale segmentation appoach for object-based land-cover classification using high-resolution imagery. *Remote Sens. Lett.* **2014**, *5*, 73–82. [CrossRef]
9. Malisiewicz, T.; Efros, A.A. Improving spatial support for objects via multiple segmentations. In Proceedings of the 18th British Machine Vision Conference, Warwick, UK, 10–13 September 2007.
10. Russell, B.C.; Freeman, W.T.; Efros, A.A.; Sivic, J.; Zisserman, A. Using multiple segmentations to discover objects and their extent in image collections. In Proceedings of the 2006 IEEE Computer Society Conference on Computer Vision and Pattern Recognition, New York, NY, USA, 17–22 June 2006.
11. Pantofaru, C.; Schmid, C.; Hebert, M. Object recognition by integrating multiple image segmentations. In Proceedings of the 10th European Conference on Computer Vision, Marseille, France, 12–28 October 2008; pp. 481–494.
12. Karadağ, Ö.Ö.; Senaras, C.; Vural, F.T.Y. Segmentation fusion for building detection using domain-specific information. *IEEE J. Sel. Top. Appl. Earth Obs. Remote Sens.* **2015**, *8*, 3305–3315. [CrossRef]
13. Akcay, H.G.; Aksoy, S. Automatic detection of geospatial objects using multiple hierarchical segmentations. *IEEE Trans. Geosci. Remote Sens.* **2008**, *46*, 2097–2111. [CrossRef]
14. Santos, J.A.d.; Gosselin, P.-H.; Philipp-Foliguet, S.; Torres, R.d.S.; Falao, A.X. Multiscale classification of remote sensing images. *IEEE Trans. Geosci. Remote Sens.* **2012**, *50*, 3764–3775. [CrossRef]
15. Syu, J.-H.; Wang, S.-J.; Wang, L.-C. Hierarchical image segmentation based on iterative contraction and merging. *IEEE Trans. Image Process.* **2017**, *26*, 2246–2260. [CrossRef] [PubMed]
16. Blei, D.M.; Ng, A.Y.; Jordan, M.I. Latent dirichlet allocation. *J. Mach. Learn. Res.* **2003**, *3*, 993–1022.
17. Hofmann, T. Probabilistic latent semantic indexing. In Proceedings of the 22nd Annual International SIGIR Conference on Research and Development in Information, Retrieval Berkeley, CA, USA, 15–19 August 1999; pp. 50–57.
18. Teh, Y.W.; Jordan, M.I.; Beal, M.J.; Blei, D.M. Hierarchical dirichlet processes. *J. Am. Stat. Assoc.* **2006**, *101*, 1566–1581. [CrossRef]
19. Lienou, M.; Maitre, H.; Datcu, M. Semantic annotation of satellite images using latent dirichlet allocation. *IEEE Geosci. Remote Sens. Lett.* **2010**, *7*, 28–32. [CrossRef]
20. Bratasanu, D.; Nedelcu, I.; Datcu, M. Bridging the semantic gap for satellite image annotation and automatic mapping applications. *IEEE J. Sel. Top. Appl. Earth Obs. Remote Sens.* **2011**, *4*, 193–204. [CrossRef]
21. Luo, W.; Li, H.; Liu, G.; Zeng, L. Semantic annotation of satellite images using author-genre-topic model. *IEEE Trans. Geosci. Remote Sens.* **2013**, *52*, 1356–1368. [CrossRef]
22. Li, S.; Tang, H.; He, S.; Shu, Y.; Mao, T.; Li, J.; Xu, Z. Unsupervised detection of earthquake-triggered roof-holes from uav images using joint color and shape features. *IEEE Geosci. Remote Sens. Lett.* **2015**, *12*, 1823–1827.
23. Zhao, B.; Zhong, Y.; Zhang, L. Scene classification via latent dirichlet allocation using a hybrid generative/discriminative strategy for high spatial resolution remote sensing imagery. *Remote Sens. Lett.* **2013**, *4*, 1204–1213. [CrossRef]
24. Zhong, Y.; Zhu, Q.; Zhang, L. Scene classification based on the multifeature fusion probabilistic topic model for high spatial resolution remote sensing imagery. *IEEE Trans. Geosci. Remote Sens.* **2015**, *53*, 6207–6222. [CrossRef]
25. Zhao, B.; Zhong, Y.; Xia, G.-S.; Zhang, L. Dirichlet-derived multiple topic scene classification model for high spatial resolution remote sensing imagery. *IEEE Trans. Geosci. Remote Sens.* **2016**, *54*, 2108–2123. [CrossRef]
26. Yi, W.; Tang, H.; Chen, Y. An object-oriented semantic clustering algorithm for high-resolution remote sensing images using the aspect model. *IEEE Geosci. Remote Sens. Lett.* **2011**, *8*, 522–526. [CrossRef]
27. Tang, H.; Shen, L.; Qi, Y.; Chen, Y.; Shu, Y.; Li, J.; Clausi, D.A. A multiscale latent dirichlet allocation model for object-oriented clustering of vhr panchromatic satellite images. *IEEE Trans. Geosci. Remote Sens.* **2013**, *51*, 1680–1692. [CrossRef]
28. Xu, K.; Yang, W.; Liu, G.; Sun, H. Unsupervised satellite image classification using markov field topic model. *IEEE Geosci. Remote Sens. Lett.* **2013**, *10*, 130–134. [CrossRef]

29. Shen, L.; Tang, H.; Chen, Y.; Gong, A.; Li, J.; Yi, W. A semisupervised latent dirichlet allocation model for object-based classification of vhr panchromatic satellite images. *IEEE Geosci. Remote Sens. Lett.* **2014**, *11*, 863–867. [CrossRef]
30. Shu, Y.; Tang, H.; Li, J.; Mao, T.; He, S.; Gong, A.; Chen, Y.; Du, H. Object-based unsupervised classification of vhr panchromatic satellite images by combining the hdp and ibp on multiple scenes. *IEEE Trans. Geosci. Remote Sens.* **2015**, *53*, 6148–6162. [CrossRef]
31. Shen, L.; Wu, L.; Li, Z. Topic modelling for object-based classification of VHR satellite images based on multiscale segmentations. *Int. Arch. Photogramm. Remote Sens. Spat. Inf. Sci.* **2016**, *41*, 359–363. [CrossRef]
32. Barnard, K.; Duygulu, P.; Forsyth, D.; de Freitas, N.; Blei, D.M.; Jordan, M.I. Matching words and pictures. *J. Mach. Learn. Res.* **2003**, *3*, 1107–1135.
33. Heinrich, G. *Parameter Estimation for Text Analysis*; Technical Note; Vsonix GmbH: Darmstadt, Germany; University of Leipzig: Leipzig, Germany, 2008.
34. Tarabalka, Y.; Benediktsson, J.A.; Chanussot, J. Spectral-spatial classification of hyperspectral imagery based on partitional clustering techniques. *IEEE Trans. Geosci. Remote Sens.* **2009**, *47*, 2973–2987. [CrossRef]
35. Halkidi, M.; Batistakis, Y.; Vazirgiannis, M. On clustering validation techniques. *J. Intell. Inf. Syst.* **2001**, *17*, 107–145. [CrossRef]
36. Liu, M.-Y.; Tuzel, O.; Ramalingam, S.; Chellappa, R. Entropy rate superpixel segmentation. In Proceedings of the 13rd IEEE Conference on Computer Vision and Pattern Recognition, Colorado Springs, CO, USA, 20–25 June 2011; pp. 2097–2104.

5

Saliency Analysis via Hyperparameter Sparse Representation and Energy Distribution Optimization for Remote Sensing Images

Libao Zhang [1,2,*], Xinran Lv [1] and Xu Liang [1]

[1] The College of Information Science and Technology, Beijing Normal University, Beijing 100875, China; 201211211009@mail.bnu.edu.cn (X.L.); 201321210018@mail.bnu.edu.cn (X.L.)
[2] The State Key Laboratory of Remote Sensing Science, Beijing Normal University, Beijing 100875, China
* Correspondence: libaozhang@163.com.

Academic Editors: Qi Wang, Nicolas H. Younan, Carlos López-Martínez and Prasad S. Thenkabail

Abstract: In an effort to detect the region-of-interest (ROI) of remote sensing images with complex data distributions, sparse representation based on dictionary learning has been utilized, and has proved able to process high dimensional data adaptively and efficiently. In this paper, a visual attention model uniting hyperparameter sparse representation with energy distribution optimization is proposed for analyzing saliency and detecting ROIs in remote sensing images. A dictionary learning algorithm based on biological plausibility is adopted to generate the sparse feature space. This method only focuses on finite features, instead of various considerations of feature complexity and massive parameter tuning in other dictionary learning algorithms. In another portion of the model, aimed at obtaining the saliency map, the contribution of each feature is evaluated in a sparse feature space and the coding length of each feature is accumulated. Finally, we calculate the segmentation threshold using the saliency map and obtain the binary mask to separate the ROI from the original images. Experimental results show that the proposed model achieves better performance in saliency analysis and ROI detection for remote sensing images.

Keywords: saliency analysis; remote sensing; ROI detection; hyperparameter sparse representation; dictionary learning; energy distribution optimizing

1. Introduction

With the rapid progress of remote sensing technology, it is becoming easier to acquire high spatial resolution remote sensing images from various satellites and sensors. However, the analysis and processing of high spatial resolution images in more effective and efficient ways still remains a great challenge, particularly in images with complicated spatial information, clear details, and well-defined geographical objects [1–4].

The detection of the region of interest (ROI) has become a popular research topic, with valuable applications in many fields, such as object segmentation [5,6], image compression [7,8], video summarization [9], and photo collage [10,11]. Introducing ROI detection into remote sensing image processing has raised great concern among some scholars.

The human visual system serves as a filter for selecting a certain subset of visual information, based on visual saliency, while ignoring irrelevant information for further processing [12,13]. The region that draws human attention in an image is called ROI. There has been a lot of work done on saliency analysis and ROI extraction based on visual saliency, which is generally constructed based on low-level visual features, pure computation or a combination of these.

Itti et al. [14] developed a biologically-based model ITTI, which was named after the presenter, using "Difference of Gaussians" across multiple scales to implement "center-surround" contrast in color, intensity, and orientation features. Li et al. [15] presented a model based on Itti's method and additionally extracted GIST features trained by a support vector machine (SVM). Klein et al. [16] extracted ROIs with the knowledge of information theory. Although the models calculated visual saliency based on biological plausibility, the computing of center-surround involved the tuning of many parameters that determined the final performance.

In addition, pure computation based algorithms for ROI extraction have also been developed. Saliency analysis based on frequency domain has been shown in [17–19]. Imamoglu et al. [20] utilized the lower-level features produced by wavelet transform (WT). The above methods based on pure computing improve the efficiency of saliency processing. However, problems related to the complexity of modeling catering to different feature distributions and the lack of sufficient plausibility of biological visual saliency mechanisms are still unsolved.

With regard to mixed models, the Graph-based visual saliency (GBVS) model proposed by Harel et al. [21] applied the principles of Markov Chain theory to normalize activation maps on each extracted feature under the ITTI model. In 2012, Borji and Itti [22] utilized the sparse representation of the image and used local and global contrast in combination to detect saliency. Goferman et al. [23] combined local underlying clues and visual organization rules with methods of local contrast to highlight significant objects, and proposed a different model based on context-aware (CA) salient information. The CA model can detect the salient object in certain scenes, but the inevitably high false detection rate affects the accuracy. Another drawback of the model is that the time complexity is much higher than for other spatial-based saliency models. Wang et al. [24] proposed a visual saliency model based on selective contrast. Additionally, methods utilizing learning have also attracted attention in recent years, such as the model for saliency detection by multiple-instance learning [25].

In terms of the application of saliency analysis in remote sensing images, some have employed support vector machines (SVM) to extract bridges and airport runways from remote sensing images [26,27]. Some have constructed parameterized models to extract roads and airports from remote sensing images with prior information of targets [28–30]. Zhang et al. [31] proposed a frequency domain analysis (FDA) model based on the principle of Quaternion Fourier Transform to attain better experimental results compared with those that only used the information of amplitude spectrum or phase spectrum in the frequency domain. Zhang et al. also adopted multi-scale feature fusion (MFF) based on integer wavelet transform (IWT) to extract residential areas along the feature channels of intensity and orientation [32]. For some remote sensing images corrupted by noise, the saliency analysis of co-occurrence histogram (SACH) model uses a co-occurrence histogram to improve robustness against Gaussian and Salt and Pepper noises [33]. In addition, global clustering methods for image pre-classification or ROI detection are also introduced in remote sensing images [34–36]. For example, Lu et al. [36] first produced an initial clustering map, and then utilized a multiscale cluster histogram to analyze the spatial information around each pixel.

It is noticeable that the data sets of remote sensing images have a high volume of dimensional information, which is usually too large to handle effectively. Aiming at this problem, sparse codes have been introduced into image processing. Sparse codes learned from image patches are similar to the receptive fields of simple-cells in the primary visual cortex (V1) [37], which shows that the mechanism of human visual saliency is consistent with sparse representation. Sparse representation has also been shown to be a quite effective technique for wiping out non-essential or irrelevant information in order to reduce the dimensions. Furthermore, it has greater flexibility for data structure capture, and better stability against perturbations of the signal, which suggests that we can obtain the sparse coefficients produced by those basic functions with good robustness against noise or corruption.

Researchers have proposed a number of methods for dictionary learning. Independent Component Analysis (ICA) is a good method for learning a dictionary in order to obtain compact basic functions. Thus, ICA is mainly utilized for the learning of basic functions based on a large number of randomly selected image patches. In addition, there are also some other methods, such as DCT [38], DWT [39], K-SVD [40], and FOCUSS [41], which also perform well at forming sparse representation of datasets.

However, these methods are difficult to use when faced with different data modalities requiring specific extensive hyper-parameter tuning on each modality when learning a dictionary in remote sensing images. For DCT and DWT, there are three parameters that need to be considered: the number of extracted features; the sparsity penalty, which is used to balance sparsity and distortion during the learning process; and the size of mini-batch, which helps improve processing efficiency. For K-SVD, sparsity and dictionary size of the target should also be considered. For FOCUSS, the calculation of the final results needs a posteriori information. Therefore, the efficiency of these dictionary learning algorithms may run into a bottleneck when applied to remote sensing images.

Considering the problems mentioned above, we propose a model based on the integration of hyperparameter sparse representation and energy distribution optimization for saliency analysis. In this study, we focus on the ROI in optical remote sensing images. As a whole, the combination has full biological plausibility in terms of the human visual mechanism. In terms of sparse representation of remote sensing images, we adopt a novel feature learning algorithm—hyperparameter sparse representation—to train a dictionary. This algorithm is simple, clear and can be quickly implemented with high effectiveness, as well as being almost parameter-free, as the feature number is the only item to be decided. As for the measure of saliency, we use an energy distribution optimization algorithm to define saliency as entropy gain. Similarly, computation of this algorithm does not involve any parameter tuning, and is computationally efficient.

In the experimental process, we first transform the image from the RGB color space to the HSI color space as a preprocessing step. Subsequently, the input remote sensing images are divided into overlapping patches, and the patches are further decomposed over the learned dictionary. Then, an algorithm is utilized to maximize the entropy of visual saliency features for energy redistribution, so as to generate a final saliency map. Finally, Otsu's threshold segmentation method is implemented in the acquisition of binary masks from saliency maps, and the masks are then used for ROI extraction from the original remote sensing images. Experimental results show that the proposed model achieves better performance than other traditional models for saliency analysis of and ROI detection in remote sensing images.

There are three major contributions in our paper: (1) we introduce hyperparameter sparse representation into dictionary learning for remote sensing images. The algorithm converges faster and has fewer parameters; (2) while training the dictionary, we define every single pixel as a feature. Thus, the sparse representation of an image is equal to the optimal features used for further saliency analysis; and (3) hyperparameter sparse representation and energy distribution optimization of features are integrated to compute the saliency map. This method is biologically rational, and consistent with cortical visual information processing.

The work in this paper is organized as followed: the proposed model is thoroughly illustrated in Section 2, Section 3 focuses on the experimental results and discussion, Sections 4 and 5 provide the applications and conclusion, respectively.

2. Methodology

In the proposed model, the whole process of ROIs detection for remote sensing images can be divided into three parts: (1) obtain sparse representation of the image feature; (2) compute saliency contribution of all sparse features; (3) extract the ROIs from saliency maps. Figure 1 illustrates the framework of the proposed model. As we can see, in the first part, an unsupervised feature learning algorithm—Hyperparameter Sparse Representation—is utilized to create a dictionary for sparse representation of remote sensing images. We define every single pixel as a feature. Thus, the sparse representation of an image is equal to the optimal features that are used for further saliency analysis. The second part measures the entropy gain of each feature. On the basis of the general principle of predictive coding [42], the rarity of features can be seen as their average energy, which is redistributed to features in terms of their code length: frequently activated features receive less energy. The final saliency map is generated by summing up the activity of all features. Finally, we segment ROI from the original remote sensing image with the mask of saliency map based on the threshold segmentation algorithm [43].

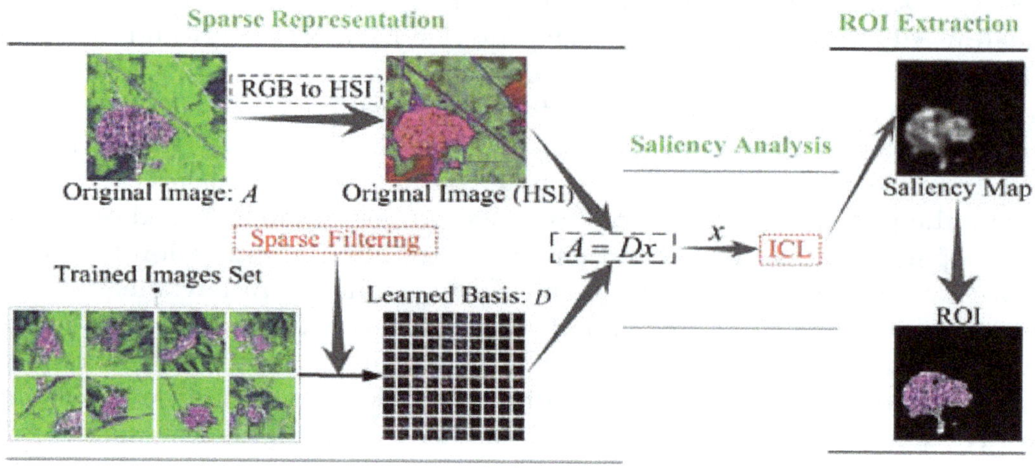

Figure 1. The framework of the proposed model.

Due to the characters of the simple computation, time efficiency and consistency in terms of the human color perception system of an HSI-based model [44], we preprocess images from RGB to HSI color space. Then the represented image is divided into overlapping patches and each patch is vectored as a column where all the pixel features were columned to form a feature matrix. Section 2.2, Section 2.3, Section 2.4 separately introduce the details of the three parts of our proposed model.

2.1. The Inadequacy of Traditional Algorithms

As we mentioned in Section 1, traditional visual saliency analysis methods have played an increasingly important role in the field of remote sensing image processing. Remote sensing images generally have high resolution and complex structure, which means that it is difficult to process directly. Visual attention models are first proposed for natural scene images. This kind of image is mostly obtained by different types of cameras, which means that we can highlight the significant targets by adjusting the aperture and the shutter. Targets will contain more information than background by selecting artificially. However, in remote sensing images, all objects have the same clarity. In other words, there is no difference in terms of clarity between the residential areas and the mountains, the roads and the ponds. Because of the clear and complex background, the problem of background interference is serious, which makes the saliency analysis hard.

The traditional methods need to combine the difference of the data distribution characteristics to select the effective calculation method for analysis, which will undoubtedly increase the diversity and complexity of the analysis. Moreover, the primary visual cortex shows that the receptive field of the single cell is similar to the sparse coding of the natural image block [45]. The human visual system also exhibits the characteristics of multilayer sparse representation of the image data. It shows that the sparse representation is consistent with the principle of human visual saliency mechanism, and can well explain the visual significance, which is biologically rational.

As shown in Figure 2, the ITTI model always mistakenly detects the background and sometimes misses the target region. The results of the frequency domain based model, Frequency-tuned (FT) model, contain a lot of debris and holes. The algorithms, which are designed specifically for ROI detection of remote sensing images, FDA and our model, obtain acceptable results. However, our results are clearly more accurate. In general, the ITTI and FT model are likely to get more inaccurate results, the FDA model makes some relative progress, and our model works best.

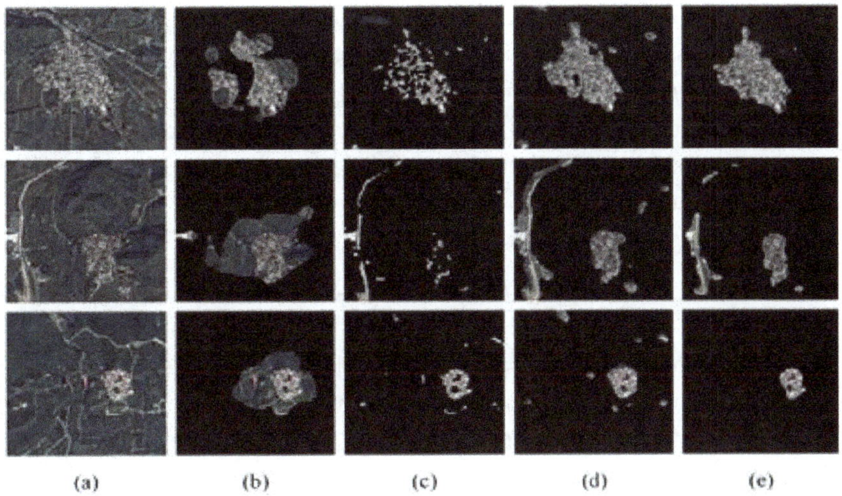

Figure 2. Region-of-interest (ROI) detection results produced by our model and the other 3 models. (**a**) origin images; (**b**) ITTI; (**c**) FT; (**d**) frequency domain analysis (FDA) and (**e**) our model.

2.2. Hyperparameter Sparse Representation

The method of dictionary learning can be considered as the generation of a particular feature distribution. For example, sparse representations are designed to use several nonzero coefficients to represent each sample, which highlight the main features of the sample. To achieve this goal, the ideal characteristics of the feature distribution should be optimized.

The desirable properties of feature distribution should meet with and include the three criteria [46]: population sparsity, lifetime sparsity and high dispersal. Population sparsity means that for each column in the feature matrix, there should be finite active (non-zero) elements. Moreover, it provides an effective coding method which is a theoretical basis for early visual cortex studies. Lifetime sparsity refers to that each row of feature matrix having only a small number of non-zero elements. This is because the features which are needed for further calculation ought to be characteristic of discrimination. High dispersal indicates that all features should have similar contributions, and the activity value of each row is supposed to be the same for every feature. Under certain circumstances, high dispersal is not completely necessary for good feature representation, on account of the same features which may be active and can prevent feature degeneration [46].

According to the characteristics that the sparse features should have, we apply a simple algorithm—hyperparameter sparse representation—which can optimize the three properties of features. Specifically, we illustrate these properties with a feature matrix of each sample. Figure 3 shows the structure of this algorithm.

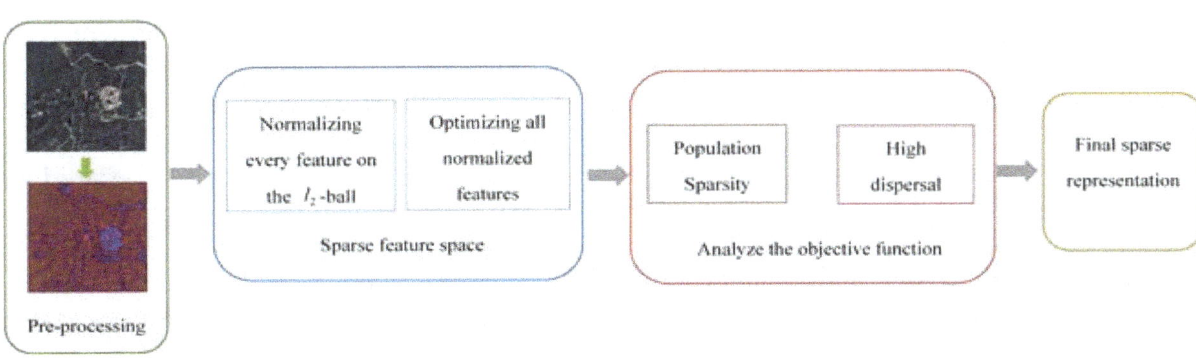

Figure 3. The structure of Hyperparameter Sparse Representation algorithm.

Each pixel column is viewed as a feature in our model. A feature matrix will be obtained after remote sensing image preprocessing. Each row of the matrix represents a feature and each column is a patch divided from the image. $f_j^{(i)}$ represents the jth feature value (rows) for the ith patch (columns). This sparse representation method aims to optimize and normalize the feature matrix by rows (feature values), then by columns (vectored image patch) and finally sums up the absolute value of all entries.

Firstly, by dividing each feature by its l_2-norm across all patches, each feature is normalized to be equally active:

$$\overset{\frown}{f_j} = f_j / \|f_j\|_2 \tag{1}$$

Then, analogously, by computing $\hat{f}^{(i)} = \overset{\frown}{f}^{(i)} / \|\overset{\frown}{f}^{(i)}\|_2$, all these features are normalized by each patch to put them on the l_2-norm ball. All normalized features are further optimized for sparsity by l_1 penalty. If there are M patches, then the sparse filtering objective function can be written as follows:

$$\min \sum_{i=1}^{M} \|\hat{f}^{(i)}\|_1 = \sum_{i=1}^{M} \left\| \frac{\overset{\frown}{f}^{(i)}}{\|\overset{\frown}{f}^{(i)}\|_2} \right\|_1 \tag{2}$$

Now it is essential to analyze whether the objective function meets with the three properties of desire features. First, population sparsity of features on the ith patch is measured by the equation as follows:

$$\|\hat{f}^{(i)}\|_1 = \left\| \frac{\overset{\frown}{f}^{(i)}}{\|\overset{\frown}{f}^{(i)}\|_2} \right\|_1 \tag{3}$$

when the features are sparse, an objective function can reach a minimum for the constraint of $\hat{f}^{(i)}$ in the l_2-norm ball. Contrarily, a patch that has similar values for each feature would incur a high penalty. Normalization of all features would cause competition between features: if only one element of $f^{(i)}$ increases, all the other elements in $f^{(i)}$ will decrease in the normalization, and vice versa. Minimal optimization of the objective function aims to make the normalization features sparse and mostly close to zero. With the principle of the competition between features, some features in $f^{(i)}$ have to be of large values while most of the rest of them are very small. To sum up, the objective function has been optimized for population sparsity.

Meanwhile, to satisfy the quality of high dispersion, each feature should equally active. As mentioned above, each feature is divided by its l_2-norm across all patches and normalized to be equally active by Equation (1). This is equal to constraining each feature to have the same expected squared value, thus contributing high dispersion. In the work of Ngiam et al. [47], they found that we can obtain over-complete sparse representation when realizing population sparsity and high dispersion in feature optimization, which also means that it is sufficient to learn good features as long as the condition of population sparsity and high dispersion are satisfied.

Therefore, obviously, the sparse filtering satisfies the three properties of desirable feature distribution and at the same time is also proved to be a fast and easy algorithm to implement. The entire optimization can be seen as the process of dictionary learning. When the objective function is optimized to reach a minimum under constraints, a dictionary D for sparse representation of the original image would appear to be the natural next-step before going on to process the image.

Notably, the entire optimization process of the feature matrix is automatically operated with the only tunable parameter: the number of the features. We can change the number of features by resizing the row number of the feature matrix to satisfy different requirements in image and signal processing. We can also learn that the dictionary learning process of the proposed model is approximately similar to the multi-layer sparsity by which the human vision system reacts to an image with the salient region from its surroundings.

2.3. Energy Distribution Optimizing

In this part, we describe the saliency of images with the optimized energy distribution (Algorithm 1), where different feature responses should have different energy intensity based on the principle of predictive coding. Therefore, incremental coding length is introduced to measure the distribution of energy on different features [48], which implies that different features have different rarity. The energy of the jth feature is defined as the ensemble's entropy gain during the activity of the jth feature. So the rarity of a dictionary feature is computed as its average energy. That is to say, rarely activated features will receive higher energy than activated ones. Then the final visually saliency is obtained by energy measurement, which shows that saliency computation by energy distribution conforms to the mechanism of human visual saliency in some degree.

Algorithm 1. Energy Distribution Optimizing

Input: A remote sensing image $A = [a_1, a_2, \cdots a_k, \cdots]$ and the liner filter $W = [w_1, w_2, \cdots w_k, \cdots]$.
  Vectorize the image patch a_k
for each feature **do**
  compute the activity ratio of the jth feature p_j.
  maximize the entropy $H(\mathbf{p})$.
  when a new excitation add a variation ε to p_i
    if $i = j$ $\hat{p_i} = (p_i + \varepsilon)/(1 + \varepsilon)$
    else $\hat{p_i} = p_i/(1 + \varepsilon)$
  end
  calculate the change of entropy of the jth feature $COE(p_j)$.
  get the salient features group $G = \{i|COE(p_i) > 0\}$
  compute the energy of the jth feature d_j
end
  obtain the saliency map m_k of image patch a_k

With the dictionary D for sparse representation mentioned above, the spare feature matrix X of image A on D can be acquired by $X = WA$, where $W = D^{-1}$. Then we can compute the activity ration p_j as follows:

$$p_j = \frac{\sum_k |w_j a_k|}{\sum_j \sum_k |w_j a_k|} \tag{4}$$

To fully consider the reaction degree of each feature in the sparse code and achieve optimality, maximizing the entropy $H(\mathbf{p})$ of the probability function $\mathbf{p}$ is a key principle to efficient coding. The probability function $\mathbf{p}$ varies at different points of time, depending upon whether there is a new perturbation on a feature, which means a variation ε will be added to p_i and further change the whole probability distribution.

This variation will change the entropy of the feature activities. We define the change of entropy of the jth feature $COE(p_j)$ as the following equation:

$$COE(p_j) = \frac{\partial H(\mathbf{p})}{\partial p_j} = -H(\mathbf{p}) - p_j - \log p_j - p_j \log p_j \tag{5}$$

The features with COE value above zero are viewed as salient and a salient feature set is obtained as G. Then the energy among features are redistributed according to their COE values. Denote the amount of energy that every sparse feature obtains d_j is computed as follows:

$$d_j = \begin{cases} \frac{COE(p_j)}{\sum_{j \in G} COE(p_j)} & j \in G \\ 0 & j \notin G \end{cases} \tag{6}$$

Finally, the saliency map $M = [m_1, m_2, \cdots m_k, \cdots]$ of image A can be obtained as the equation below:

$$m_k = \sum_{j \in G} d_j w_j a_k \tag{7}$$

The final saliency map can be obtained by restoring all the vectorization image patches to the whole original remote sensing image.

2.4. Threshold Segmentation

To further evaluate the performance of the proposed model, we segment the saliency maps from the original images and obtain masks of the ROIs with the threshold algorithm proposed by Otsu [43].

Assume that the total number of pixels in an image is N, gray values of the image range from 1 to L, and the number of pixels with gray value i in the entire image is n_i. The occurrence ratio of pixels is computed as follows:

$$\begin{gathered} p_i = n_i / N \ (i = 1, 2, \ldots L) \\ \sum_{i=1}^{L} p_i = 1 \end{gathered} \tag{8}$$

Suppose that the gray threshold value is k, pixels of the whole image is thus divided into two classes: A and B. Values in class A range from 1 to k, and values in class B from k + 1 to L. Their respective ratio is:

$$\begin{gathered} \omega_A = \sum_{i=1}^{k} p_i = \omega(k) \\ \omega_B = \sum_{i=k+1}^{k} p_i = 1 - \omega(k) \end{gathered} \tag{9}$$

Then, the average gray value of each cluster is:

$$\begin{gathered} \lambda_A = \sum_{i=1}^{k} i p_i / \omega_A = \frac{\lambda(k)}{\omega(k)} \\ \lambda_B = \sum_{i=1+k}^{L} i p_i / \omega_B = \frac{\lambda_T - \lambda(k)}{1 - \omega(k)} \end{gathered} \tag{10}$$

where $\lambda(k) = \sum_{i=1}^{k} i p_i$ and $\lambda_T = \sum_{i=1}^{L} i p_i$. λ_T is the average gray value of the whole image. The variance between A and B are calculated as follows:

$$\sigma^2(k) = \frac{[\lambda_T \omega(k) - \lambda(k)]^2}{\omega(k)[1 - \omega(k)]} \tag{11}$$

Then, the optimal segmentation threshold can be obtained by:

$$k^* = \underset{1 \le k \le L}{\text{argmax}} \sigma^2(k) \tag{12}$$

The segmentation threshold value varies for different saliency maps. With the image binary segmentation, the masks of the ROIs are produced, and the masks are overlaid onto the original images to extract the final ROI in the next step.

3. Experimental Results and Discussion

To evaluate the performance of the proposed model, we used 300 remote sensing images of two different kinds as the experimental data. One is the remote sensing images from the SPOT 5 satellite with a spatial resolution of 2.5 m; the other is the remote sensing images from Google Earth with a higher spatial resolution of 1.0 m. The size of the experimental data are all 512 × 512 pixels. Among experiment images, we define the rural residential regions as ROIs, which should be detected primarily. As we have presented before, these regions typically include rich texture, irregular boundary, the area of brightness and color highlighting.

For the proposed model, the size of all these images used for learning a dictionary is down-sampled to 128 × 128 pixels, considering that we chose each pixel as a feature for saliency detection and ROI extraction. Therefore, the time consumed will be unbelievably excessive if we directly process images of original size. For remote sensing images of each kind, we randomly selected 60 images of to train the dictionary for sparse representation and all the 150 images were demonstrated for saliency analysis and ROIs extraction. The performance of the proposed model was compared qualitatively and quantitatively with other nine models including the Itti's model (ITTI) [14], the frequency-tuned (FT) model [17], the spectral residual (SR) model [18], the Graph-based visual saliency (GBVS) model [21], the Wavelet-transform-based (WT) model [20], the context aware (CA) model [23], the multiscale feature fusion (MFF) model [32], the frequency domain analysis (FDA) model [31] and the saliency analysis of co-occurrence histogram (SACH) model [33]. These nine models are selected for the following reasons:

- high citation rate: The classic model ITTI and SR have been widely cited;
- variety: ITTI is biologically motivated; FT, SR, and WT model all are the purely computational based models and estimate saliency in the frequency domain; GBVS and CA both belong to biological models and partly to the computational model;
- affinity: MFF, FDA and SACH model all are specially designed for saliency analysis in remote sensing images.

Notably, we use resized original images of 128 × 128 pixels to test their respective performance on different models. Finally, we resized the saliency maps of all models uniformly to the size of 128 × 128 pixels for fair comparison. Here, in each kind of image, we choose eight out of all the 150 images to make up the display figures for our experimental results.

After the transformation from RGB to HSI color space, we divide all the input remote sensing images used for dictionary training into overlapped patches of the size of 8 × 8 pixels with 192-dimension and further form an up to 130,000 large set of vectorization image patches.

Here, what we should pay attention to is the selecting feature number which is the only tunable parameter in the process of dictionary learning. Generally, a greater numbers of features correlates to a better performance. For consistency with the input dimension of the vectorization image set to form a square matrix, we choose 192 features for dictionary learning and saliency analysis. In our experiments, we adopted the off-the-shelf L-BFGS [49] package to optimize the sparse filtering objective until convergence with a maximum iteration number of 100. The learned dictionary we have obtained is shown in Figure 4.

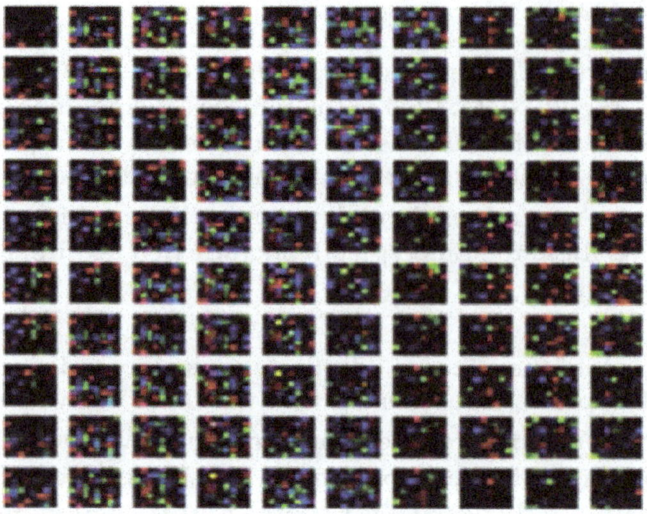

Figure 4. The learned dictionary.

3.1. Qualitative Experiment

As shown in Figures 5 and 6, the comparison among saliency maps generated by the proposed model and the other nine competing models on remote sensing images from SPOT 5 satellite and Google Earth, respectively. We can see that the saliency maps obtained by the proposed method focus on the residential areas and hardly have any background information. In contrast to the original images, the results of our model detected almost all salient objects. However, the other nine models detected some redundant information from the original images and cannot accurately locate the salient region. Although the CA model detects a clear boundary, it also includes the non-residential areas, thus enlarging the fall-out ratio and meanwhile is quite time-consuming.

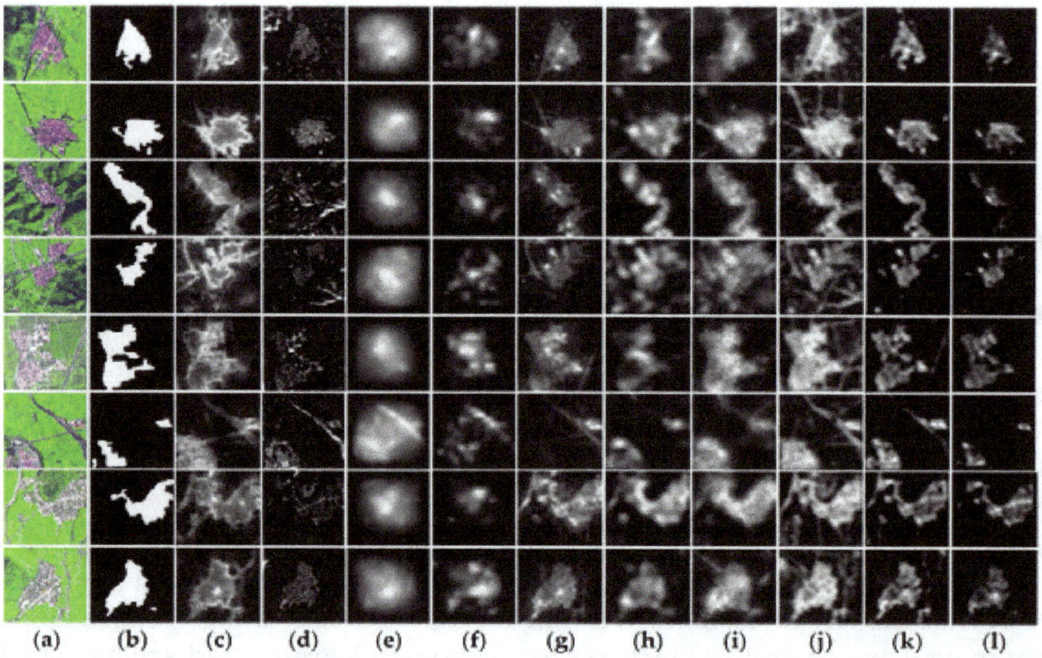

Figure 5. Saliency maps by our proposed model and nine competing models on SPOT 5 images. (**a**) Origin images; (**b**) Ground truth; (**c**) CA; (**d**) FT; (**e**) GBVS; (**f**) ITTI; (**g**) WT; (**h**) SR; (**i**) MFF; (**j**) SACH; (**k**) FDA and (**l**) Ours.

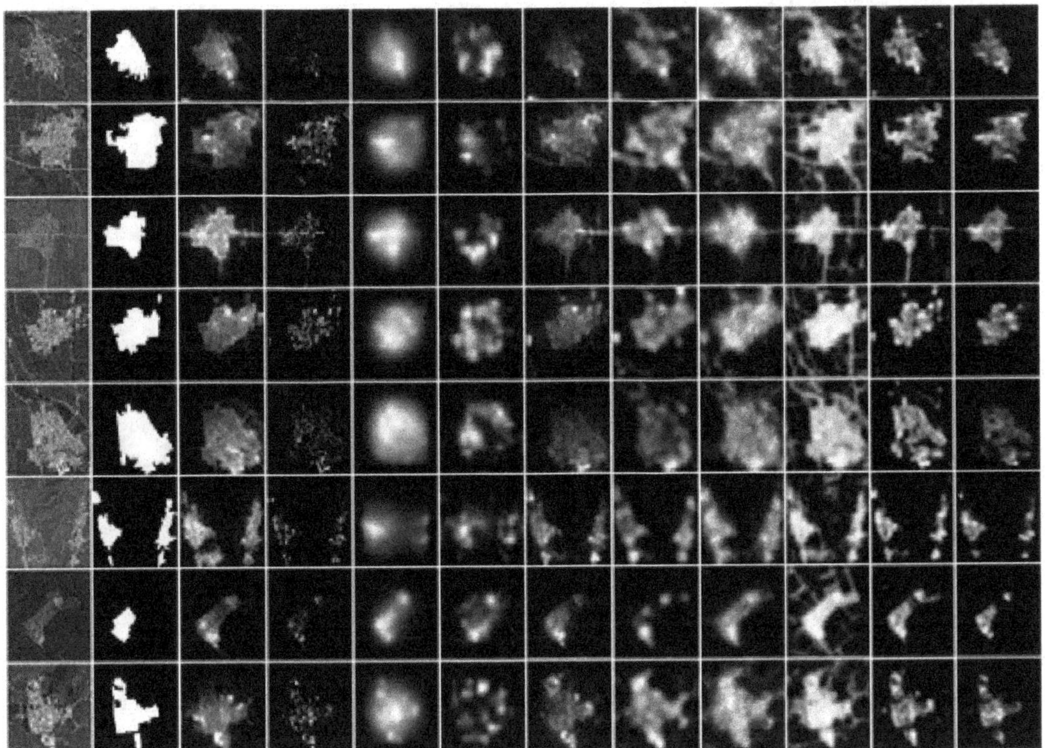

(a) (b) (c) (d) (e) (f) (g) (h) (i) (j) (k) (l)

Figure 6. Saliency maps by our proposed model and nine competing models on Google Earth images. (**a**) Origin images; (**b**) Ground truth; (**c**) CA; (**d**) FT; (**e**) GBVS; (**f**) ITTI; (**g**) WT; (**h**) SR; (**i**) MFF; (**j**) SACH; (**k**) FDA and (**l**) Ours.

For SPOT 5 images, the experimental results of FDA model seem close to ours but we can see that there are still some little non-salient regions such as roads contained in the last four saliency maps in Figure 5. The MFF and SACH model can also obtain saliency maps which are not bad, but they are not accurate enough. Other models such as the ITTI, GBVS, and SR generate the final saliency maps of low resolution with blurred boundaries, which do not contribute to further ROI extraction. The CA and WT model always get acceptable results, but the inevitable needless background information can always be highlighted, too. Conversely, FT model fails to highlight the entire salient area, which results in the so-called hole effect that is the incomplete description of the salient area's interior. Meanwhile, for Google Earth images, although the performance of all the other models on saliency details such as border information is a little worse than that on SPOT images because of the higher spatial resolution, the proposed model still performs better intuitively.

Similarly, we can see the ROIs extraction results for two kinds of images from Figures 7 and 8 after Otsu's threshold segmentation. For the other nine models, some extracted ROIs are not able to completely contain the residential areas while some ROIs include excessively large redundant background information such as roads, especially in the ROI extraction results of the ITTI model and the GBVS model. In contrast, the proposed model exactly extracts the ROIs with clear boundaries and also has a good performance for remote sensing images with complex background, especially for the images with non-salient regions inside the outline of the residential areas and those with more than one salient region, as is shown in the ROI extraction result on the fifth and sixth images in Figure 7.

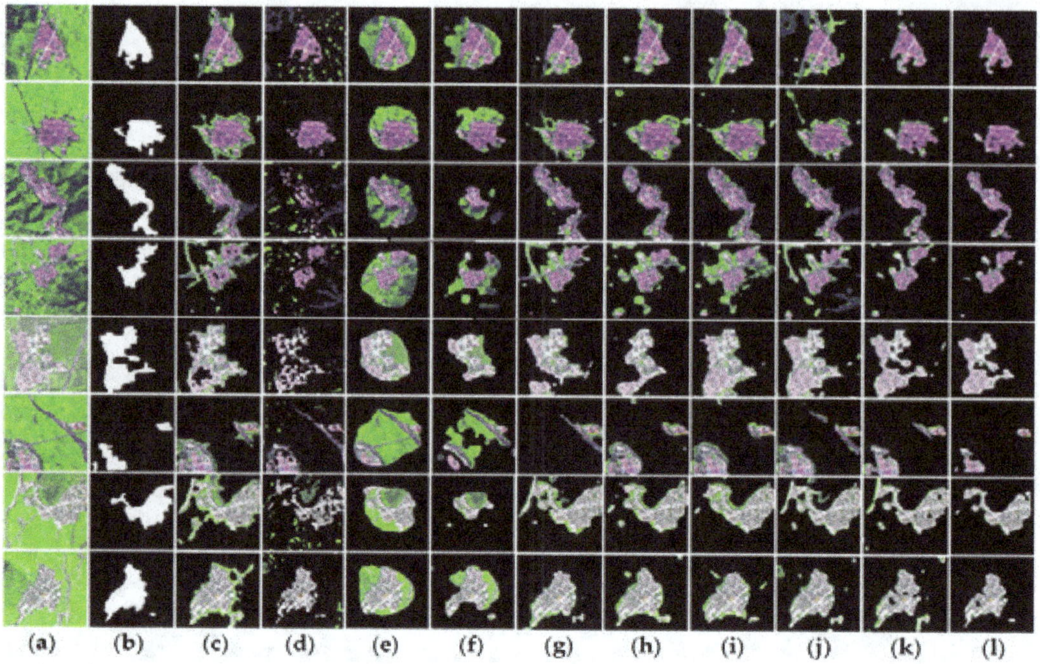

Figure 7. ROIs extracted by our proposed model and nine competing models on SPOT 5 images. (**a**) Origin images; (**b**) Ground truth; (**c**) CA; (**d**) FT; (**e**) GBVS; (**f**) ITTI; (**g**) WT; (**h**) SR; (**i**) MFF; (**j**) SACH; (**k**) FDA and (**l**) ours.

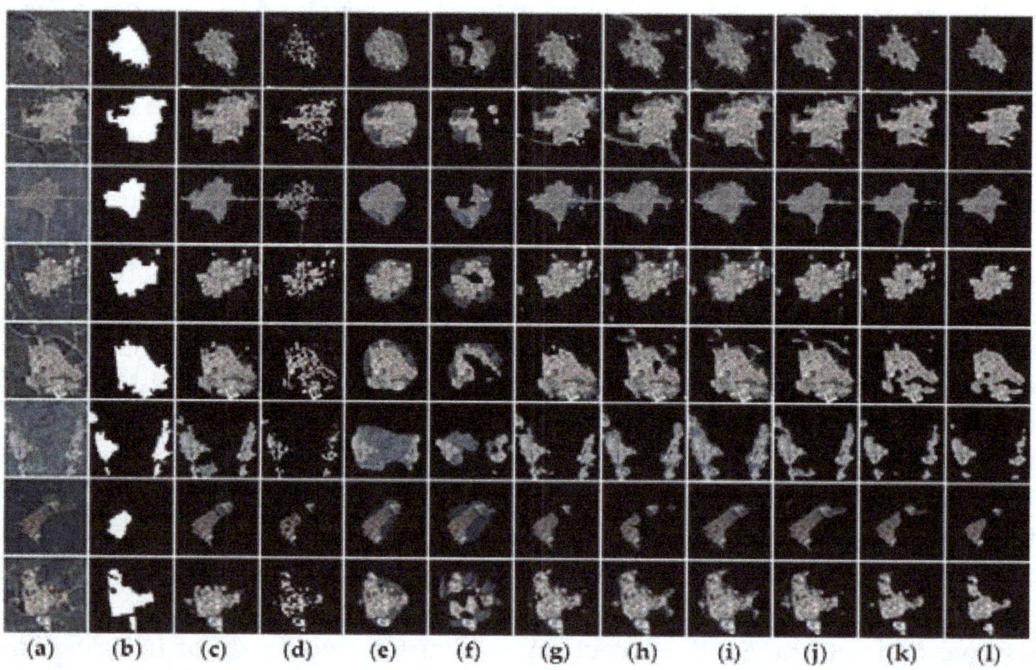

Figure 8. ROIs extracted by our proposed model and nine competing models on Google Earth images. (**a**) Origin images; (**b**) Ground truth; (**c**) CA; (**d**) FT; (**e**) GBVS; (**f**) ITTI; (**g**) WT; (**h**) SR; (**i**) MFF; (**j**) SACH; (**k**) FDA and (**l**) ours.

On a qualitative level, the experimental results show that the proposed model can not only generate saliency maps with a clear boundary with no excessive redundant background information, but also extracts exactly the ROIs with irregular shape and multi-saliency.

3.2. Quantitative Experiment

In the quantitative analysis of the experiment results, the ROC (Receiver Operator Characteristic) curve is adopted to measure the performance of different models. The ROC curve is derived by thresholding a saliency map at the threshold within the range [0, 255] and further classifying the saliency map into the ROIs and the background. The True Positive Rate (TPR) and the False Positive Rate (FPR) are two dimensions for spanning the ROC curve and respectively denote the percentage of the ROIs from the ground truth intersecting with the ROI from the saliency map and the percentage of the remaining background except for the ROIs. They are both computed as follows:

$$TPR = \frac{\sum_{i=1}^{M}\sum_{j=1}^{N} g(i,j)s(i,j)}{\sum_{i=1}^{M}\sum_{j=1}^{N} g(i,j)} \tag{13}$$

$$FPR = \frac{\sum_{i=1}^{M}\sum_{j=1}^{N} [1-g(i,j)]s(i,j)}{\sum_{i=1}^{M}\sum_{j=1}^{N} [1-g(i,j)]} \tag{14}$$

where, for an $M \times N$ image, g denotes the ground truth, s denotes the saliency map after the binary image, and (i,j) denotes the coordinate of the images. A higher TPR value indicates a better performance when the FPR value is the same and, conversely, better performance depends on a smaller FPR value at the same TPR value. The area beneath the curve is called the Area Under the Curve (AUC). Thus, a larger AUC indicates better performance. The AUCs of all the models are shown in Tables 1 and 2. From the Tables we can see that our model obtains the largest value of AUC compared to the other nine competing models, thus achieving better performance.

Table 1. The Area Under the Curve (AUC)s of our proposed model and nine competing models on SPOT 5 images.

Model	CA	FT	GBVS	ITTI	WT	SR	MFF	SACH	FDA	OURS
AUC	0.8832	0.9008	0.8216	0.7973	0.8934	0.9107	0.9278	0.9350	0.9408	0.9629

Table 2. The AUCs of our proposed model and nine competing models on Google Earth images.

Model	CA	FT	GBVS	ITTI	WT	SR	MFF	SACH	FDA	OURS
AUC	0.9274	0.9227	0.9267	0.8634	0.9531	0.9354	0.9639	0.9889	0.9789	0.9887

Similarly, we used two kinds of resized remote sensing images of 128×128 pixel size to test our model's performance. For each image, a manually segmented binary map using graphic software was generated as the ground truth. The average TPR and FPR values of every model are computed, and their ROCs on two kinds of images are shown in Figure 9a,b, respectively. From Figure 9a, we can conclude that the ROC curve that our model generated seems to show better performance than the others. However, we can see from Figure 9b that the performance of the SACH model is slightly better than our model whose ROC trace almost coincides with the other one. Therefore, we can know that the same model may have different performance for different kinds of remote sensing images, such as the FDA model and SACH model. The AUC comparison in Figure 10a,b further verifies our conclusion exactly, meanwhile, the Tables 1 and 2 also show the clear value of AUC.

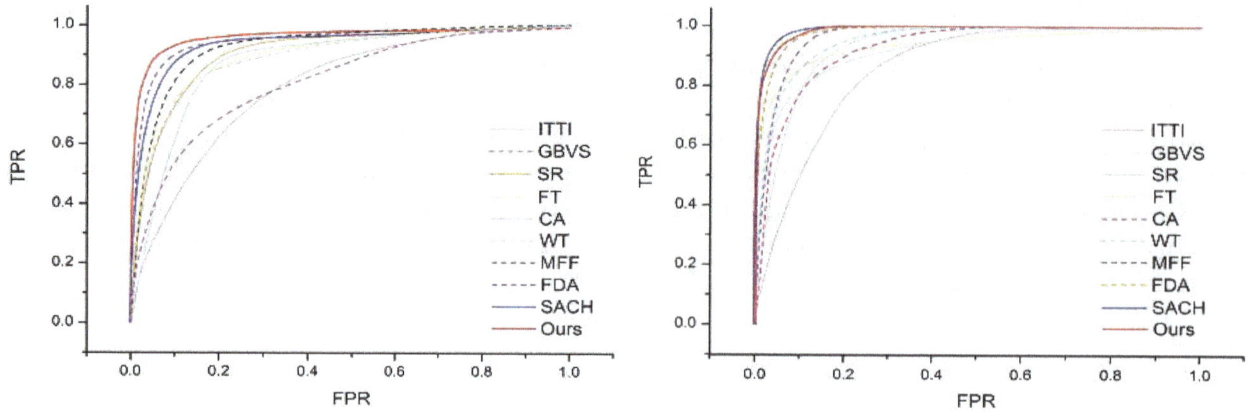

(a) (b)

Figure 9. ROC curves of our proposed model and nine competing models on (**a**) SPOT 5 and (**b**) Google Earth images.

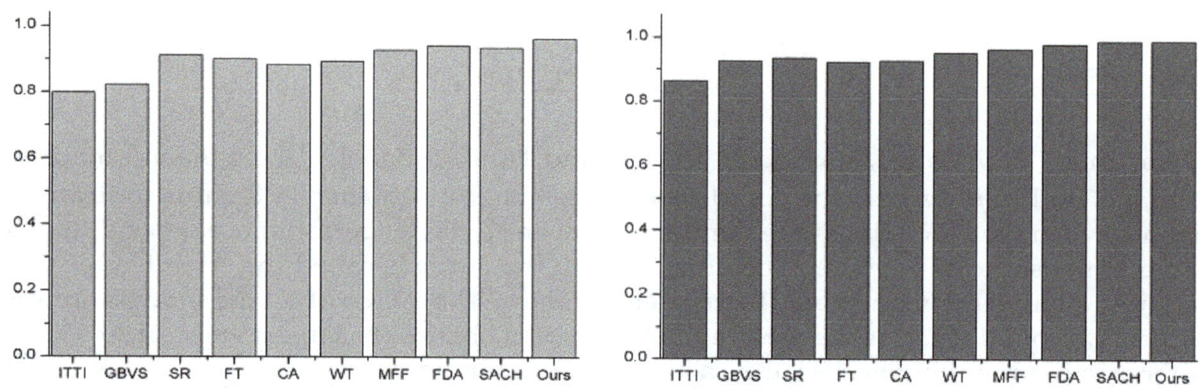

(a) (b)

Figure 10. AUC of ROC curves of our proposed model and nine competing models on (**a**) SPOT 5 and (**b**) Google Earth images.

Another method based on Precision, Recall and the F-Measure which are denoted as P, R and F is also adopted to further evaluate the model's performance. They are computed as follows and the comparison of different models is shown in Figure 11a,b.

$$P = \frac{\sum_{x=1}^{M} \sum_{y=1}^{N} t(x,y)s(x,y)}{\sum_{x=1}^{M} \sum_{y=1}^{N} s(x,y)} \tag{15}$$

$$R = \frac{\sum_{x=1}^{M} \sum_{y=1}^{N} t(x,y)s(x,y)}{\sum_{x=1}^{M} \sum_{y=1}^{N} t(x,y)} \tag{16}$$

$$F_\beta = (1+\beta^2)\frac{P \cdot R}{\beta^2 \cdot P + R} \tag{17}$$

where, for an image with size of $M \times N$, $t(x,y)$ denotes the ground truth, and $s(x,y)$ denotes the saliency map. The β serves as an indicator for the relative importance between precision and recall.

The larger the value of β, the more emphasis we put on recall than precision and vice versa. We choose $\beta = 1$ to equally balance the weight in our experiment.

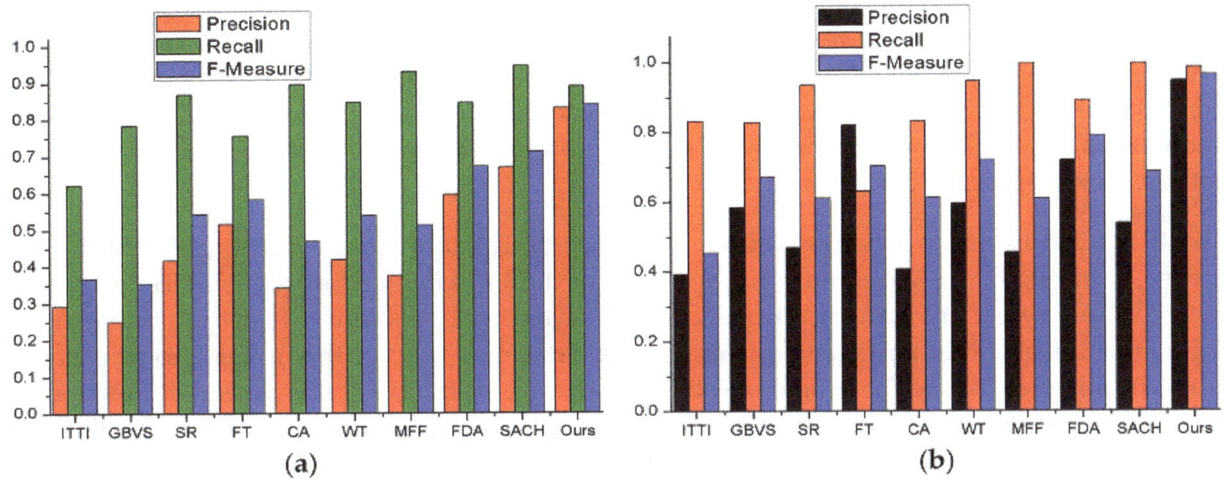

Figure 11. Precision, Recall and F-Measure of ROIs by our proposed model and nine competing models on (**a**) SPOT 5 and (**b**) Google Earth images.

From Figure 11a,b the precision of our model is obviously much higher than the other nine competing models, which means our model returns substantially more salient regions than background regions. Based on the previous qualitative analysis, the CA, WT, SR, MFF, SACH and FDA models achieve higher recall than the proposed model, probably because these models capture not only salient areas but some little non-salient regions with blurred boundaries. Meanwhile, this can be obtained clearly and reasonably according to Equation (17). Although the Recall is not the highest among these models, and in Google Earth dataset our ROC curve is slightly worse than SACH, our model still achieves the highest F-measure, thus showing better performance than others on different kinds of remote sensing images.

Additionally, we have compared the computational time for each method using matlab on a PC with 8 G RAM, Intel Core i3-4170 CPU @ 3.70 GHz. For the proposed model, the size of all these images used for learning a dictionary is down-sampled to 128 $\times$ 128 pixels. Here, we resized all images to the size of 128 $\times$ 128 pixels for fair comparison. From the Table 3 we can see that the run time of our proposed model is in the middle of the ten methods.

The FDA, FT, SR, ITTI and SACH model have a shorter run time than our model. The ITTI, FT and SR model are not proposed for remote sensing images. They do not take into account the complex background of remote sensing images, and use only a few simple features for analysis. The models FDA and SACH are specially designed for remote sensing images. For the former, there remain some holes in ROIs and the latter is not as high as our F-measure evaluation.

The MFF, GBVS, WT and CA model have a longer run time than our model. GBVS generates the final saliency maps of low resolution with blurred boundaries. WT and CA can always get acceptable results some non-salient regions were still extracted. Although MFF does not perform badly, it is not accurate enough.

Table 3. Running time comparisons for 10 models.

Model	FDA	FT	SR	ITTI	SACH	MFF	GBVS	WT	CA	OURS
AUC	0.85	1.72	2.04	3.68	4.83	6.81	18.43	106.51	1664	5.72

4. Applications

Because of the development of remote sensing technology, remote sensing image registration and fusion have been paid more and more attention in this field. Some researchers have applied region based image fusion algorithms to remote sensing images [50]. In the previous section, our experiments show that our model can extract ROI accurately from high resolution remote sensing images. Therefore, according to the region information provided by our model and the Gauss Pyramid decomposition, we can obtain more details from different scales of the original images, and then carry out image fusion to construct a clearer and accurate map.

The JPEG 2000 standard demonstrates many attractive features, including the ROI definition. In this case, ROI needs to encode with higher quality than the background [51]. However, knowing how to accurately select investment returns is still a prominent problem. Therefore, the results of our model can also be applied to image compression. The saliency map of the image can be detected and the visual importance [52] of the image pixels is measured, so that ROI can be considered as a step in the process of image compression priority encoding. According to Figure 12, the ROI still has a high subjective quality even at low bit rates (e.g., 0.5 bpp).

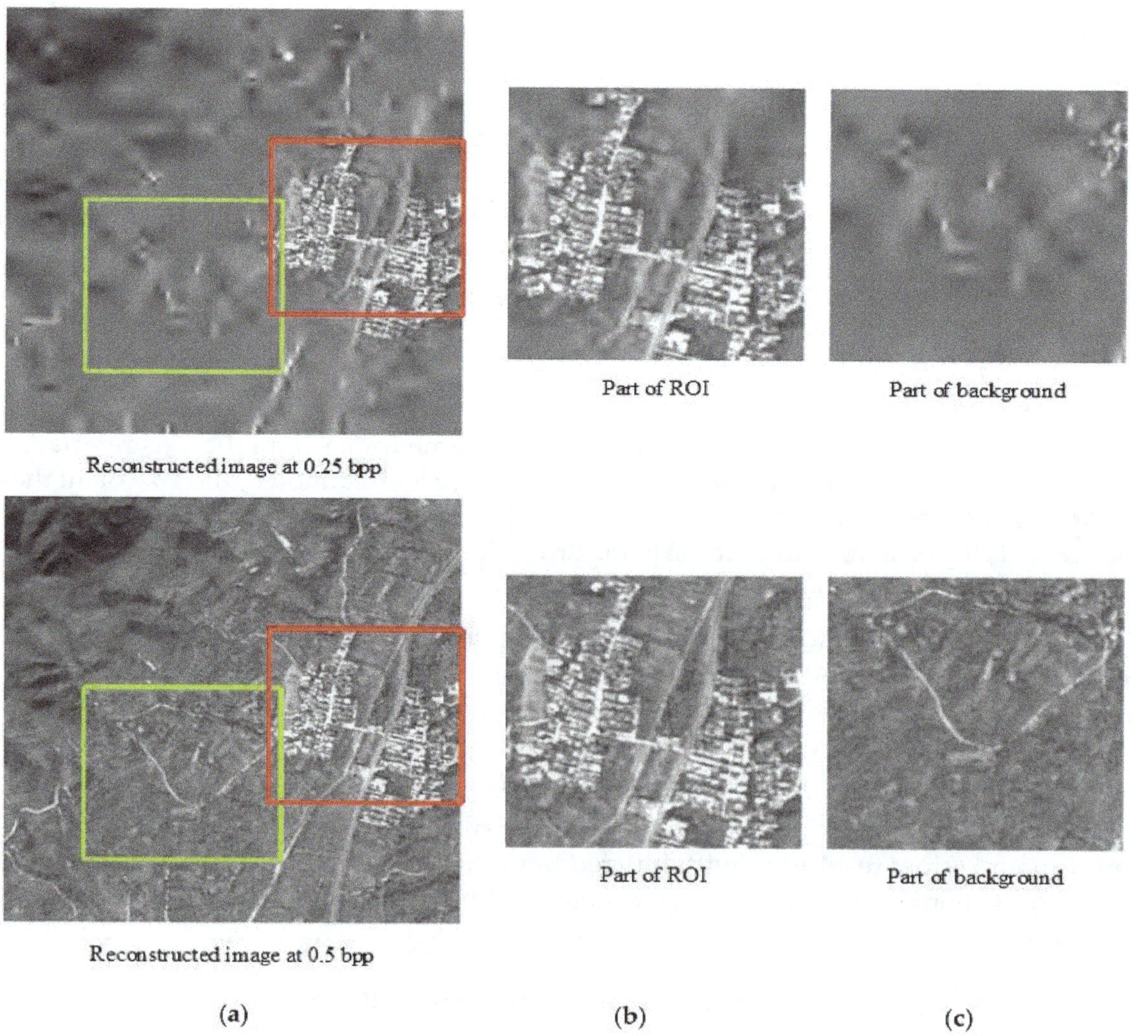

Figure 12. ROI compression example of remote sensing image. (**a**) reconstructed image; (**b**) part of ROI; and (**c**) part of background region. From top to bottom: reconstructed images are 0.5 bpp and 2.0 bpp, respectively.

5. Conclusions

This paper proposes a novel model based on hyperparameter sparse representation and energy distribution optimizing for saliency analysis and ROI detection in remote sensing images. The proposed model is simple to use and makes up the deficiency of biological plausibility as well as achieving better performance on saliency analysis and ROI detection. In this model, we firstly down-sample the original images and then transform them to HSI color space to increase the efficiency for further processing. After the overlapped patches segmentation and vectorization, a feature learning algorithm is adopted to train the dictionary for sparse representation. Then, energy distribution optimizing based on the principle of predictive coding is used to maximize the entropy of the feature of visual saliency, thereby generating the final saliency map. Finally, ROIs are extracted from original images with Otsu's segmentation method implemented in the obtained saliency map. Experimental results in two different kinds of remote sensing images demonstrate that the proposed model outperforms the other nine models in ROI extraction, qualitatively and quantitatively. In our experiments, each pixel is simply used as feature and only the number of features need to be chosen. Thus, there is no need to consider the specific structural information of different remote sensing images, which may provide a new unified method for feature extraction for image processing areas such as object compression, segmentation and recognition in the future.

Acknowledgments: This work was supported by the National Natural Science Foundation of China under grant numbers 61571050; the Beijing Natural Science Foundation under grant number 4162033; and the Open Fund of State Key Laboratory of Remote Sensing Science under grant number OFSLRSS201621.

Author Contributions: Libao Zhang, Xinran Lv and Xu Liang had the original idea for the study; Libao Zhang supervised the research and contributed to the article's organization; Libao Zhang and Xinran Lv conceived and designed the experiments; Xinran Lv performed the experiments; Xinran Lv and Xu Liang analyzed the data; Libao Zhang and Xinran Lv wrote the paper. All of the authors read and approved the submitted manuscript.

References

1. Zhang, L.; Li, A.; Zhang, Z.; Yang, K. Global and local saliency analysis for the extraction of residential areas in high-spatial-resolution remote sensing image. *IEEE Trans. Geosci. Remote Sens.* **2016**, *54*, 3750–3763. [CrossRef]
2. Sedaghat, A.; Mokhtarzade, M.; Ebadi, H. Uniform robust scale-invariant feature matching for optical remote sensing images. *IEEE Trans. Geosci. Remote Sens.* **2011**, *49*, 4516–4527. [CrossRef]
3. Liu, Z.; Dezert, J.; Mercier, G.; Pan, Q. Dynamic evidential reasoning for change detection in remote sensing images. *IEEE Trans. Geosci. Remote Sens.* **2012**, *50*, 1955–1967. [CrossRef]
4. Yi, L.; Zhang, G.; Wu, Z. A scale-synthesis method for high spatial resolution remote sensing image segmentation. *IEEE Trans. Geosci. Remote Sens.* **2012**, *50*, 4062–4070. [CrossRef]
5. Zhang, L.; Li, A.; Li, X.; Xu, S.; Yang, X. Remote Sensing Image Segmentation Based on an Improved 2-D Gradient Histogram and MMAD Model. *IEEE Geosci. Remote Sens. Lett.* **2015**, *12*, 58–62. [CrossRef]
6. Faur, D.; Gavat, I.; Datcu, M. Salient remote sensing image segmentation based on rate-distortion measure. *IEEE Geosci. Remote Sens. Lett.* **2009**, *6*, 855–859. [CrossRef]
7. Giusto, D.; Murroni, M.; Petrou, M. Region-based remote sensing image compression in wavelet domain using free angle segmentation model. *Electron. Lett.* **2002**, *38*, 1335–1337. [CrossRef]
8. Ancis, M.; Murroni, M.; Giusto, D.; Petrou, M. Region-based remote-sensing image compression in the wavelet domain. In Proceedings of the IEEE Conference on Geoscience and Remote Sensing Symposium, Hamburg, Germany, 28 June–2 July 1999; pp. 2054–2056.
9. Ma, Y.; Hua, X.; Lu, L.; Zhang, H. A generic framework of user attention model and its application in video summarization. *IEEE Trans. Multimedia* **2005**, *7*, 907–919.
10. Wang, J.; Sun, J.; Quan, L.; Tang, X.; Shum, H.Y. Picture collage. In Proceedings of the IEEE Conference on Computer Vision and Pattern Recognition, New York, NY, USA, 17–22 June 2006; pp. 347–354.
11. Goferman, S.; Tal, A.; Zelnik-Manor, L. Puzzle-like collage. *Comput. Gr. Forum* **2010**, *29*, 459–468. [CrossRef]

12. Maunsell, J.; Treue, S. Feature-based attention in visual cortex. *Trends Neurosci.* **2006**, *29*, 317–322. [CrossRef] [PubMed]
13. Najemnik, J.; Geisler, W. Optimal eye movement strategies in visual search. *Nature* **2005**, *434*, 387–391. [CrossRef] [PubMed]
14. Itti, L.; Koch, C.; Niebur, E. A Model of Saliency-Based Visual Attention for Rapid Scene Analysis. *IEEE Trans. Pattern Anal. Mach. Intell.* **1998**, *20*, 1254–1259. [CrossRef]
15. Li, Z.; Itti, L. Saliency and gist features for target detection in satellite images. *IEEE Trans. Image Process.* **2011**, *20*, 2017–2029. [PubMed]
16. Klein, D.; Frintrop, S. Center-surround Divergence of Feature Statistics for Salient Object Detection. In Proceedings of the IEEE Conference on Computer Vision and Pattern Recognition, Colorado Springs, CO, USA, 20–25 June 2011; pp. 2204–2219.
17. Achanta, R.; Hemami, S.; Estrada, F.; Susstrunk, S. Frequency-tuned salient region detection. In Proceedings of the IEEE Conference on Computer Vision and Pattern Recognition, Miami, FL, USA, 20–25 June 2009; pp. 1597–1604.
18. Hou, X.; Zhang, L. Saliency detection: A spectral residual approach. In Proceedings of the IEEE Conference on Computer Vision and Pattern Recognition, Los Angeles, CA, USA, 17–22 June 2007; pp. 1–8.
19. Guo, C.; Ma, Q.; Zhang, L. Spatio-temporal saliency detection using phase spectrum of quaternion fourier transform. In Proceedings of the IEEE Conference on Computer Vision and Pattern Recognition, Anchorage, AK, USA, 23–28 June 2008; pp. 1–8.
20. Imamoglu, N.; Lin, W.; Fang, Y. A saliency detection model using low-level features based on wavelet transform. *IEEE Trans. Multimedia* **2013**, *15*, 96–105. [CrossRef]
21. Harel, J.; Koch, C.; Perona, P. Graph-based visual saliency. *Neural Inf. Process. Syst.* **2006**, *19*, 545–552.
22. Borji, A.; Itti, L. Exploiting local and global patch rarities for saliency detection. In Proceedings of the IEEE Conference on Computer Vision and Pattern Recognition, Providence, RI, USA, 16–21 June 2012.
23. Goferman, S.; Zelnik-Manor, L.; Tal, A. Context-aware saliency detection. *IEEE Trans. Pattern Anal. Mach. Intell.* **2012**, *34*, 1915–1926. [CrossRef] [PubMed]
24. Wang, Q.; Yuan, Y.; Yan, P. Visual saliency by selective contrast. *IEEE Trans. Circuits Syst. Video Technol.* **2013**, *23*, 1150–1155. [CrossRef]
25. Wang, Q.; Yuan, Y.; Yan, P.; Li, X. Saliency detection by multiple-instance learning. *IEEE Trans. Cybern.* **2013**, *43*, 660–672. [CrossRef] [PubMed]
26. Tong, X.; Xie, H.; Weng, Q. Urban Land Cover Classification with Airborne Hyperspectral Data: What Features to Use? *IEEE J. Sel. Top. Appl. Earth Obs. Remote Sens.* **2014**, *7*, 3998–4009. [CrossRef]
27. Li, L.; Wang, C.; Chen, J.; Ma, J. Refinement of Hyperspectral Image Classification with Segment-Tree Filtering. *Remote Sens.* **2017**, *9*, 69. [CrossRef]
28. Valero, S.; Chanussot, J.; Benediktsson, J.A.; Talbot, H.; Waske, B. Directional mathematical morphology for the detection of the road network in very high resolution remote sensing images. *Pattern Recogn. Lett.* **2010**, *31*, 1120–1127. [CrossRef]
29. Chao, T.; Tan, Y.; Cai, H.; Tian, J. Airport detection from large IKONOS images using clustered SIFT keypoints and region information. *IEEE Geosci. Remote Sens. Lett.* **2011**, *8*, 128–132.
30. Lyu, C.; Jiang, J. Remote Sensing Image Registration with Line Segments and Their Intersections. *Remote Sens.* **2017**, *9*, 439. [CrossRef]
31. Zhang, L.; Yang, K. Region-of-interest extraction based on frequency domain analysis and salient region detection for remote sensing image. *IEEE Geosci. Remote Sens. Lett.* **2014**, *11*, 916–920. [CrossRef]
32. Zhang, L.; Yang, K.; Li, H. Regions of interest detection in panchromatic remote sensing images based on multiscale feature fusion. *IEEE J. Sel. Top. Appl. Earth Obs. Remote Sens.* **2014**, *7*, 4704–4716. [CrossRef]
33. Zhang, L.; Li, A. Region-of-Interest Extraction Based on Saliency Analysis of Co-occurrence Histogram in High Spatial Resolution Remote Sensing Images. *IEEE J. Sel. Top. Appl. Earth Obs. Remote Sens.* **2015**, *8*, 2111–2124. [CrossRef]
34. Martinez-Uso, A.; Pla, F.; Sotoca, J.M.; Garcia-Sevilla, P. Clustering-based hyperspectral band selection using information measures. *IEEE Trans. Geosci. Remote Sens.* **2007**, *45*, 4158–4171. [CrossRef]
35. Chen, J.; Zhang, L. Joint Multi-Image Saliency Analysis for Region of Interest Detection in Optical Multispectral Remote Sensing Images. *Remote Sens.* **2016**, *8*, 461. [CrossRef]

36. Lu, Q.; Huang, X.; Zhang, L. A Novel Clustering-Based Feature Representation for the Classification of Hyperspectral Imagery. *Remote Sens.* **2014**, *6*, 5732–5753. [CrossRef]
37. Huang, K.; Aviyente, S. Sparse representation for signal classification. *Adv. Neural Inf. Process. Syst.* **2006**, *19*, 609–616.
38. Lam, E.Y.; Goodman, J.W. A Mathematical Analysis of the DCT Coefficient Distributions for Images. *IEEE Trans. Image Process.* **2000**, *9*, 1661–1666. [CrossRef] [PubMed]
39. Bruce, L.M.; Koger, C.H.; Jiang, L. Dimensionality reduction of hyperspectral data using discrete wavelet transform feature extraction. *IEEE Trans. Geosci. Remote Sens.* **2002**, *40*, 2331–2338. [CrossRef]
40. Elad, M.; Aharon, M. Image denoising via sparse and redundant representations over learned dictionary. *IEEE Trans. Image Process.* **2006**, *15*, 3736–3745. [CrossRef] [PubMed]
41. Gorodnitsky, I.; Rao, B. Sparse signal reconstruction from limited data using FOCUSS: A re-weighted minimum norm algorithm. *IEEE Trans. Signal Process.* **1997**, *45*, 600–616. [CrossRef]
42. Rao, R.; Ballard, D. Predictive coding in the visual cortex: A functional interpretation of some extra-classical receptive-field effects. *Nat. Neurosci.* **1999**, *2*, 79–87. [CrossRef] [PubMed]
43. Otsu, N. A threshold selection algorithm from gray-level histograms. *IEEE Trans. Syst. Man Cybern.* **1979**, *9*, 62–66. [CrossRef]
44. Rahmani, S.; Strait, M.; Merkurjev, D.; Moeller, M.; Wittman, T. An adaptive IHS pan-sharpening method. *IEEE Geosci. Remote Sens. Lett.* **2010**, *7*, 746–750. [CrossRef]
45. Han, J.; He, S.; Qian, X.; Wang, D.; Guo, L.; Liu, T. An object-oriented visual saliency detection framework based on sparse coding representations. *IEEE Trans. Circuits Syst. Video Technol.* **2013**, *23*, 2009–2021. [CrossRef]
46. Willmore, B.; Tolhurst, D.J. Characterizing the sparseness of neural codes. *Network* **2001**, *12*, 255–270. [CrossRef] [PubMed]
47. Ngiam, J.; Koh, P.W.; Chen, Z.; Bhaskar, S.; Ng, A.Y. Sparse filtering. *Proc. Neural Inf. Process. Syst.* **2011**, *11*, 1125–1133.
48. Hou, X.; Zhang, L. Dynamic Visual Attention: Searching for coding length increments. *Adv. Neural Inf. Process. Syst.* **2008**, *21*, 681–688.
49. Schmidt, M. minFunc: Unconstrained Differentiable Multivariate Optimization in Matlab. Available online: http://www.cs.ubc.ca/~schmidtm/Software/minFunc.html (accessed on 1 January 2005).
50. Younggi, B.; Jaewan, C.; Youkyung, H. An Area-Based Image Fusion Scheme for the Integration of SAR and Optical Satellite Imagery. *IEEE J. Sel. Top. Appl. Earth Obs. Remote Sens.* **2013**, *6*, 2212–2220. [CrossRef]
51. Skodras, A.; Christopoulos, C.; Ebrahimi, T. The JPEG 2000 still image compression standard. *IEEE Signal Process. Mag.* **2001**, *18*, 36–58. [CrossRef]
52. Zhang, L.; Chen, J.; Qiu, B. Region-of-interest coding based on saliency detection and directional wavelet for remote sensing images. *IEEE Geosci. Remote Sens. Lett.* **2017**, *14*, 23–27. [CrossRef]

6

A Fuzzy-GA Based Decision Making System for Detecting Damaged Buildings from High-Spatial Resolution Optical Images

Milad Janalipour and Ali Mohammadzadeh *

Faculty of Geodesy & Geomatics Engineering, K. N. Toosi University of Technology, Tehran 1996715433, Iran; m_janalipour89@yahoo.com

* Correspondence: almoh2@gmail.com.

Academic Editors: Qi Wang, Nicolas H. Younan, Carlos López-Martínez, Soe Myint and Prasad S. Thenkabail

Abstract: In this research, a semi-automated building damage detection system is addressed under the umbrella of high-spatial resolution remotely sensed images. The aim of this study was to develop a semi-automated fuzzy decision making system using Genetic Algorithm (GA). Our proposed system contains four main stages. In the first stage, post-event optical images were pre-processed. In the second stage, textural features were extracted from the pre-processed post-event optical images using Haralick texture extraction method. Afterwards, in the third stage, a semi-automated Fuzzy-GA (Fuzzy Genetic Algorithm) decision making system was used to identify damaged buildings from the extracted texture features. In the fourth stage, a comprehensive sensitivity analysis was performed to achieve parameters of GA leading to more accurate results. Finally, the accuracy of results was assessed using check and test samples. The proposed system was tested over the 2010 Haiti earthquake (Area 1 and Area 2) and the 2003 Bam earthquake (Area 3). The proposed system resulted in overall accuracies of 76.88 $\pm$ 1.22%, 65.43 $\pm$ 0.29%, and 90.96 $\pm$ 0.15% over Area 1, Area 2, and Area 3, respectively. On the one hand, based on the concept of the proposed Fuzzy-GA decision making system, the automation level of this system is higher than other existing systems. On the other hand, based on the accuracy of our proposed system and four advanced machine learning techniques, i.e., bagging, boosting, random forests, and support vector machine, in the detection of damaged buildings, it seems that our proposed system is robust and efficient.

Keywords: building damage detection; Fuzzy-GA decision making system; machine learning techniques; optical remotely sensed images; sensitivity analysis; texture analysis

1. Introduction

Detecting damaged buildings after a massive disaster in a robust manner is a critical task, because it helps relief and rescue teams to manage related works accurately and precisely and then may reduce losses. Hence, the production of accurate building damage maps after disasters would help relief and rescue teams in emergency situations. Remote sensing (RS) data is one of the sources which can be used for generating building damage maps. Due to specific characteristics of the RS data such as its high temporal frequency and the availability of various sensors with different spatial and spectral resolutions, it plays an important role in producing building damage maps. Satellite optical images, as a source of the RS data, have been frequently used to produce damage maps [1]. In this study, we present a novel semi-automated decision making system based on the fuzzy theory and genetic algorithm (GA) in order to produce the building damage maps. Our proposed system can be used as a knowledge extraction tool in future works. Knowledge extraction is a necessary stage in order to convert a semi-structured problem into a structured one [2].

1.1. Literature Review

The existing damage detection methods can be discussed from viewpoints of data used and methodology. From the perspective of data used, researchers employed different sources of RS data, including Light Detection and Ranging (LiDAR) [3], Synthetic Aperture Radar (SAR) [4–9], and optical imagery [10–12]. Optical satellite imagery is one of the useful sources in building damage detection process. Disasters may damage components of a building. A building's roof is one of these components. Optical satellite sensors usually observe the roof of buildings. The spectral signature is the outcome of an optical satellite sensor. The spectral signature of each phenomenon is unique. Therefore, spectral signatures of a damaged roof and an intact roof are different from each other. For this reason, spectral signatures obtained from optical satellite sensors or their extracted features are suitable for detecting damaged buildings [2,13].

From the viewpoint of the methodology, presenting a comprehensive analysis about existing methods used in damage detection application is a difficult task, because there are many research studies. Therefore, we attempt to mention the related works to our research. In this study, we propose a fuzzy decision making system in order to detect damaged buildings from textural features extracted from post-event optical images. Hence, the literature review is presented in three parts: (1) the use of optical images for detecting damaged areas, (2) the role of textural features for detecting damaged areas, and (3) fuzzy systems used for detecting damaged areas.

Some researchers attempted to specify the role of optical images in damage detection application. Eguchi and Mansouri (2005) focused on investigating and categorizing papers that used RS technology for detecting buildings damaged after the 2003 Bam earthquake. It was concluded that detecting regional damages by RS technology is possible [14]. Voigt et al. (2011) presented results concerning actions of the German Aerospace Center (DLR) after the 2010 Haiti earthquake. It was deduced that extracting building damage maps even with high spatial-resolution optical satellite images is not an easy task and needs several human experts [15]. Lu et al. (2012) implemented a building damage detection method based on mono-temporal very high-spatial resolution optical images. In this paper, integrating manual and automatic interpretations resulted in a robust building damage map [16]. Tiede et al. (2011) used shadow information extracted from pre- and post-event optical images for generating a damage map after the 2010 Haiti earthquake. The proposed method was able to create the damage map of the Carrefour area after 12 h [17]. Lemoine et al. (2013) used aerial optical data for providing a realistic estimate from damaged buildings. Using the aerial optical data instead of satellite data was the key objective of the presented study to obtain more accurate results [18]. Based on the presented research works, it appears that optical data is a suitable source for detecting damaged buildings. However, owing to the complexity of this problem (i.e., damage detection), the role of experts is important and undeniable.

Many researches have benefited from textural features extracted from the RS data for identifying areas damaged after disasters. The ability of textural features in measuring spectral and height variations in the spatial domain over RS data is the main reason for use of them in the damage detection problem. Table 1 briefly depicts researches that used textural features in damage detection applications, especially for building damage detection [1] and road damage detection [19]. The existing researches can be discussed from three viewpoints. The used feature extraction method is the first viewpoint. From this perspective, Laws mask [1], Haralick [20], Multivariate variogram [21], 1st statistical [19], and Gabor filter [22] feature extraction methods were frequently employed to produce textural features in previous research works. Moreover, from the second viewpoint, textural features were extracted from different remotely sensed data including optical images, light detection and ranging (LiDAR) data, and synthetic aperture radar (SAR) data. Based on the literature, textural features extracted from pre- and/or post-event optical images were widely employed for detecting damaged areas. From the third viewpoint, researchers utilized textural features for improving the accuracy of the final damage map. In fact, it seems that textural features positively affect the performance of machine learning techniques and decision making systems in identifying damaged areas.

Table 1. A brief presentation from previous research studies that used textural features in damage detection applications.

Reference	Textural Features Used	Remotely Sensed Data
[1]	Laws mask	Pre- and post-event optical images
[23]	Haralick	Pre- and post-event optical images
[20]	Haralick	Pre- and post-event optical images
[21]	Multivariate variogram	Pre- and post-event optical images and post-event LiDAR data
[22]	Gabor Filter	Pre- and post-event optical images
[19]	1st statistical features, Gabor features and Haralick features	Pre- and post-event optical images
[24]	Multivariate variogram	Pre- and post-event optical images
[25]	Haralick	Pre- and post-event digital elevation models
[26]	Haralick	Post-event SAR data
[27]	1st statistical features, Gabor features and Haralick features	Pre- and post-event optical images
[28]	Haralick	Pre- and post-event optical images

After extracting features, it is necessary to use a classifier or machine learning technique or decision making system for creating a relation between the extracted features and the damage extent of buildings. To this end, some researchers used advanced and non-parametric classifiers. Chesnel et al. (2008) utilized the Support Vector Machine (SVM) classifier to partition the feature space for detecting damaged and undamaged buildings [22]. Li et al. (2010) used the One-Class Support Vector Machine (OCSVM) classifier to obtain damaged areas from high spatial-resolution optical images [21]. Dubois and Lepage (2014) employed a multilayer backpropagation perceptron neural network to detect damaged buildings after the 2010 Haiti earthquake [1]. In addition to advanced classifiers, based on Table 2, some researchers used fuzzy-based decision making systems in the damage detection process [2,13]. The fuzzy-based decision making systems are usually employed in issues where experts want to model their knowledge. Damage detection is one of these issues. Producing damage maps using experts after disasters in a manual manner proves our claim. To the best of our knowledge, in the damage detection application, the fuzzy-based decision making systems can be used for two main procedures including: (1) land use/cover classification [29,30] and (2) modeling the damage extent of buildings from the extracted features [2,13]. Ural et al. (2011) employed a fuzzy classifier in order to map buildings and their rubble after the 2010 Haiti earthquake in a robust manner [30]. Moreover, researches have used Mamdani fuzzy inference systems (MFISs) as a decision making system for modeling the damage extent of buildings [19,27,28]. In these researches, parameters of fuzzy inference systems were manually adjusted in a trial and error manner that is a time consuming task. In these cases, the accuracy of results completely relies on the selected parameters. For this reason, Janalipour, M. et al. [2,13] used semi-automated Sugeno fuzzy decision making systems in order to detect damage and changed areas. The use of these systems was a good solution for improving the automation level of fuzzy systems, but it is a difficult task to extract knowledge from a Sugeno fuzzy system due to the structure of its rules [31]. Knowledge extraction is an important stage for converting a semi-structured problem (i.e., the damage detection) into a structured one. For further study on damage detection methods, we encourage readers to refer to [32–34].

Table 2. A brief presentation from previous research works which used fuzzy inference systems in damage detection application.

Reference	Type of Fuzzy System	Automation Level
[19]	Mamdani	Manually
[28]	Mamdani	Manually
[27]	Mamdani	Manually
[35]	Mamdani	Manually
[13]	Sugeno	Semi-automated
[2]	Sugeno	Semi-automated

1.2. Research Aims

In this study, three important objectives are satisfied. Based on the previous works, researchers widely used Mamdani fuzzy decision making systems for detecting damaged areas whose parameters of these systems were manually chosen in a trial and error basis which is a time consuming task. Moreover, the robustness of outcomes relies on the selected parameters. For this reason, Janalipour, M. et al. [2,13] proposed a semi-automated Sugeno fuzzy decision making system. However, for knowledge extraction, the Sugeno fuzzy system is not appropriate due to the structure of its rules. To this end, it is necessary to employ a semi-automated Mamdani fuzzy system. However, there is no semi-automated Mamdani fuzzy system. Therefore, it is essential to propose a semi-automated (or fully-automated) Mamdani fuzzy decision making system to detect damaged areas. In the first and main objective, we develop a semi-automated Mamdani fuzzy decision making system using Genetic Algorithm (GA). Based on the previous researches [2,13], sensitivity analysis plays an important role in identifying the appropriate parameters of a system leading to more accurate results. In the mentioned researches, a step-by-step sensitivity analysis method was used. However, it is necessary to simultaneously test all parameters of a system, because it permits us to consider the relationship between changes of all the parameters. For this reason, in the second objective, we study the effect of the simultaneous change of all parameters of the system on the final result. To investigate the robustness and effectiveness of our semi-automated Mamdani fuzzy decision making system, we compare results of this system with four advanced machine learning techniques including random forests (RF), bagging, boosting, and support vector machine (SVM)—that is our third objective.

2. Materials and Methods

In this section is included information about study areas and data used and description about our proposed methodology.

2.1. The First Study Area: The 2010 Haiti Earthquake

Port-au-Prince city is the first study area, where an earthquake occurred on 12 January 2010. Port-au-Prince is the capital of Haiti. Two areas including Area 1 and Area 2 were chosen over Port-au-Prince.

The previous research studies proved that ancillary information such as a pre-event vector map improves the accuracy of damage detection methods [36–38]. On the other hand, the use of the pre-event map in our proposed system is necessary, because it is difficult to find the footprint of buildings on post-earthquake optical images. For these reasons, a pre-event building map was injected into our methodology. There are some old-vector maps in the Haiti area such as [39]. To update the old-vector map, pre-event Geoeye-1 and IKONOS-2 images were employed. To this end, the old-vector map was updated as much as possible by an expert. Moreover, in the first study area, an ortho-rectified, pansharpened and georeferenced post-event Geoeye-1 image acquired on 13 January 2010 with a spatial resolution of 50 cm and three spectral bands (blue, green, and red) was employed.

2.2. The Second Study Area: The 2003 Bam Earthquake

Bam city is the second study area, a city located in southwestern Iran, where an earthquake occurred on 26 December 2003. The post- earthquake pansharpened QuickBird image and pre-event digital vector map of the Bam area were used in this study. The post-earthquake image was acquired on 3 January 2004 and also has 61 cm spatial resolution and four spectral bands (red, green, blue, and near infrared). The second study area has about 400 buildings with different damage extent. The pre-event digital vector map with a scale of 1:500 was produced by the National Cartographic Center (NCC) of Iran in 1994, which was updated using an expert according to [13].

2.3. Methodology

Our proposed damage detection system is presented in four main stages according to Figure 1. In stage "1", post-event optical images are pre-processed. In stage "2", Haralick texture features are extracted from the pre-processed optical images using the pre-event map and related equations. In stage "3", a Fuzzy- GA (Genetic Algorithm) based decision making system is developed to estimate the damage extent of buildings from the extracted texture features. In stage "4", a comprehensive sensitivity analysis is performed to achieve the best parameters leading to more accurate results. Finally, the accuracy of results is firstly assessed and then the building damage map is obtained. The aforementioned stages are presented in more detail below.

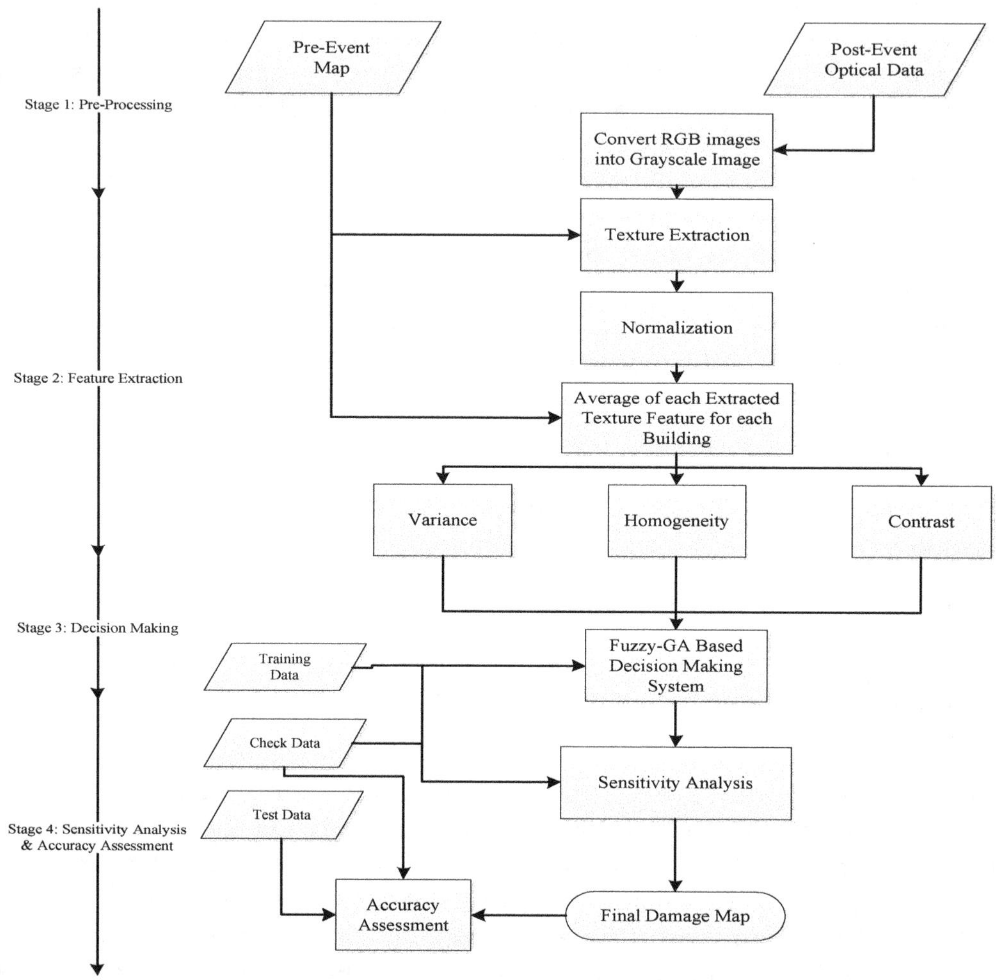

Figure 1. The workflow of our semi-automated damage detection system in this study.

2.3.1. Stage 1: Pre-Processing

Pre-processing is one of the important stages in building damage detection. Geo-rectification and pansharpening are two of the important pre-processes, which should be performed on the post-event optical images. Georeferencing and pansharpening were performed on the post-event optical image of the Haiti earthquake. Moreover, the mentioned pre-processes were performed on the post-event optical image of the Bam earthquake according to [13].

As another pre-process, based on previous research works [2], the correlation between corresponding textural features extracted from spectral bands (i.e., red, blue, and green) is high. Hence, numerous and correlated features result, which increase the computational cost of the proposed method. For this reason, a grayscale image is produced from red, blue, and green bands using

Equation (1) (please see [40] for further study about Equation (1)). Textural features are extracted from the grayscale band.

$$E = 0.2989 \times R_r + 0.5870 \times R_g + 0.1140 \times R_b \tag{1}$$

where, R_r, R_g, R_b, and E are the reflectance value of the red, green, blue, and resulting grayscale bands, respectively.

2.3.2. Stage 2: Feature Extraction

Extracting textural features is the main process of the feature extraction stage. The potential of textural features in measuring variations of digital numbers in the spatial domain enables us to use them in satellite image processing, especially after natural hazards. Natural hazards suddenly cause damage of objects of the earth's surface leading to reflectance changes in the spatial domain. Therefore, the textural features extracted from remotely sensed optical data are widely used for detecting damaged areas. Based on previous works presented in the literature review section, different texture extraction methods have been used in damage detection applications. In most of the previous works [20,23], Haralick features were widely chosen for extracting textural features. For this reason, these features are also used in our study. For further study on the Haralick texture extraction method, please see [41,42].

In this study, in order to detect damaged buildings from textural features, variance, homogeneity, and contrast features were chosen, which can be calculated from Equation (2), Equation (3), and Equation (4), respectively. The selection of these features was based on three reasons. First of all, three texture features with three linguistic terms and Gaussian membership functions (MFs) generate 24 unknown parameters regarding MFs in a Mamdani fuzzy inference system which should be simultaneously set. It seems that the number of unknown parameters is sufficient to test an optimization algorithm and the selection of them by an expert is a difficult task. For the second reason, based on equations of the mentioned features, it appears that correlation among the selected features is low. Finally, the performance of our decision making system and advanced machine learning techniques would be investigated in similar conditions (i.e., with three texture features). Hence, the selection of the mentioned texture features is within the path of objectives of this study.

$$Variance = \sum_{i=0}^{G-1} (i - \mu)^2 P(i,j) \tag{2}$$

$$Homogenity = \sum_{i=0}^{G-1} \sum_{j=0}^{G-1} \frac{P(i,j)}{1 + (i-j)^2} \tag{3}$$

$$Contrast = \sum_{i=0}^{G-1} \sum_{j=0}^{G-1} (i-j)^2 \times P(i,j) \tag{4}$$

where, μ is the mean value of gray-levels in an area selected for producing texture features. Moreover, P and G are the probability matrix and the number of image gray-levels, respectively.

There are two important points about extracting textural features in this study. For the first point, for preserving the negative effects of non-building pixels on extracted textural features, building pixels are specified by the pre-event map and are only used to extract textural features. For the second point, to compare the three mentioned texture features, they should be standardized. Hence, Equation (5) is employed to standardize the extracted textural features.

$$x_i = \frac{D_i - D_i^{\min}}{D_i^{\max} - D_i^{\min}}, \; i = 1:3 \tag{5}$$

where, x_i is the ith standardized texture feature; $D_i^{\min}$ and $D_i^{\max}$ are the minimum and maximum values of the ith texture feature; and D_i is an arbitrary value of the ith texture feature.

2.3.3. Stage 3: Decision Making

After extracting textural features, it is necessary to employ a decision making system [43] or a machine learning technique to provide a relation between the extracted features and the damage extent of buildings. In this study, a decision making system based on MFIS and GA is used to provide the mentioned relation. MFIS was firstly proposed by Zadeh [44]. In a MFIS, initially, crisp input values are converted into fuzzy values by input MFs. This is called "fuzzification". In fact, fuzzification is a mapping process that is performed using membership functions. In this process, membership functions act as connectors among crisp and fuzzy spaces. Then, using fuzzy values, the inference system and existing rules in the fuzzy rule base, fuzzy output values are generated. Finally, the fuzzy output values are transformed into crisp output values by a defuzzification method [31].

In general, a MFIS is a function of three main parameters according to Equation (6) including: parameters of membership functions (C_1), rules (C_2), and parameters of the inference system (C_3). Therefore, in designing a MFIS, two essential tasks must be performed by an expert: (1) designing rules of the fuzzy rule base and parameters of the inference system, and (2) designing and selecting type and parameters of input and output MFs. In this study, we focus on the second task, because the number of rules in our study is minor and can be easily selected and also an expert could select the small parameters of the inference system.

$$\textit{Mamdani Fuzzy System} = F(C_1, C_2, C_3) \tag{6}$$

For expressing the importance of the second mentioned task, an example is employed here. Please note that this example is also used as a MFIS in our damage detection method. A MFIS with three inputs and one output (its MF type is Gaussian) is presented in Figure 2. Suppose three rules similar to Equations (7) to (9) have been designed by an expert. Based on these equations, we can express that C_1 is a function of some unknown parameters (Equation (10)).

$$\begin{gathered} \textit{If}\left(X_1 \textit{ is} \exp\left(\frac{-(X_1-m_1)^2}{2\sigma_1^2}\right) \textit{and } X_2 \textit{ is} \exp\left(\frac{-(X_2-m_4)^2}{2\sigma_4^2}\right) \textit{ and } X_3 \textit{ is} \exp\left(\frac{-(X_3-m_7)^2}{2\sigma_7^2}\right)\right) \\ \textit{Then } Z \textit{ is} \exp\left(\frac{-(Z-m_{10})^2}{2\sigma_{10}^2}\right) \end{gathered} \tag{7}$$

$$\begin{gathered} \textit{If}\left(X_1 \textit{ is} \exp\left(\frac{-(X_1-m_2)^2}{2\sigma_2^2}\right) \textit{and } X_2 \textit{ is} \exp\left(\frac{-(X_2-m_5)^2}{2\sigma_5^2}\right) \textit{ and } X_3 \textit{ is} \exp\left(\frac{-(X_3-m_8)^2}{2\sigma_8^2}\right)\right) \\ \textit{Then } Z \textit{ is} \exp\left(\frac{-(Z-m_{11})^2}{2\sigma_{11}^2}\right) \end{gathered} \tag{8}$$

$$\begin{gathered} \textit{If}\left(X_1 \textit{ is} \exp\left(\frac{-(X_1-m_3)^2}{2\sigma_3^2}\right) \textit{and } X_2 \textit{ is} \exp\left(\frac{-(X_2-m_6)^2}{2\sigma_6^2}\right) \textit{ and } X_3 \textit{ is} \exp\left(\frac{-(X_3-m_9)^2}{2\sigma_9^2}\right)\right) \\ \textit{Then } Z \textit{ is} \exp\left(\frac{-(Z-m_{12})^2}{2\sigma_{12}^2}\right) \end{gathered} \tag{9}$$

$$C_1 = F(m_1, \sigma_1, \ldots, m_{12}, \sigma_{12}) \tag{10}$$

where, X_1, X_2, and X_3 are input linguistic variables 1 to 3, respectively and Z is the output linguistic variable. Moreover, m_i and σ_i are the mean and standard deviation of a Gaussian membership function.

According to Figure 2 and the designed rules, 24 unknown parameters of MFs (Equation (11)) must be set by an expert. From our viewpoint, the procedure of selection of 24 unknown parameters in a continuous-space is an optimization problem. For this reason, in this study, Genetic Algorithm (GA), as an optimization algorithm, is employed to select unknown parameters [45,46]. The ability of GA in selecting optimum answers was the main reason for the selection of this algorithm. In fact, in this research, MFIS and GA are integrated to select appropriate parameters of MFs leading to the best result. In the following, concepts of GA and its integration with the MFIS are presented.

$$\textit{Unknown Parameters} = \begin{bmatrix} m_1 & \sigma_1 & \cdots & m_{12} & \sigma_{12} \end{bmatrix}_{1\times 24} \tag{11}$$

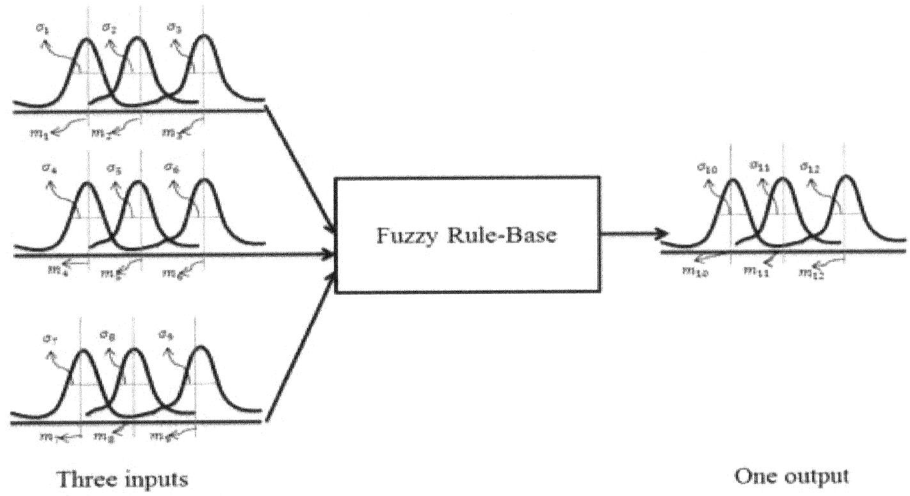

Figure 2. A schematic presentation of a MFIS with three inputs and one output and its MFs.

Genetic Algorithm (GA) is based on the mechanism exhibited by nature incorporating the robustness of biological systems as presented by Charles Darwin [47]. This algorithm is one of the powerful artificial intelligence algorithms, which selects the optimal answer using a random searching method in the search-space.

To find the optimal unknown parameters regarding MFs of a fuzzy inference system using GA, five main steps are employed in GA. The employed steps are presented below:

The first step: initial parameters regarding MFs of the fuzzy system are generated in a random manner. Suppose the number of the population is equal to k and the generated parameters are:

$$positions = \begin{bmatrix} m_1^1 & \sigma_1^1 & \cdots & m_{12}^1 & \sigma_{12}^1 \\ \vdots & \vdots & \ddots & \vdots & \vdots \\ m_1^k & \sigma_1^k & \cdots & m_{12}^k & \sigma_{12}^k \end{bmatrix}_{k\times 24} = \begin{bmatrix} the\ 1\text{st}\ population \\ \vdots \\ the\ k\text{th}\ population \end{bmatrix} \quad (12)$$

After generating the initial parameters, it is necessary to use a cost function to calculate the efficiency of each population. It should be considered that the fitness function and objective function terms are also used instead of the cost function one. The cost function is at the heart of our proposed system. The integration of the fuzzy system and GA is the main task of this function. Moreover, the cost function is responsible for assessing population. In this study, to obtain the cost of population, some training and check samples according to Equations (13) and (14) are considered. The duty of training samples is to learn the fuzzy-GA system. Furthermore, check samples are employed to prevent an over-learning problem.

$$training\ samples = \begin{bmatrix} x_{t1}^1 & x_{t2}^1 & x_{t3}^1 & z_t{}^1 \\ \vdots & \vdots & \vdots & \vdots \\ x_{t1}^s & x_{t2}^s & x_{t3}^s & z_t{}^s \end{bmatrix}_{s\times 4} = \left[X_1^{train}\ X_2^{train}\ X_3^{train}\ Z^{train}\right] \quad (13)$$

$$check\ samples = \begin{bmatrix} x_{c1}^1 & x_{c2}^1 & x_{c3}^1 & z_c{}^1 \\ \vdots & \vdots & \vdots & \vdots \\ x_{c1}^u & x_{c2}^u & x_{c3}^u & z_c{}^u \end{bmatrix}_{u\times 4} = \left[X_1^{check}\ X_2^{check}\ X_3^{check}\ Z^{check}\right] \quad (14)$$

where, x_{t1}^h, x_{t2}^h, x_{t3}^h are values of textural features regarding the hth training sample. Moreover, $z_t{}^h$ is the damage extent of the hth training sample. x_{c1}^h, x_{c2}^h, x_{c3}^h are values of textural features regarding the hth check sample. Moreover, $z_c{}^h$ is the damage extent of the hth check sample. Furthermore, s and u are the number of training samples and the number of check samples, respectively.

To estimate the cost of each population for training samples (Equation (15)), at first, the parameters of the MFs are updated using the population (Equation (16)). For population r, the updated MFIS is represented by Equation (17). Then, using the updated MFIS and Equation (18), the damage extent of training samples is obtained from Equation (19). Finally, the cost of population r ($cost^r_{train}$) obtained from the training samples is calculated from Equation (20). The cost of check samples is also estimated in the same way.

$$Costs_{train} = \begin{bmatrix} cost^1_{train} \\ \vdots \\ cost^k_{train} \end{bmatrix}_{k\times 1} \tag{15}$$

$$C^r_1 = F(m^r_1, \sigma^r_1, \ldots, m^r_{12}, \sigma^r_{12}) \; ; \; 1 \le r \le k \tag{16}$$

$$Mamdani\ Fuzzy\ System^r = F(C^r_1, C_2, C_3) \tag{17}$$

$$\hat{Z} = Mamdani\ Fuzzy\ System\ (X^{train}_1, X^{train}_2,\ X^{train}_3) \tag{18}$$

$$\begin{bmatrix} \hat{Z}^1 \\ \vdots \\ \hat{Z}^s \end{bmatrix} = \begin{bmatrix} Mamdani\ Fuzzy\ System^r\ (x^1_{t1}, x^1_{t2},\ x^1_{t3}) \\ \vdots \\ Mamdani\ Fuzzy\ System^r\ (x^s_{t1}, x^s_{t2},\ x^s_{t3}) \end{bmatrix}_{s\times 1} \tag{19}$$

$$cost^r_{train} = \tfrac{1}{s} \times \sqrt{\sum_{i=1}^{s} \left(z_t{}^s - \hat{Z}^s\right)^2},\ r = 1, \ldots, k \tag{20}$$

where, C^r_1 and $Mamdani\ Fuzzy\ System^r$ are parameters of MFs obtained from population r and the fuzzy inference system updated from C^r_1, respectively.

The second step: In this step, new MF parameters are obtained from crossover function and population generated by the previous step. Crossover is one of the important functions in GA, which is responsible for generating new children (parameters of fuzzy systems) from their parents. To generate new children, the number of uses of crossover function ($n_{crossover}$) should be specified. To this end, at first, parameter α is calculated from the crossover rate and the number of population (*population*) using Equation (21). Then, the number of uses of crossover function is obtained from Equation (22). For generating new MF parameters from the crossover function, first, two random parents (like m_i and n_i) are chosen. Afterwards, new children are achieved from Equation (23). Since unknown parameters were defined in specific ranges, there are two conditions according to Equations (24) and (25) for undefined values. Finally, by using Equations (15) to (20), costs of the new children are calculated and inserted into Equation (26) [47].

$$\alpha = \frac{crossover\ rate \times population}{2} \tag{21}$$

$$n_{crossover} = \max\{n \in \mathbb{Z} | n \le \alpha\} \tag{22}$$

$$\begin{aligned} & positions_{crossover} \\ & = \begin{bmatrix} \gamma_1 \times positions(m_1, :) + (1-\gamma_1) \times positions(n_1, :) \\ \gamma_1 \times positions(n_1, :) + (1-\gamma_1) \times positions(m_1, :) \\ \vdots \\ \gamma_{n_{crossover}} \times positions(m_{n_{crossover}}, :) + (1-\gamma_{n_{crossover}}) \times positions(n_{n_{crossover}}, :) \\ \gamma_{n_{crossover}} \times positions(n_{n_{crossover}}, :) + (1-\gamma_{n_{crossover}}) \times positions(m_{n_{crossover}}, :) \end{bmatrix}_{(2\times n_{crossover})\times 24} \\ & = \begin{bmatrix} \overline{m}^1_1 & \overline{\sigma}^1_1 & \cdots & \overline{m}^1_{12} & \overline{\sigma}^1_{12} \\ \overline{m}^2_1 & \overline{\sigma}^2_1 & \cdots & \overline{m}^2_{12} & \overline{\sigma}^2_{12} \\ \vdots & \vdots & \ddots & \vdots & \vdots \\ \overline{m}^{2\times n_{crossover}-1}_1 & \overline{\sigma}^{2\times n_{crossover}-1}_1 & \cdots & \overline{m}^{2\times n_{crossover}-1}_{12} & \overline{\sigma}^{2\times n_{crossover}-1}_{12} \\ \overline{m}^{2\times n_{crossover}}_1 & \overline{\sigma}^{2\times n_{crossover}}_1 & \cdots & \overline{m}^{2\times n_{crossover}}_{12} & \overline{\sigma}^{2\times n_{crossover}}_{12} \end{bmatrix} \end{aligned} \tag{23}$$

$$\overline{\sigma}_i = \begin{cases} 0.01\ if\ \overline{\sigma}_i < 0 \\ 0.5\ if\ \overline{\sigma}_i > 0.5 \end{cases} \tag{24}$$

$$\overline{m}_i = \begin{cases} 0\ if\ \overline{m}_i < 0 \\ 1\ if\ \overline{m}_i > 1 \end{cases} \tag{25}$$

$$Costs_{crossover} = \begin{bmatrix} cost^1 \\ \vdots \\ cost^{2\times n_{crossover}} \end{bmatrix}_{(2\times n_{crossover})\times 1} \tag{26}$$

where, γ_1 is a random number between -0.1 and 1.1. $\overline{\sigma}_i$ and $\overline{m}_i$ are the variance and mean of the ith variable achieved from the crossover function, respectively.

The third step: In this step, one of the parameters of a population is changed using the mutation function. The mutation is another important function in GA. The mutation function has an undeniable role in solving the local minimum problem in GA. To generate new children by the mutation function, the number of uses of the mutation function ($n_{mutation}$) should be specified. To this end, at first, parameter β is calculated from the mutation rate and the number of population using Equation (27). Then, the number of uses of the mutation function is obtained from Equation (28). For each use of the mutation function, one random population is selected. Afterwards, a new child is achieved. Equation (29) depicts all new children achieved from the mutation function. Undefined values of variables are corrected using Equations (24) and (25). Finally, using Equations (15) to (20), costs of the new children are calculated and inserted into Equation (30).

$$\beta = \frac{mutation\ rate \times population}{2} \tag{27}$$

$$n_{mutation} = \max\{n \in \mathbb{Z} | n \leq \beta\} \tag{28}$$

$$positions_{mutation} = \begin{bmatrix} m_1 & \sigma_1 + sigma \times rand_1 & \cdots & m_{12} & \sigma_{12} \\ \vdots & \vdots & \ddots & \vdots & \vdots \\ m_1 + sigma \times rand_{n_{mutation}} & \sigma_1 & \cdots & m_{12} & \sigma_{12} \end{bmatrix}_{n_{mutation}\times 24} \tag{29}$$

$$Costs_{mutation} = \begin{bmatrix} cost^1 \\ \vdots \\ cost^{n_{mutation}} \end{bmatrix}_{(n_{mutation})\times 1} \tag{30}$$

where, $rand_i$ is a random number. Moreover, $sigma$ is calculated from Equation (31). Variables σ_i and m_i are defined respectively in a range of [0.01–0.5] and [0–1]. Because the range of the variable σ_i is lower than the variable m_i, sigma values of Equation (31) regarding these variables are different:

$$sigma = \begin{cases} 0.049\ for\ variable\ \sigma_i \\ 0.1\ for\ variable\ m_i \end{cases} \tag{31}$$

The fourth step: In this step, using elitism operator, GA is able to preserve the best answer of iterations. To this end, at first, all population and costs obtained from the previous steps are inserted in two pools according to Equations (32) and (33). Afterwards, they are sorted in descending order by Equation (34) (i.e., a population with the minimum cost value is the best answer). Finally, we select k population with minimum cost values as the best answers from the sorted population (Equation (35)). With the elitism operator, it is possible to preserve the best solutions and GA can be converged on the best solution. In fact, with deep insight into this step, we can conclude that the artificial intelligence of GA exists in the fourth step.

$$all\ poistions = [positions; positions_{crossover}; positions_{mutation}]_{(n_{mutation}+2\times n_{crossover}+k)\times 24} \tag{32}$$

$$Costs = [Costs_{train}, Costs_{crossover}, Costs_{mutation}] \quad (33)$$

$$all\ poistions = sort(all\ poistions, Costs) \quad (34)$$

$$positions = all\ poistions(1 : population, :) \quad (35)$$

The fifth step: The second, third and fourth steps should be repeated to obtain the best solution. The first row of Equation (35) is the best solution with the minimum cost value.

2.3.4. Stage 4: Damage Map and Accuracy Assessment

In order to assess the accuracy of the final damage maps, the confusion matrix and statistical descriptors extracted from this matrix are considered. To generate a confusion matrix, it is necessary to employ training, test, and check samples. To this end, the damage extent of some buildings was specified as training, test, and check samples. More information about these samples is presented in Section 3.2. After generating the confusion matrix, statistical descriptors are employed to specify the accuracy of results. For this purpose, some statistical descriptors, including overall, user, and producer accuracies extracted from the confusion matrix, are used to display the accuracy of the proposed system. Based on a sample confusion matrix presented in Table 3, the overall, user, and producer accuracies are calculated from Equations (36) to (38), respectively.

$$Overall\_Accuracy = \frac{a+d}{a+b+c+d} \quad (36)$$

$$User\_accuracy\_class1 = \frac{a}{a+c} \quad (37)$$

$$Producer\_accuracy\_class1 = \frac{a}{a+b} \quad (38)$$

Table 3. A sample confusion matrix.

	References	
	Class 1	Class 2
Class 1	a	b
Class 2	c	d

Sensitivity Analysis

A sensitivity analysis should be carried out to confirm the stability and the reliability of the proposed system's results with respect to changes of its parameters [48–50]. In this study, the sensitivity of the Fuzzy-GA is assessed against any changes in GA's parameters: (a) the number of iterations; (b) the number of population; (c) the mutation rate; and (d) the crossover rate. In the previous works [2,13], a step-by-step sensitivity analysis method was used. However, step-by-step sensitivity analysis methods are unable to consider simultaneous changes of parameters of a system [51]. For this reason, in this study, a grid-partitioning based sensitivity analysis method was performed to study the effect of change of parameters on the accuracy of the final results [52]. In this method, the mentioned parameters of GA are changed in a limited and meaningful range to achieve accurate results. The range of parameters was adjusted based on experimental results and our knowledge about GA.

3. Results

3.1. Feature Extraction Considerations

In the feature extraction stage, variance, homogeneity, and contrast features were extracted by producing the probability matrix (P) for a distance of 1 pixel, orientation angles of 0, 45, 90, and 135 degrees and a window size of 3 × 3. Our experimental results showed that increasing the window size and the distance parameter negatively affects the accuracy of the building damage detection methods. For this reason, the mentioned parameters were chosen for producing textural features. The textural features were finally calculated from the average of features obtained on all orientation angles.

3.2. The Number of Training, Test and Check Samples Used over Area 1, Area 2, and Area 3

Training, check, and test samples must be used in modeling and validation processes of the Mamdani fuzzy system. To this end, two different sources were employed to extract these samples from the Haiti and Bam areas. The Haiti building damage atlas (the first source) was used to collect the mentioned samples over the Haiti area (Area 1 and Area 2). In this atlas, buildings were classified as undamaged (D1), substantial to heavy damage (D3), very heavy damage (D4), and destruction (D5). Moreover, according to [13], the damage extent of buildings over the Bam area was specified using the visual interpretation of an expert on pre- and post-event high-spatial resolution images (the second source). Table 4 shows the number of training, check, and test samples selected over Area 1, Area 2, and Area 3.

Table 4. The number of training, check and test samples selected over Area 1, Area 2, and Area 3.

	Training Samples				Check Samples				Test Samples			
	D1	D3	D4	D5	D1	D3	D4	D5	D1	D3	D4	D5
Area 1	172	15	30	83	66	5	13	40	185	8	36	100
Area 2	380	22	196	102	118	10	61	35	530	43	268	142
Area 3	25	-	56	115	14	-	30	74	23	-	38	93

3.3. Considerations for Implementing the Fuzzy System

For implementing the fuzzy system, in this study, its parameters were adjusted according to Table 5. Based on the opinion of an expert about the number of linguistic terms and results presented to [2,13], it seems that a number of three MFs for each input or output variable are appropriate for building damage detection. Moreover, according to the previous works [2,13], the Gaussian MF is efficient and robust for damage detection application, therefore this function was also utilized in this study. Furthermore, 12 fuzzy rules were designed by an expert, which were employed over three selected areas. In addition, Min, Max, Min and Max operators were chosen for "and", "or", "aggregation" and "implication" methods, respectively.

Table 5. Parameters of Mamdani fuzzy inference system selected in this study.

Parameter Name	Fuzzy Parameters Selected
Number of MF for each input	3
Number of MF for each output	3
Type of input MF	Gaussian Function
Type of output MF	Gaussian Function
Number of iterations	Flexible
Defuzzification method	Centroid
"and" method	Min
"or" method	Max
Implication method	Min
Aggregation method	Max
Number of Rules	12

3.4. Sensitivity Analysis on Fuzzy-GA Parameters

In order to perform the sensitivity analysis of results with respect to GA parameters, including the number of iteration (*it*), the number of population (*pop*), the mutation rate (P_m) and the crossover rate (P_c), these parameters were varied in a limited range. Tables 6–8 depict the overall accuracy of our proposed system obtained from variations of GA parameters over Area 1, Area 2, and Area 3, respectively. The range of values of parameters was selected based on experimental results and our knowledge about GA. Here, our description is presented on the range of the selected values. In general, low mutation rates are selected in GA, because increasing the mutation rate leads GA into a random search method. For this reason, mutation rates 0.1, 0.2, and 0.3 were chosen in this study. In contrast with the mutation rate, the selection of high values for the crossover rate guaranties optimized solutions to be achieved in a speedy manner. Hence, crossover rates 0.7, 0.8, and 0.9 were selected. Furthermore, based on the diagram of convergence of GA (see Section 3.5), it seems that GA is approximately converged after the 100th iteration. Therefore, the sensitivity of results with respect to iterations 100, 200, and 300 was tested. Finally, based on the previous works [53], the effect of population 50, 150, and 250 on the accuracy of the proposed damage detection system was investigated.

Table 6. Overall accuracies achieved from variations of GA parameters over Area 1.

pop →	50	50	50	150	150	150	250	250	250	*it* ↓
0.1	76%	73%	73%	77%	79%	79%	77%	79%	79%	100
0.2	76%	77%	74%	79%	77%	72%	79%	78%	77%	100
0.3	76%	75%	74%	76%	80%	77%	79%	73%	79%	100
0.1	73%	75%	75%	77%	76%	78%	77%	79%	79%	200
0.2	74%	77%	79%	77%	79%	77%	79%	79%	79%	200
0.3	77%	77%	78%	77%	73%	77%	79%	79%	77%	200
0.1	79%	78%	70%	75%	77%	77%	75%	73%	80%	300
0.2	77%	79%	77%	78%	79%	78%	78%	79%	77%	300
0.3	77%	79%	77%	77%	80%	78%	71%	71%	79%	300
P_m ↑	0.7	0.8	0.9	0.7	0.8	0.9	0.7	0.8	0.9	← P_c

Table 7. Overall accuracies achieved from variations of GA parameters over Area 2.

pop →	50	50	50	150	150	150	250	250	250	*it* ↓
0.1	65%	65%	65%	64%	66%	66%	64%	64%	64%	100
0.2	64%	67%	66%	64%	66%	66%	66%	66%	64%	100
0.3	65%	67%	66%	66%	66%	66%	66%	66%	66%	100
0.1	65%	65%	66%	65%	67%	66%	65%	65%	63%	200
0.2	65%	64%	67%	65%	65%	66%	66%	65%	66%	200
0.3	63%	65%	64%	67%	66%	66%	65%	66%	65%	200
0.1	66%	67%	66%	66%	65%	64%	66%	66%	65%	300
0.2	65%	67%	67%	65%	65%	65%	64%	65%	66%	300
0.3	65%	65%	66%	67%	67%	65%	66%	65%	65%	300
P_m ↑	0.7	0.8	0.9	0.7	0.8	0.9	0.7	0.8	0.9	← P_c

Table 8. Overall accuracies achieved from variations of GA parameters over Area 3.

pop →	50	50	50	150	150	150	250	250	250	*it* ↓
0.1	91%	91%	92%	90%	89%	90%	91%	90%	91%	100
0.2	91%	90%	90%	90%	91%	90%	91%	92%	92%	100
0.3	92%	90%	90%	89%	90%	92%	92%	90%	92%	100
0.1	90%	90%	90%	91%	91%	91%	90%	91%	91%	200
0.2	93%	92%	91%	91%	92%	92%	91%	91%	92%	200
0.3	92%	90%	90%	91%	92%	92%	91%	92%	91%	200
0.1	92%	90%	90%	91%	91%	90%	91%	91%	91%	300
0.2	92%	91%	90%	92%	92%	92%	91%	92%	90%	300
0.3	92%	91%	89%	91%	92%	90%	92%	91%	91%	300
P_m ↑	0.7	0.8	0.9	0.7	0.8	0.9	0.7	0.8	0.9	← P_c

3.5. Results of Optimized Fuzzy System

In this section, results of optimized fuzzy systems, including the diagram of convergence of GA and optimized parameters regarding MFs, are presented. Figure 3 depicts the cost value of the best population with the minimum cost at 300 iterations over Area 1 and Area 3. Figure 3 shows that 300 iterations are appropriate for converging GA in this research. Moreover, in order to illustrate the importance of GA in selecting parameters of the fuzzy system, changes of MFs of input "2" and input "3" in an experiment are presented in Figure 4. Based on Figure 4, it is easily observed that vast variations must be applied on MF parameters to achieve optimized results. Therefore, it seems that selecting these parameters by an expert is not an easy task.

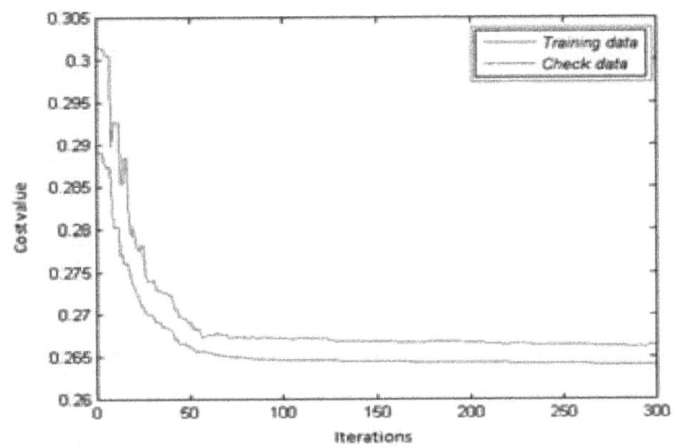

(**a**)

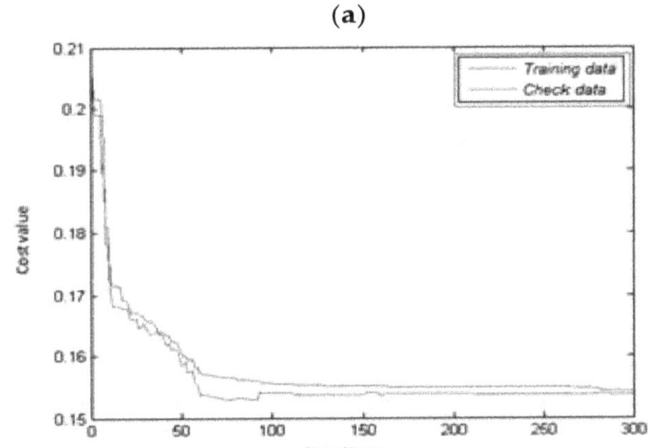

(**b**)

Figure 3. The diagram of convergence of GA over (**a**) Area 1, (**b**) Area 3.

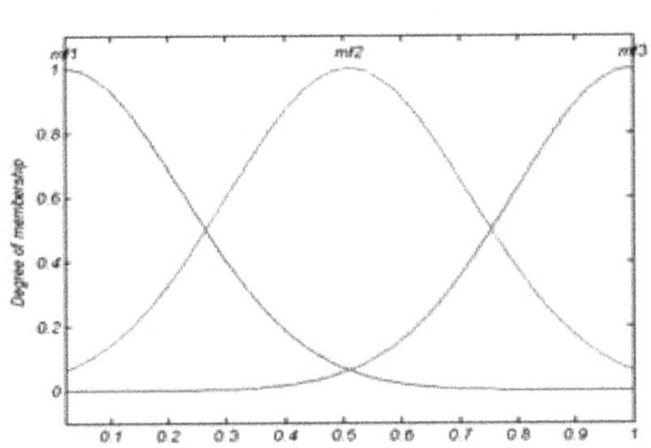

(**a**)

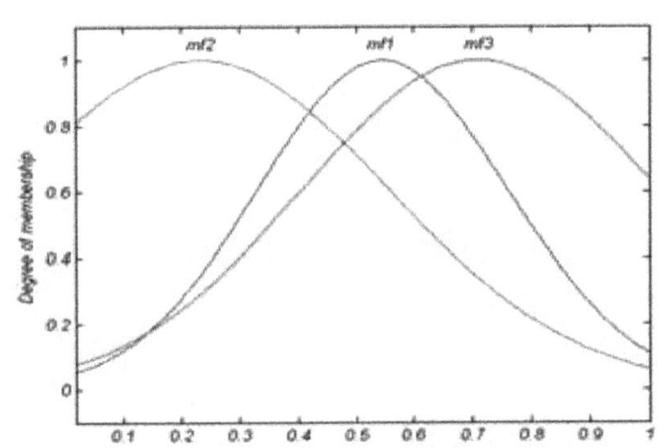

(**b**)

Figure 4. *Cont.*

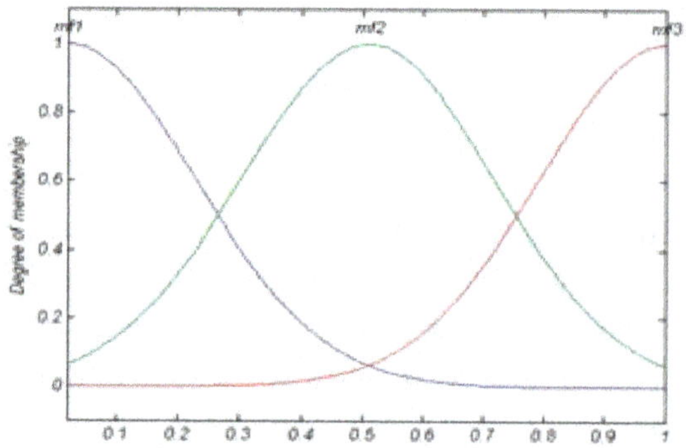

(c)

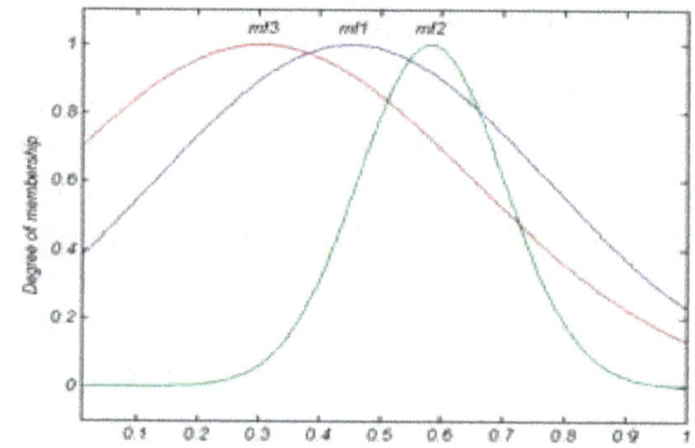

(d)

Figure 4. The presentation of preliminary MFs and optimized MFs in an experiment for input 2 and input 3: (**a**) preliminary MFs for input 2, (**b**) optimized MFs for input 2, (**c**) preliminary MFs for input 3, (**d**) optimized MFs for input 3.

3.6. Accuracy Assessment by Confusion Matrix

According to Section 3.5, to produce a confusion matrix, it is necessary to specify the number of classes and their definitions. In this study, we consider two damage classes including "damaged" and "undamaged" regarding each building. The "damaged" class includes very heavy damage (D4) and destruction (D5) whose definitions of D4 and D5 were presented in European Macroseismic Scale 1998 (EMS 98) [54]. To the best of our knowledge, substantial to heavy damage class (D3) which is related to cracks on the buildings facades would not be detected by our optical data used [54]. Therefore, "undamaged" class includes negligible to slight damage (D1) and D3. The confusion matrix of training, test and check samples as well as some statistical descriptors such as user and producer accuracies are presented to Table 9. Moreover, the range of overall accuracy of the generated damage maps with 90% confidence level is presented in Table 10.

3.7. Damage Map Resulting

In this subsection, final building damage maps regarding Area 1, Area 2, and Area 3 extracted from the proposed method are presented. Figure 5 shows the building damage maps of the mentioned areas.

Table 9. Confusion matrix of training, check, and test samples over Area 1, Area 2, and Area 3.

		Area 1				**Area 2**				**Area 3**			
		undamaged	damaged	User acc. (%)	Producer acc. (%)	undamaged	damaged	User acc. (%)	Producer acc. (%)	undamaged	damaged	User acc. (%)	Producer acc. (%)
Test samples	undamaged	153	40	79	79	444	195	77	69	20	9	87	69
	damaged	40	96	70	70	129	215	52	63	3	122	93	98
Check samples	undamaged	51	11	72	82	97	45	76	68	10	7	71	59
	damaged	20	42	79	68	31	51	53	62	4	97	93	96
Training samples	undamaged	142	34	76	81	310	126	77	71	21	4	84	84
	damaged	45	79	70	60	92	172	58	65	4	167	98	98

Table 10. Overall accuracy of the proposed method obtained on Area 1, Area 2, and Area 3.

	Area 1 (with 90% Confidence Level)	Area 2 (with 90% Confidence Level)	Area 3 (with 90% Confidence Level)
Overall accuracy (%)	76.88 ± 1.22	65.43 ± 0.29	90.96 ± 0.15

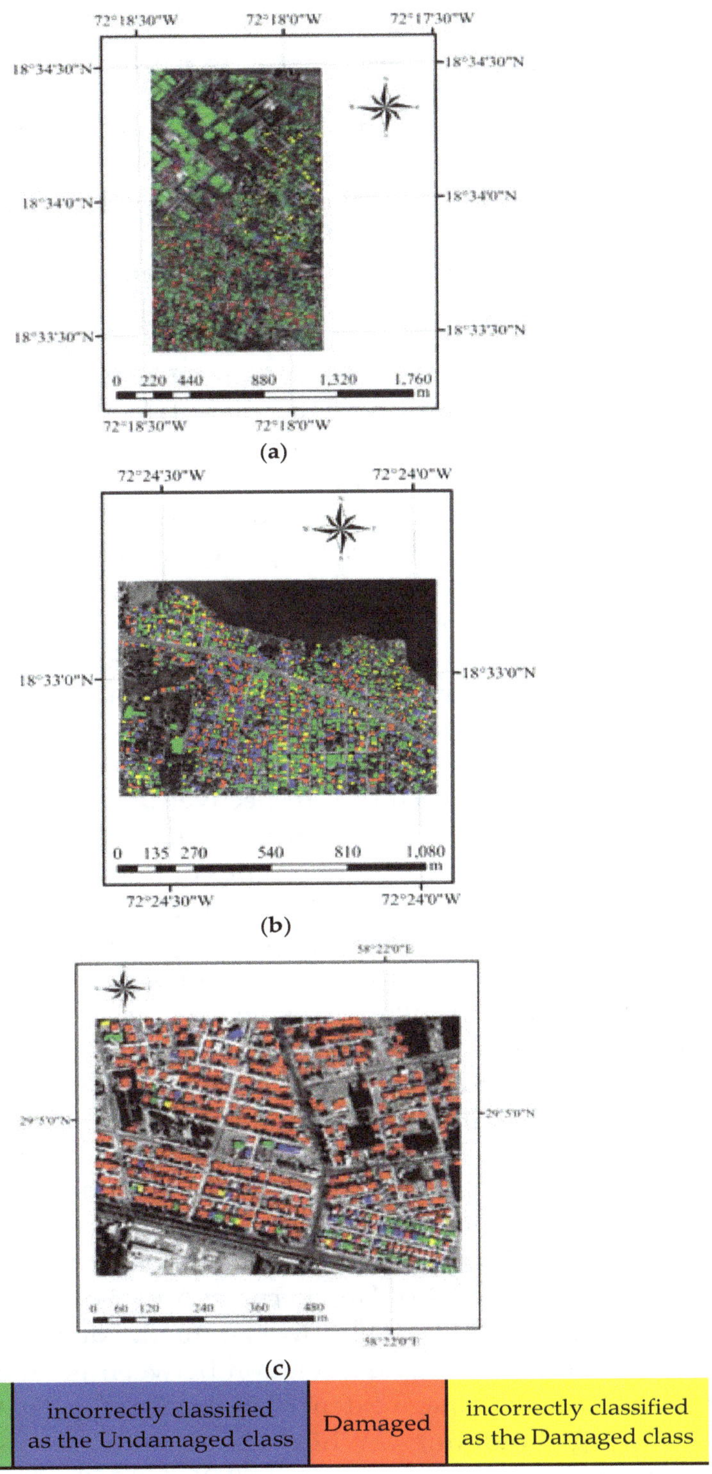

Figure 5. Building damage maps resulting from the proposed method on (**a**) Area 1, (**b**) Area 2, (**c**) Area 3.

4. Discussion

In this section, a comprehensive discussion about the obtained results is presented. Moreover, sources of error of the proposed method are specified. Furthermore, the accuracy and precision of Fuzzy-GA are compared with advanced machine learning techniques.

Based on the convergence diagram of GA presented to Figure 3, it is concluded that GA has been fully converged. Figure 3 shows that the selected parameters regarding GA were appropriate for providing a robust fuzzy system. Moreover, based on Figure 3, Fuzzy-GA succeeded in decreasing the cost value of training and check samples over the selected areas. It should be considered that the reduction of cost value directly improves the accuracy of the final results. These results show that the change of parameters of MFs in a fuzzy system is very important in obtaining an accurate result. Furthermore, based on Figure 4, in order to obtain more accurate results, vast variations should be performed on parameters of MFs (i.e., mean (m_i) and variance (σ_i) of Gaussian MF). Owing to employing an optimization algorithm (i.e., GA) in our study, the selection of the optimized parameters was carried out in a semi-automated way. However, it seems that adjusting these parameters in a manual way is not an easy task.

In this part, conclusions obtained from the sensitivity analysis stage are presented. Based on sensitivity analysis results (i.e., Tables 6–8), it was found that the selected parameters are appropriate for extracting an accurate damage map. Moreover, based on the outcomes, the accuracy of the final damage maps varies over a specific range. For example, overall accuracies of 90%, 91%, and 92% were achieved a lot over Area 3. In fact, in this study, it seems that the optimum answer is defined for a limited continuous range of variables. For this reason, increasing or decreasing the parameters of GA in the defined ranges does not follow a predictable procedure. Hence, based on these results, it appears that step-by-step sensitivity analysis methods presented to [2,13] are not proper for performing the sensitivity analysis in optimization-based decision making systems.

Based on the description presented in the previous paragraph on the range of optimum answers, it is better to present the overall accuracy of the obtained damaged maps in a $(a \pm b)$% form. In this form, a is the most probable overall accuracy and $[a - b, a + b]$ is the most probable range of overall accuracy. Based on Table 10, overall accuracies obtained on Area 1, Area 2, and Area 3 were equal to $76.88 \pm 1.22\%$, $65.43 \pm 0.29\%$ and $90.96 \pm 0.15\%$, respectively. Based on these results, the study area and its characteristics completely affect the accuracy of the damage detection method. For example, Area 2 is a complex urban region. In this area, different types of buildings, including buildings connected with trees, buildings with small area, and gabled roof buildings, are observed. It should be considered that a complex area could spread many uncertainties and errors over the final damage maps (we will present the source of errors in the following). For this reason, the overall accuracy of the proposed system over Area 2 was lower than the other areas. In general, based on our visual observations from high-spatial resolution satellite images, Area 1 and Area 2 are more complex than Area 3. For this reason, the accuracy of our proposed system over Area 3 was higher than Area 1 and Area 2.

There have been many damage detection methods implemented in the Haiti and Bam areas [4,13,55,56]. However, it is a difficult issue to present a fair and comprehensive judgment between the existing methods over study areas and our proposed method, because based on our experimental results, study area and its characteristics affect the accuracy of damage detection methods. Results achieved on Area 1, Area 2, and Area 3 completely confirm our claim. In this part, we compare our achieved results with the outcomes of [13], because Area 3 of our study is similar to the selected area of the mentioned paper. Janalipour and Mohammadzadeh (2016) employed a Nuero-fuzzy decision making system in order to create a relation between geometrical features obtained from post-event satellite image and the damage extent of buildings [13]. They reported an overall accuracy of 76% for detecting damaged buildings. While, in our study, the use of Fuzzy-GA and Haralick textural features resulted in an overall accuracy of $90.96 \pm 0.15\%$. Therefore, as a result, Fuzzy-GA and Haralick textural features outperformed Nuero-fuzzy and the geometrical features in detecting damaged buildings.

In this part, outcomes of Fuzzy-GA decision making system are compared with four advanced machine learning techniques, including bagging, boosting, random forests (RF) [57] and support vector machine (SVM), from viewpoints of accuracy and precision. It should be considered that results of the mentioned techniques and our decision making system were achieved with similar textural features. Hence, we can compare them with respect to each other. Based on [58], precision measures the degree of consistency among overall accuracies obtained in the selected areas (Tables 6–8) and accuracy shows the degree of closeness of measurements to true value. Based on the presented definitions and the form of presentation of overall accuracy (i.e., $(a \pm b)\%$), a and b can be used respectively as an accuracy measure and a precision creation. Based on Table 11, from the perspective of the accuracy, the Fuzzy-GA decision making system was more successful than bagging, boosting, and RF and SVM machine learning techniques. For example, over Area 3, overall accuracies of Fuzzy-GA, bagging, boosting, RF and SVM were respectively equal to 90.96%, 89.05%, 87.47%, 88.03% and 89.03%. To compare results of machine learning techniques with our proposed system from the viewpoint of precision, it should be considered that b is a function of precision. A machine learning technique with the lowest b is more precise than other ones. Because values of b for our proposed system are lower than the corresponding values for the advanced machine learning techniques, it appears that Fuzzy-GA decision making system is more precise than bagging, boosting, RF and SVM machine learning techniques.

Table 11. Overall accuracies of Fuzzy-GA, bagging, boosting, RF, and SVM machine learning techniques obtained on Area 1, Area 2, and Area 3.

	Area 1 (%)	**Area 2 (%)**	**Area 3 (%)**
Fuzzy-GA	76.88 ± 1.22	65.43 ± 0.29	90.96 ± 0.15
Bagging	74.5 ± 2.0	56.03 ± 1.2	89.05 ± 0.6
Boosting	71.33 ± 2.2	62.79 ± 2.9	87.47 ± 1.0
RF	73.07 ± 1.4	55.92 ± 1.4	88.03 ± 0.5
SVM with a radial basis function (RBF) kernel	72.53 ± 1.6	60.52 ± 1.2	89.03 ± 0.6

As mentioned, an urban area with a complex structure could spread many uncertainties and errors on outcomes. In this part, reasons of some misclassifications that occurred in the study areas are discussed. Table 12 depicts post-event optical image, variance feature, and damage map of nine miss-classified buildings. Building No. 1 is connected to a tall building. According to Table 12, parts from building No. 1 were covered with the shadow of the tall building. This is sufficient for texture extraction methods to produce high texture values in the roof of building No. 1. For this reason, this building was incorrectly classified as damaged class. The root of error of building No. 2 is similar to building No. 1 but with the difference that the overlapped shadow with this building resulted from a tall tree. This type error was frequently observed over Area 2 which is a complex urban area. Moreover, based on the Haiti building damage atlas, building No. 3 was classified as undamaged class by experts. While our visual interpretation showed that this is a damaged building. Hence, the proposed method correctly performed its task. Building No. 4 is an inclined damaged one. Based on the definition of inclined buildings [59] and Table 12, textural features are unable to detect these damage types. Therefore, the class of this building was incorrectly assigned as unchanged category. The use of LiDAR data may be an appropriate solution for detecting inclined damaged buildings.

Table 12. The presentation of pre-event image, variance feature, and damage map of nine mis-classified buildings.

No.	Post-Event Image	Variance Feature	Damage Map
1		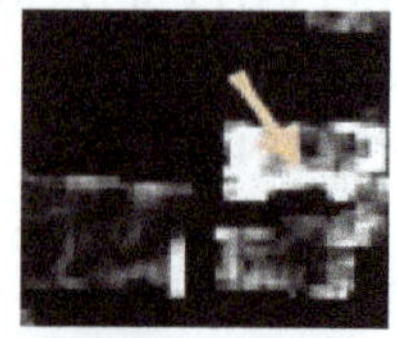	
2		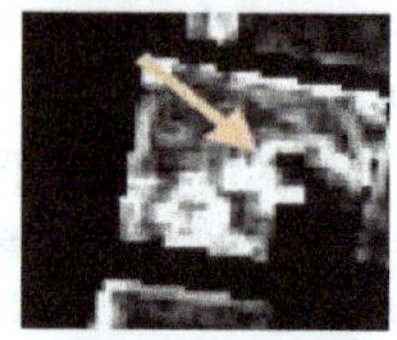	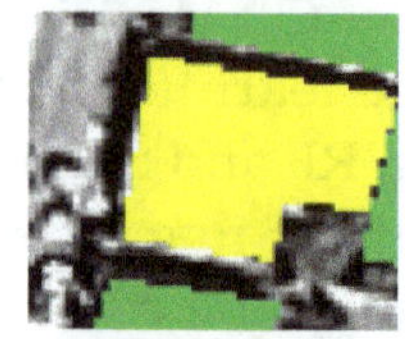
3			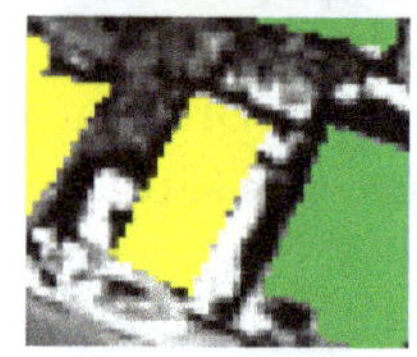
4			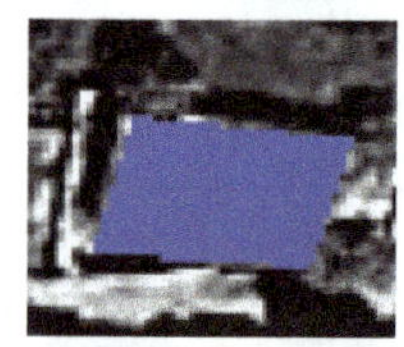
5			
6		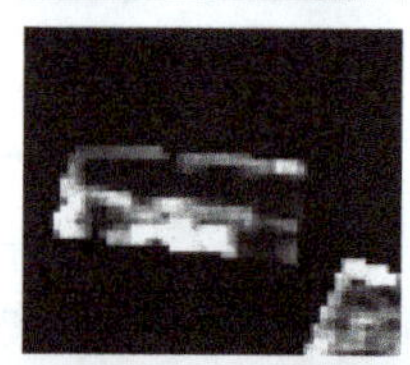	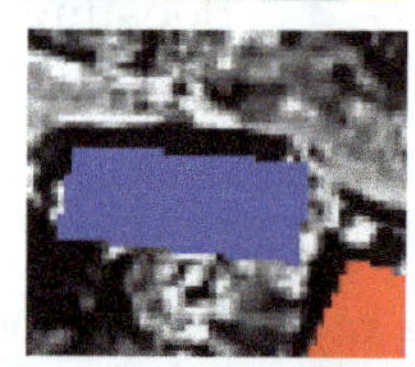
7		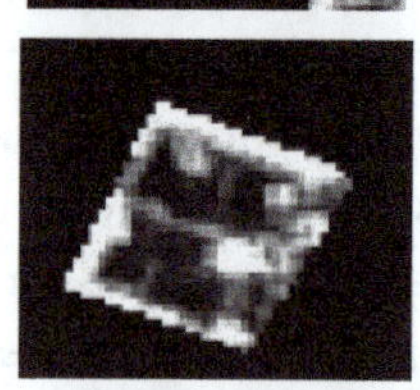	
8	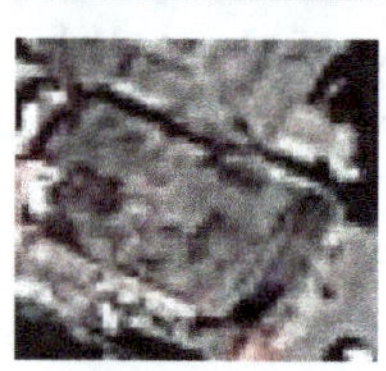	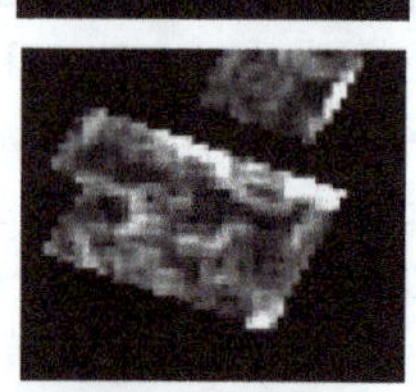	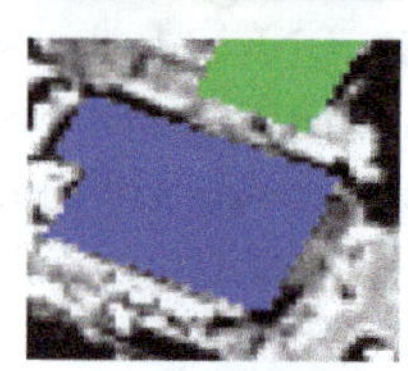
9		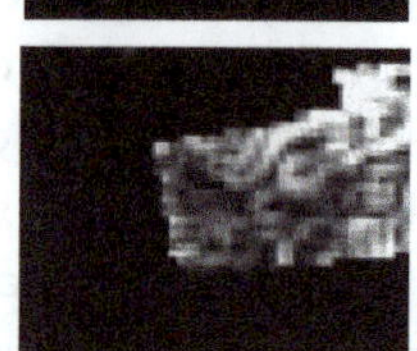	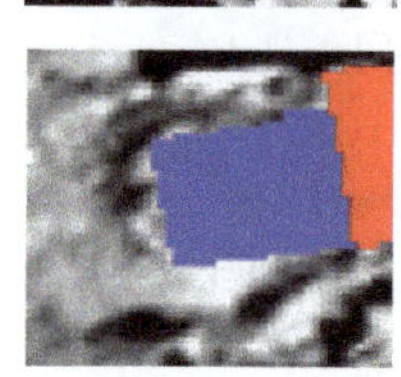

The high spectral variation of pixels of undamaged building roofs over satellite optical images is a critical issue for texture-based damage detection methods, because the behavior of texture features of these buildings is similar to damaged ones. This issue was the main reason for classifying building No. 5 as a damaged one. Building No. 6 is a gabled roof one, in which parts of the roof of this building were destroyed. The texture extraction method correctly identified the damaged parts. However, it seems that the Fuzzy-GA decision making system was the main reason for classifying this building as undamaged category incorrectly. The class of building No. 7 is undamaged. The spectral variation of pixels of building No. 7 is very high. For this reason, our proposed system classified building No. 7 as damaged category. Based on our interpretation, the texture extraction method and Fuzzy-GA decision making system are two sources of error for the misclassification of building No. 8. Finally, the damage extent of building No. 9 is totally pancaked. The proposed system was unsuitable in the detection of the damage class of this building, because the variation of the digital numbers of the building roof was not high enough for assigning it to the damaged class. The use of a normalized digital surface model may be an appropriate way for identifying totally pancaked damaged buildings.

5. Conclusions

In this study, a semi-automated Mamdani based fuzzy decision making system was developed in order to identify damaged buildings using their textural features. For improving the automation level of the Mamdani fuzzy system, a genetic algorithm was used to find its optimized parameters concerning membership functions. The proposed system was tested over two areas of the 2010 Haiti earthquake and one area of the 2003 Bam earthquake. Based on the concept of our proposed system, its automation level is higher than other existing decision making systems [19,27,28]. Moreover, based on the statistical descriptors and results of bagging, boosting, RF, and SVM machine learning techniques, it seems that Fuzzy-GA decision making system is more accurate and precise than the mentioned techniques for building damage detection. Furthermore, based on outcomes of the sensitivity analysis stage, it seems that results of our proposed system are robust enough for building damage detection. Based on the results of the sensitivity analysis stage, the overall accuracy of $76.88 \pm 1.22\%$, $65.43 \pm 0.29\%$ and $90.96 \pm 0.15\%$ was obtained on Area 1 (the Haiti earthquake), Area 2 (the Haiti earthquake), and Area 3 (the Bam earthquake), respectively. According to these results, the study area and its characteristics directly affect the accuracy achieved from the proposed method.

Based on the presented discussions, there are some major error sources: (1) high spectral variation of digital numbers over the roofs of undamaged buildings, (2) the shadow of tall buildings and trees connected with undamaged buildings, (3) inability of optical data in detecting inclined and pancaked damaged buildings.

As a future work, based on limitations of this study, it will be necessary to propose a damage detection method for integrating optical and LiDAR data. Moreover, due to the importance of the automation level of the damage detection method, it is important to present an automatic and accurate fuzzy decision making system in these future works. As another future work, the robustness of the Fuzzy-GA decision making system can be assessed over SAR data for detecting damaged areas. Moreover, the Fuzzy-GA decision making system can be adapted for applications that use hyperspectral bands. Because we are dealing with a high dimensional feature space, it is very important to select the appropriate bands using efficient feature selection methods such as [60,61]. Finally, owing to our proposed method which depends on a pre-event updated map, proposing an efficient building extraction method such as [62,63] for obtaining building footprints from pre-event optical images is very important to consider as a research study.

Acknowledgments: We would like to express our sincere thanks to the National Cartographic Center of Iran (NCC) who provided the pre-event map of BAM city. The author thanks DigitalGlobe for providing the pre- and post-event high spatial resolution images of the study area. The Haiti building damage atlas was produced by the United Nations' Institute for Training and Research (UNITAR) Operational Satellite Applications Programme (UNOSAT), the European Commission's Joint Research Centre (JRC), and the World Bank in support to the Post Disaster Needs Assessment (PDNA) process led by the Government of Haiti. The authors are grateful to the anonymous reviewers for their time and effort spent in reviewing the paper and for their constructive comments. We also like to thank Qi Wang, Nicolas H. Younan, and Carlos López-Martínez for the rapid peer reviewing process.

Author Contributions: Milad Janalipour proposed and implemented the method, also wrote the manuscript. Ali Mohammadzadeh supervised the research and also reviewed and revised the manuscript.

Abbreviations

The following abbreviations are used in this manuscript:

LiDAR	Light Detection and Ranging
SAR	Synthetic Aperture Radar
OA	Overall Accuracy
EMS 98	European Macroseismic Scale 1998
GA	Genetic Algorithm
Fuzzy-GA	Fuzzy Genetic Algorithm
RS	Remote Sensing
SVM	Support Vector Machine
RF	Rnadom Forests
OCSVM	One-Class Support Vector Machine
MF	Membership Function
MFIS	Mamdani Fuzzy Inference System

References

1. Dubois, D.; Lepage, R. Fast and efficient evaluation of building damage from very high resolution optical satellite images. *IEEE J. Sel. Top. Appl. Earth Obs. Remote Sens.* **2014**, *7*, 4167–4176. [CrossRef]
2. Janalipour, M.; Taleai, M. Building change detection after earthquake using multi-criteria decision analysis based on extracted information from high spatial resolution satellite images. *Int. J. Remote Sens.* **2017**, *38*, 82–99. [CrossRef]
3. Khoshelham, K.; Oude Elberink, S. Role of dimensionality reduction in segment-based classification of damaged building roofs in airborne laser scanning data. In Proceedings of the International Conference on Geographic Object Based Image Analysis, Rio de Janeiro, Brazil, 7–9 May 2012; pp. 372–377.
4. Dell'Acqua, F.; Gamba, P.; Polli, D.A. Earthquake damage assessment from post-event VHR radar data: From Sichuan, 2008 to Haiti, 2010. In Proceedings of the IEEE 2011 Joint Urban Remote Sensing Event, Munich, Germany, 11–13 April 2011; pp. 201–204.
5. Balz, T.; Liao, M. Building-damage detection using post-seismic high-resolution SAR satellite data. *Int. J. Remote Sens.* **2010**, *31*, 3369–3391. [CrossRef]
6. Matsuoka, M.; Yamazaki, F. Use of satellite SAR intensity imagery for detecting building areas damaged due to earthquakes. *Earthq. Spectra* **2004**, *20*, 975–994. [CrossRef]
7. Matsuoka, M.; Yamazaki, F. Building damage mapping of the 2003 bam, Iran, earthquake using ENVISAT/ASAR intensity imagery. *Earthq. Spectra* **2005**, *21*, 285–294. [CrossRef]
8. Matsuoka, M.; Yamazaki, F. Comparative analysis for detecting areas with building damage from several destructive earthquakes using satellite synthetic aperture radar images. *J. Appl. Remote Sens.* **2010**, *4*. [CrossRef]
9. Dell'Acqua, F.; Polli, D.A. Post-event only VHR radar satellite data for automated damage assessment. *Photogramm. Eng. Remote Sens.* **2011**, *77*, 1037–1043. [CrossRef]
10. Sirmacek, B.; Unsalan, C. Damaged building detection in aerial images using shadow information. In Proceedings of the IEEE 2009 4th International Conference on Recent Advances in Space Technologies, Istanbul, Turkey, 11–13 June 2009; pp. 249–252.
11. Geiß, C.; Pelizari, P.A.; Marconcini, M.; Sengara, W.; Edwards, M.; Lakes, T.; Taubenböck, H. Estimation of seismic building structural types using multi-sensor remote sensing and machine learning techniques. *ISPRS J. Photogramm. Remote Sens.* **2015**, *104*, 175–188. [CrossRef]

12. Geiß, C.; Taubenböck, H.; Tyagunov, S.; Tisch, A.; Post, J.; Lakes, T. Assessment of seismic building vulnerability from space. *Earthq. Spectra* **2014**, *30*, 1553–1583. [CrossRef]
13. Janalipour, M.; Mohammadzadeh, A. Building damage detection using object-based image analysis and anfis from high-resolution image (case study: Bam earthquake, Iran). *IEEE J. Sel. Top. Appl. Earth Obs. Remote Sens.* **2016**, *9*, 1937–1945. [CrossRef]
14. Eguchi, R.T.; Mansouri, B. Use of remote sensing technologies for building damage assessment after the 2003 Bam, Iran, earthquake—Preface to remote sensing papers. *Earthq. Spectra* **2005**, *21*, 207–212. [CrossRef]
15. Voigt, S.; Schneiderhan, T.; Twele, A.; Gähler, M.; Stein, E.; Mehl, H. Rapid damage assessment and situation mapping: Learning from the 2010 Haiti earthquake. *Photogramm. Eng. Remote Sens.* **2011**, *77*, 923–931. [CrossRef]
16. Lu, L.; Guo, H.; Corbane, C.; Pesaresi, M.; Ehrlich, D. Rapid damage assessment of buildings with VHR optical airborne images in Yushu earthquake. In Proceedings of the 2012 2nd International Conference on Remote Sensing, Environment and Transportation Engineering, Nanjing, China, 1–3 June 2012.
17. Tiede, D.; Lang, S.; Füreder, P.; Hölbling, D.; Hoffmann, C.; Zeil, P. Automated damage indication for rapid geospatial reporting. *Photogramm. Eng. Remote Sens.* **2011**, *77*, 933–942. [CrossRef]
18. Lemoine, G.; Corbane, C.; Louvrier, C.; Kauffmann, M. Intercomparison and validation of building damage assessments based on post-Haiti 2010 earthquake imagery using multi-source reference data. *Nat. Hazards Earth Syst. Sci. Discuss.* **2013**, *1*, 1445–1486. [CrossRef]
19. Haghighattalab, A.; JAVAD Valadan zoej, M.; Mohammadzadeh, A.; Taleai, M.; Kalantari, M. Détection de dommages et évaluation des dégâts du réseau routier après un séisme, en utilisant des images quickbird haute résolution. *XYZ J.* **2010**, *32*, 41–47.
20. Vu, T.T.; Matsuoka, M.; Yamazaki, F. Multilevel detection of damaged buildings from high-resolution optical satellite images. In Proceedings of the 2006 International Society for Optics and Photonics Asia-Pacific Remote Sensing Symposium, Goa, India, 13 November 2006.
21. Wang, X.; Li, P. Urban building collapse detection using very high resolution imagery and airborne Lidar data. *ISPRS Int. Arch. Photogramm. Remote Sens. Spat. Inf. Sci.* **2013**, *40*, 127–132. [CrossRef]
22. Chesnel, A.-L.; Binet, R.; Wald, L. Urban damage assessment using multimodal Quickbird images and ancillary data: The Bam and the Boumerdes earthquakes. In Proceedings of the 6th International Workshop on Remote Sensing for Disaster Management Applications, Pavia, Italy, 11–12 September 2008.
23. Miura, H.; Modorikawa, S.; Chen, S.H. Texture characteristics of high-resolution satellite images in damaged areas of the 2010 Haiti earthquake. In Proceedings of the 9th International Workshop on Remote Sensing for Disaster Response, Stanford, CA, USA, 15–16 September 2011.
24. Li, P.; Xu, H.; Guo, J. Urban building damage detection from very high resolution imagery using OCSVM and spatial features. *Int. J. Remote Sens.* **2010**, *31*, 3393–3409. [CrossRef]
25. Tian, J.; Nielsen, A.A.; Reinartz, P. Building damage assessment after the earthquake in Haiti using two post-event satellite stereo imagery and DSMS. *Int. J. Image Data Fusion* **2015**, *6*, 155–169. [CrossRef]
26. Polli, D.; Dell'Acqua, F.; Gamba, P.; Lisini, G. Earthquake damage assessment from post-event only radar satellite data. In Proceedings of the Eighth International Workshop on Remote Sensing for Disaster Response, Tokyo, Japan, 1 October 2010.
27. Samadzadegan, F.; Rastiveisi, H. Automatic detection and classification of damaged buildings, using high resolution satellite imagery and vector data. *ISPRS Int. Arch. Photogramm. Remote Sens. Spat. Inf. Sci.* **2008**, *37*, 415–420.
28. Mansouri, B.; Hamednia, Y. A soft computing method for damage mapping using VHR optical satellite imagery. *IEEE J. Sel. Top. Appl. Earth Obs. Remote Sens.* **2015**, *8*, 4935–4941. [CrossRef]
29. Chen, P.; Wu, J.; Liu, Y.; Wang, J. Extraction method for earthquake-collapsed building information based on high-resolution remote sensing. In Proceedings of the IOP Conference Series: Earth and Environmental Science, Beijing, China, 22–26 April 2014; p. 012096.
30. Ural, S.; Hussain, E.; Kim, K.; Fu, C.-S.; Shan, J. Building extraction and rubble mapping for city port-au-prince post-2010 earthquake with GeoEye-1 imagery and Lidar data. *Photogramm. Eng. Remote Sens.* **2011**, *77*, 1011–1023. [CrossRef]
31. Jang, J.-S.R.; Sun, C.-T.; Mizutani, E. Neuro-fuzzy and soft computing; a computational approach to learning and machine intelligence. *IEEE Trans. Autom. Control* **1997**, *42*, 1482–1484. [CrossRef]

32. Dong, L.; Shan, J. A comprehensive review of earthquake-induced building damage detection with remote sensing techniques. *ISPRS J. Photogramm. Remote Sens.* **2013**, *84*, 85–99. [CrossRef]
33. Dell'Acqua, F.; Gamba, P. Remote sensing and earthquake damage assessment: Experiences, limits, and perspectives. *IEEE Proc.* **2012**, *100*, 2876–2890. [CrossRef]
34. Geiß, C.; Taubenböck, H. Remote sensing contributing to assess earthquake risk: From a literature review towards a roadmap. *Nat. Hazards* **2013**, *68*, 7–48. [CrossRef]
35. Rehor, M.; Bähr, H.P.; Tarsha-Kurdi, F.; Landes, T.; Grussenmeyer, P. Contribution of two plane detection algorithms to recognition of intact and damaged buildings in Lidar data. *Photogramm. Rec.* **2008**, *23*, 441–456. [CrossRef]
36. Gamba, P.; Dell'Acqua, F.; Trianni, G. Rapid damage detection in the Bam area using multitemporal SAR and exploiting ancillary data. *IEEE Trans. Geosci. Remote Sens.* **2007**, *45*, 1582–1589. [CrossRef]
37. Dell'Acqua, F.; Lisini, G.; Gamba, P. Experiences in optical and SAR imagery analysis for damage assessment in the Wuhan, May 2008 Earthquake. *IEEE Int. Geosci. Remote Sens. Symp.* **2009**, *4*, 37–40.
38. Trianni, G.; Gamba, P. Fast damage mapping in case of earthquakes using multitemporal SAR data. *J. Real-Time Image Process.* **2009**, *4*, 195–203. [CrossRef]
39. Berman, L. Haiti Earthquake Data (Vectors). *Harv. Dataverse* **2015**. [CrossRef]
40. Pascale, D. *A Review of RGB Color Spaces*; Technical Report; The Babel Color Company: Montreal, QC, Canada, 2003.
41. Theodoridis, S.; Koutroumbas, K. Chapter 7–feature generation II. In *Pattern Recognition*; Academic Press: San Diego, CA, USA, 2009; pp. 411–479.
42. Haralick, R.M.; Shanmugam, K.; Dinstein, I.H. Textural features for image classification. *IEEE Trans. Syst. Man Cybern.* **1973**, *3*, 610–621. [CrossRef]
43. Mollalo, A.; Khodabandehloo, E. Zoonotic cutaneous leishmaniasis in northeastern Iran: A GIS-based spatio-temporal multi-criteria decision-making approach. *Epidemiol. Infect.* **2016**, *144*, 2217–2229. [CrossRef] [PubMed]
44. Zadeh, L.A. The role of fuzzy logic in the management of uncertainty in expert systems. *Fuzzy Sets Syst.* **1983**, *11*, 199–227. [CrossRef]
45. Homaifar, A.; McCormick, E. Simultaneous design of membership functions and rule sets for fuzzy controllers using genetic algorithms. *IEEE Trans. Fuzzy Syst.* **1995**, *3*, 129–139. [CrossRef]
46. Lee, M.A.; Takagi, H. Integrating design stage of fuzzy systems using genetic algorithms. In Proceedings of the 2nd IEEE International Conference on Fuzzy Systems, San Francisco, CA, USA, 18 March–1 April 1993; pp. 612–617.
47. Haupt, R.L.; Haupt, S.E. *Practical Genetic Algorithms*; Wiley-Interscience: Hoboken, NJ, USA, 2004.
48. Hwang, D.; Karimi, H.A.; Byun, D.W. Uncertainty analysis of environmental models within GIS environments. *Comput. Geosci.* **1998**, *24*, 119–130. [CrossRef]
49. Lodwick, W.A. Developing confidence limits on errors of suitability analyses in geographical information systems. *Accuracy Spat. Databases* **1989**, 69–80.
50. Tarantola, S.; Giglioli, N.; Jesinghaus, J.; Saltelli, A. Can global sensitivity analysis steer the implementation of models for environmental assessments and decision-making? *Stoch. Environ. Res. Risk Assess.* **2002**, *16*, 63–76. [CrossRef]
51. Iooss, B.; Lemaître, P. A review on global sensitivity analysis methods. In *Uncertainty Management in Simulation-Optimization of Complex Systems*; Springer: New York, NY, USA, 2015; pp. 101–122.
52. Srinivas, C.; Reddy, B.R.; Ramji, K.; Naveen, R. Sensitivity analysis to determine the parameters of genetic algorithm for machine layout. *Procedia Mater. Sci.* **2014**, *6*, 866–876. [CrossRef]
53. Roeva, O.; Fidanova, S.; Paprzycki, M. Influence of the population size on the genetic algorithm performance in case of cultivation process modelling. In Proceedings of the 2013 Federated Conference on Computer Science and Information Systems, Krakow, Poland, 8–11 September 2013; pp. 371–376.
54. Grünthal, G. *European Macroseismic Scale 1998*; European Seismological Commission, Subcommission on Engineering Seismology, Working Group Macroseismic Scales: Luxembourg, 1998.
55. Labiak, R.C.; Van Aardt, J.A.; Bespalov, D.; Eychner, D.; Wirch, E.; Bischof, H.-P. Automated method for detection and quantification of building damage and debris using post-disaster Lidar data. *Proc. SPIE* **2011**. [CrossRef]

56. Pham, T.-T.-H.; Apparicio, P.; Gomez, C.; Weber, C.; Mathon, D. Towards a rapid automatic detection of building damage using remote sensing for disaster management: The 2010 Haiti earthquake. *Disaster Prev. Manag.* **2014**, *23*, 53–66. [CrossRef]
57. Kuncheva, L.I. *Combining Pattern Classifiers: Methods and Algorithms*; John Wiley & Sons: Hoboken, NJ, USA, 2004.
58. Wolf, P.R.; Ghilani, C.D. *Adjustment Computations: Statistics and Least Squares in Surveying and GIS*; Wiley-Interscience: Hoboken, NJ, USA, 1997.
59. Schweier, C.; Markus, M. Classification of collapsed buildings for fast damage and loss assessment. *Bull. Earthq. Eng.* **2006**, *4*, 177–192. [CrossRef]
60. Yuan, Y.; Lin, J.; Wang, Q. Dual-clustering-based hyperspectral band selection by contextual analysis. *IEEE Trans. Geosci. Remote Sens.* **2016**, *54*, 1431–1445. [CrossRef]
61. Yuan, Y.; Zhu, G.; Wang, Q. Hyperspectral band selection by multitask sparsity pursuit. *IEEE Trans. Geosci. Remote Sens.* **2015**, *53*, 631–644. [CrossRef]
62. Sirmacek, B.; Unsalan, C. Urban-area and building detection using sift keypoints and graph theory. *IEEE Trans. Geosci. Remote Sens.* **2009**, *47*, 1156–1167. [CrossRef]
63. Sirmacek, B.; Unsalan, C. Building detection from aerial images using invariant color features and shadow information. In Proceedings of the 23rd International Symposium on Computer and Information Sciences, Istanbul, Turkey, 27–29 October 2008; pp. 1–5.

7

Hyperspectral Dimensionality Reduction by Tensor Sparse and Low-Rank Graph-Based Discriminant Analysis

Lei Pan [1], Heng-Chao Li [1,*], Yang-Jun Deng [1], Fan Zhang [2], Xiang-Dong Chen [1] and Qian Du [3]

1 School of Information Science & Technology, Southwest Jiaotong University, Chengdu 610031, China; mapan.lei@163.com (L.P.); dyj2012@yeah.net (Y.-J.D.); xdchen@home.swjtu.edu.cn (X.-D.C.)

2 College of Information Science & Technology, Beijing University of Chemical Technology, Beijing 100029, China; zhangf@mail.buct.edu.cn

3 Department of Electrical & Computer Engineering, Mississippi State University, Starkville, MS 39762, USA; du@ece.msstate.edu

* Correspondence: lihengchao_78@163.com.

Academic Editors: Qi Wang, Nicolas H. Younan, Carlos López-Martínez, Lenio Soares Galvao and Prasad S. Thenkabail

Abstract: Recently, sparse and low-rank graph-based discriminant analysis (SLGDA) has yielded satisfactory results in hyperspectral image (HSI) dimensionality reduction (DR), for which sparsity and low-rankness are simultaneously imposed to capture both local and global structure of hyperspectral data. However, SLGDA fails to exploit the spatial information. To address this problem, a tensor sparse and low-rank graph-based discriminant analysis (TSLGDA) is proposed in this paper. By regarding the hyperspectral data cube as a third-order tensor, small local patches centered at the training samples are extracted for the TSLGDA framework to maintain the structural information, resulting in a more discriminative graph. Subsequently, dimensionality reduction is performed on the tensorial training and testing samples to reduce data redundancy. Experimental results of three real-world hyperspectral datasets demonstrate that the proposed TSLGDA algorithm greatly improves the classification performance in the low-dimensional space when compared to state-of-the-art DR methods.

Keywords: hyperspectral image; sparse and low-rank graph; tensor; dimensionality reduction

---

1. Introduction

A hyperspectral image contains a wealth of spectral information about different materials by collecting the reflectance of hundreds of contiguous narrow spectral bands from the visible to infrared electromagnetic spectrum [1–3]. However, the redundant information in a hyperspectral image not only increases computational complexity but also degrades classification performance when training samples are limited. Some research has demonstrated that the redundancy can be reduced without a significant loss of useful information [4–7]. As such, reducing the dimensionality of hyperspectral images is a reasonable and important preprocessing step for subsequent analysis and practical applications.

Dimensionality reduction (DR) aims to reduce the redundancy among features and simultaneously preserve the discriminative information. In general, existing DR methods may belong to one of three categories: unsupervised, supervised, and semisupervised. The unsupervised methods do not take the class label information of training samples into consideration. The most commonly used unsupervised DR algorithm is principal component analysis (PCA) [8], which is to find a linear transformation by maximizing the variance in the projected subspace. Linear discriminant analysis (LDA) [9], as a simple supervised DR method, is proposed to maximize the trace ratio of between-class and within-class scatter matrices. To address the application limitation in data distribution of LDA, local Fisher's discriminant analysis (LFDA) [10] is developed. In order to overcome the difficulty that the number of training samples is usually limited, some semisupervised DR methods in [11,12] are proposed.

The graph, as a mathematical data representation, has been successfully embedded in the framework of DR, resulting in the development of many effective DR methods. Recently, a general graph embedding (GE) framework [13] has been proposed to formulate most of the existing DR methods, in which an undirected graph is constructed to characterize the geometric information of the data. k-nearest neighbors and ε-radius ball [14] are two traditional methods to construct adjacency graphs. However, these two methods are sensitive to the noise and may lead to incorrect data representation. To construct an appropriate graph, a graph-based discriminant analysis with spectral similarity (GDA-SS) measurement was recently proposed by considering curves changing description among spectral bands in [15]. Sparse representation (SR) [16,17] has attracted much attention because of its benefits of data-adaptive neighborhoods and noise robustness. Based on this work, a sparse graph embedding (SGE) model [18] was developed by exploring the sparsity structure of the data. In [19], a sparse graph-based discriminant analysis (SGDA) model was developed for hyperspectral image dimensionality reduction and classification by exploiting the class label information, improving the performance of SGE. In [20], a weighted SGDA integrated both the locality and sparsity structure of the data. To reduce the computational cost, collaborative graph-based discriminant analysis (CGDA) [21] was introduced by imposing an l_2 regularization on sparse coefficient vector. In [22], Laplacian regularization was imposed on CGDA, resulting in the LapCGDA algorithm. SR is able to reveal the local structure but fails in capturing the global structure. To solve this problem, a sparse and low-rank graph-based discriminant analysis (SLGDA) [23] was proposed to simultaneously preserve the local and global structure of hyperspectral data.

However, the aforementioned graph-based DR methods only deal with spectral vector-based (first-order) representations, which do not take the spatial information of hyperspectral data into consideration. Aiming to overcome this shortcoming, simultaneous sparse graph embedding (SSGE) was proposed to improve the classification performance in [24]. Although SSGE has obtained enhanced performance, it still puts the spectral-spatial feature into first-order data for analysis and ignores the cubic nature of hyperspectral data that can be taken as a third-order tensor. Some researchers have verified the advantage of tensor representation when processing the hyperspectral data. For example, multilinear principal component analysis (MPCA) [25] was integrated with support vector machines (SVM) for tensor-based classification in [26]. A group based tensor model [27] by exploiting clustering technique was developed for DR and classification. In addition, a tensor discriminative locality alignment (TDLA) [28] algorithm was proposed for hyperspectral image spectral-spatial feature representation and DR, which has been extended in [29] by combining with well-known spectral-spatial feature extraction methods (such as extended morphological profiles (EMPs) [30], extended attribute profiles (EAPs) [31], and Gabors [32]) for classification. Though the previous tensor-based DR methods have achieved great improvement on performance, they do not consider the structure property from other perspectives, such as representation-based and graph-based points.

In this context, we propose a novel DR method, i.e., tensor sparse and low-rank graph-based discriminant analysis (TSLGDA), for hyperspectral data, in which the information from three perspectives (tensor representation, sparse and low-rank representation, and graph theory) is exploited to present the data structure for hyperspectral image. It is noteworthy that the proposed method aims to exploit the spatial information through tensor representation, which is different from the work in [23] only considering the spectral information. Furthermore, tensor locality preserving projection (TLPP) [33] is exploited to obtain three projection matrices for three dimensions (one spectral dimension and two spatial dimensions) in TSLGDA, while SLGDA [23] only considers one spectral projection matrix by locality preserving projection. The contributions of our work lie in the following aspects: (1) tensor representation is utilized in the framework of sparse and low-rank graph-based discriminant analysis for DR of hyperspectral image. To the best of our knowledge, this is the first time that tensor theory, sparsity, and low-rankness are combined in graph embedding framework; (2) Tensorial structure contains the spectral-spatial information, sparse and low-rank representation reveals both local and global structure and a graph preserves manifold structure. The integration of these three techniques remarkably promotes discriminative ability of reduced features in low-dimensional subspaces; (3) The proposed method can effectively deal with small training size problem, even for the class with only two labeled samples.

The rest of this paper is organized as follows. Section 2 briefly describes the tensor basics and some existing DR methods. The proposed TSLGDA algorithm for DR of hyperspectral imagery is provided in detail in Section 3. Parameters discussions and experimental results compared with some state-of-the-art methods are given in Section 4. Finally, Section 5 concludes this paper with some remarks.

2. Related Work

In this paper, if not specified otherwise, lowercase italic letters denote scalars, e.g., i, j, k, bold lowercase letters denote vectors, e.g., $\mathbf{x}$, $\mathbf{y}$, bold uppercase letters denote matrices, e.g., $\mathbf{U}$, $\mathbf{X}$, and bold uppercase letters with underline denote tensors, e.g., $\underline{\mathbf{A}}$, $\underline{\mathbf{X}}$.

2.1. Tensor Basics

A multidimensional array is defined as a tensor, which is represented as $\underline{\mathbf{A}} \in \mathbf{R}^{I_1\times\ldots I_n\times\ldots I_N}$. We regard $\underline{\mathbf{A}} \in \mathbf{R}^{I_1\times\ldots I_n\times\ldots I_N}$ as an N-order tensor, corresponding to an N-dimensional data array, with its element denoted as $\underline{\mathbf{A}}_{i_1\ldots i_n\ldots i_N}$, where $1 \le i_n \le I_n$, and $1 \le n \le N$. Some basic definitions related to tensor operation are provided as follows [28,33,34].

Definition 1. *(Frobenius norm): The Frobenius norm of a tensor* $\underline{\mathbf{A}}$ *is defined as* $\|\underline{\mathbf{A}}\|_F = (\sum_{i_1\ldots i_N}(\underline{\mathbf{A}}_{i_1\ldots i_N})^2)^{1/2}$.

Definition 2. *(Mode-n matricizing): The n-mode vector of an N-order tensor* $\underline{\mathbf{A}} \in \mathbf{R}^{I_1\times\ldots I_n\times\ldots I_N}$ *is defined as an n-dimensional vector by fixing all indices except* i_n. *The n-mode matrix is composed of all the n-mode vectors in column form, denoted as* $\mathbf{A}^n \in \mathbf{R}^{I_n\times(I_1\ldots I_{n-1}I_{n+1}\ldots I_N)}$. *The obtained n-mode matrix is also known as n-mode unfolding of a tensor* $\underline{\mathbf{A}}$.

Definition 3. *(Mode-n product): The mode-n product of a tensor* $\underline{\mathbf{A}}$ *with a matrix* $\mathbf{U} \in \mathbf{R}^{I_n'\times I_n}$ *yields* $\underline{\mathbf{C}} = \underline{\mathbf{A}} \times_n \mathbf{U}$, *and* $\underline{\mathbf{C}} \in \mathbf{R}^{I_1\ldots I_{n-1}I_n'I_{n+1}\ldots I_N}$, *whose entries are computed by*

$$\underline{\mathbf{C}}_{i_1\ldots i_{n-1}i_n'i_{n+1}\ldots i_N} = \sum_{i_n=1}^{I_n} \underline{\mathbf{A}}_{i_1\ldots i_{n-1}i_ni_{n+1}\ldots i_N}\mathbf{U}_{i_n'i_n} \tag{1}$$

where $i_k = 1, 2, \ldots, I_k$, $(k \neq n)$ *and* $i_n' = 1, 2, \ldots, I_n'$. *Note that the n-mode product can also be expressed in terms of unfolding tensor*

$$\underline{\mathbf{C}} = \underline{\mathbf{A}} \times_n \mathbf{U} \Leftrightarrow \mathbf{C}^n = \mathbf{U}\mathbf{A}^n \tag{2}$$

where $\times_n$ *denotes mode-n product between a tensor and a matrix.*

Definition 4. *(Tensor contraction): The contraction of tensors* $\underline{\mathbf{A}} \in \mathbf{R}^{I_1\times\ldots\times I_N\times I_1'\times\ldots\times I_{N'}'}$ *and* $\underline{\mathbf{B}} \in \mathbf{R}^{I_1\times\ldots\times I_N\times I_1''\times\ldots\times I_{N''}''}$ *is defined as*

$$[\underline{\mathbf{A}} \otimes \underline{\mathbf{B}}; (1:N)(1:N)]_{i_1,i_2,\ldots,i_N} = \sum_{i_1=1}^{I_1}\cdots\sum_{i_N=1}^{I_N} \underline{\mathbf{A}}_{i_1,\ldots,i_N,i_1',\ldots,i_{N'}'}\underline{\mathbf{B}}_{i_1,\ldots,i_N,i_1'',\ldots,i_{N''}''} \tag{3}$$

The condition for tensor contraction is that both two tensors should have the same size at the specific mode. For example, when the contraction is conducted on all indices except for the index n on tensors $\underline{\mathbf{A}}, \underline{\mathbf{B}} \in \mathbf{R}^{I_1 \times \ldots I_n \times \ldots I_N}$, *this operation can be denoted as* $[\underline{\mathbf{A}} \otimes \underline{\mathbf{B}}; (\overline{n})(\overline{n})]$. *According to the property of tensor contraction, we have*

$$[\underline{\mathbf{A}} \otimes \underline{\mathbf{B}}; (\overline{n})(\overline{n})] = \mathbf{A}^n \mathbf{B}^{nT} \tag{4}$$

2.2. Sparse and Low-Rank Graph-Based Discriminant Analysis

In [19], sparse graph-based discriminant analysis (SGDA), as a supervised DR method, was proposed to extract important features for hyperspectral data. Although SGDA can successfully reveal the local structure of the data, it fails to capture the global information. To address this problem, sparse and low-rank graph-based discriminant analysis (SLGDA) [23] was developed to preserve local neighborhood structure and global geometrical structure simultaneously by combining the sparse and low-rank constraints. The objective function of SLGDA can be formulated as

$$\begin{aligned} &\arg\min_{\mathbf{W}^{(l)}} \frac{1}{2}\|\mathbf{X}^{(l)} - \mathbf{X}^{(l)}\mathbf{W}^{(l)}\|_F^2 + \beta\|\mathbf{W}^{(l)}\|_* + \lambda\|\mathbf{W}^{(l)}\|_1, \\ &s.t.\ diag(\mathbf{W}^{(l)}) = 0 \end{aligned} \tag{5}$$

where β and λ are two regularization parameters to control the effect of low-rank term and sparse term, respectively, $\mathbf{X}^{(l)}$ represents samples from the lth class in a vector-based way, and $l = [1, 2, \ldots, c]$, in which c is the number of total classes. After obtaining the complete graph weight matrix $\mathbf{W} = diag(\mathbf{W}^{(1)}, \mathbf{W}^{(2)}, \ldots, \mathbf{W}^{(c)})$, the projection operator can be solved as

$$\begin{aligned} \mathbf{P}^* &= \arg \min_{\mathbf{P^T X L_p X^T P}} \sum_{i \neq j} \|\mathbf{P}^T\mathbf{x}_i - \mathbf{P}^T\mathbf{x}_j\|_2^2 \mathbf{W}_{ij} \\ &= \arg \min_{\mathbf{P^T X L_p X^T P}} tr(\mathbf{P^T X L}_s \mathbf{X^T P}) \end{aligned} \tag{6}$$

where $\mathbf{L}_s = \mathbf{D} - \mathbf{W}$ is defined as the Laplacian matrix, $\mathbf{D}$ is a diagonal matrix with the ith diagonal entry being $\mathbf{D}_{ii} = \sum_{j=1}^{N} \mathbf{W}_{ij}$, and $\mathbf{L}_p$ may be a simple scale normalization constraint [13].

The projection can be further formulated as

$$\mathbf{P}^* = \arg\min_{\mathbf{P}} \frac{|\mathbf{P}^T\mathbf{X}\mathbf{L}_s\mathbf{X}^T\mathbf{P}|}{|\mathbf{P}^T\mathbf{X}\mathbf{L}_p\mathbf{X}^T\mathbf{P}|} \tag{7}$$

which can be solved as a generalized eigendecomposition problem

$$\mathbf{X}\mathbf{L}_s\mathbf{X}^T\mathbf{p}_b = \lambda_b \mathbf{X}\mathbf{L}_p\mathbf{X}^T\mathbf{p}_b \tag{8}$$

The bth projection vector $\mathbf{p}_b$ is the eigenvector corresponding to the bth smallest nonzero eigenvalue. The projection matrix can be formed as $\mathbf{P} = [\mathbf{p}_1, \ldots, \mathbf{p}_B] \in \mathbf{R}^{d \times B}$, $B \ll d$. Finally, the reduced features are denoted as $\widehat{\mathbf{X}} = \mathbf{P}^T\mathbf{X} \in \mathbf{R}^{B \times M}$.

2.3. Multilinear Principal Component Analysis

In order to obtain a set of multilinear projections that will map the original high-order tensor data into a low-order tensor space, MPCA performs to directly maximize the total scatter matrix on the subspace $\mathbf{U}_i (i \neq n)$

$$\max_{\mathbf{U}_n\mathbf{U}_n^T = \mathbf{I}_n} tr(\mathbf{U}_n \mathbf{S}_T^n \mathbf{U}_n^T) = \max_{\mathbf{U}_n\mathbf{U}_n^T = \mathbf{I}_n} tr\Big(\mathbf{U}_n (\sum_{k=1}^{M} \mathbf{X}_k^n \mathbf{X}_k^{nT}) \mathbf{U}_n^T\Big), \tag{9}$$

where $\mathbf{S}_T^n = \sum_{k=1}^{M} \mathbf{X}_k^n \mathbf{X}_k^{nT}$ and $\mathbf{X}_k^n$ is the n-mode unfolding matrix of tensor $\underline{\mathbf{X}}_k$.

The optimal projections of MPCA can be obtained from the eigendecomposition

$$\mathbf{S}_T^n \mathbf{U}_n^T = \mathbf{U}_n^T \mathbf{D}_n \tag{10}$$

where $\mathbf{U}_n = [\mathbf{u}_n^1, \ldots, \mathbf{u}_n^{d_n}]$ is the eigenvector matrix and $\mathbf{D}_n = diag(\lambda_n^1, \ldots, \lambda_n^{d_n})$ is the eigenvalue matrix of $\mathbf{S}_T^n$, in which the eigenvalues are ranked in descending order, and λ_n^j is the eigenvalue corresponding to the eigenvector $\mathbf{u}_n^j$. The optimal projection matrix for mode-n is composed of the eigenvectors corresponding to the first B_n largest eigenvalues, e.g., $\mathbf{U}_n = [\mathbf{u}_n^1, \ldots, \mathbf{u}_n^{B_n}]$. After obtained the projection matrix for each mode, the reduced features can be formulated as

$$\widehat{\underline{\mathbf{X}}}_k = \underline{\mathbf{X}}_k \times_1 \mathbf{U}_1 \ldots \times_N \mathbf{U}_N \tag{11}$$

where $\mathbf{U}_i \in \mathbf{R}^{B_n \times I_n} (B_n \leq I_n)$.

3. Tensor Sparse and Low-Rank Graph-Based Discriminant Analysis

Consider a hyperspectral image as a third-order tensor $\underline{\mathbf{A}} \in \mathbf{R}^{I_1 \times I_2 \times I_3}$, in which I_1 and I_2 refer to the width and height of the data cube, respectively, and I_3 represents the number of spectral bands, $I_3 = d$. Assume that the kth small patch is composed of the kth training sample and its $i_1 \times i_2$ neighbors, which is denoted as $\underline{\mathbf{X}}_k \in \mathbf{R}^{i_1 \times i_2 \times d}$. M patches construct the training set $\{\underline{\mathbf{X}}_k\}_{k=1}^{M}$. The training patches belonging to the lth class are expressed as $\{\underline{\mathbf{X}}_{k,l}\}_{k=1}^{M_l}$, where M_l represents the number of patches belonging to the lth class and $l \in \{1, 2, \ldots, c\}$. For the purpose of convenient expression, a fourth-order tensor $\underline{\mathbf{X}}^{(l)} \in \mathbf{R}^{i_1 \times i_2 \times d \times M_l}$ is defined to represent these M_l patches, and $\underline{\mathbf{X}} \in \mathbf{R}^{i_1 \times i_2 \times d \times M}$ denotes all training patches for c classes, where $M = \sum_{l=1}^{c} M_l$. A visual illustration of 3-mode vectors, 3-mode unfolding, and 3-mode product is shown in Figure 1.

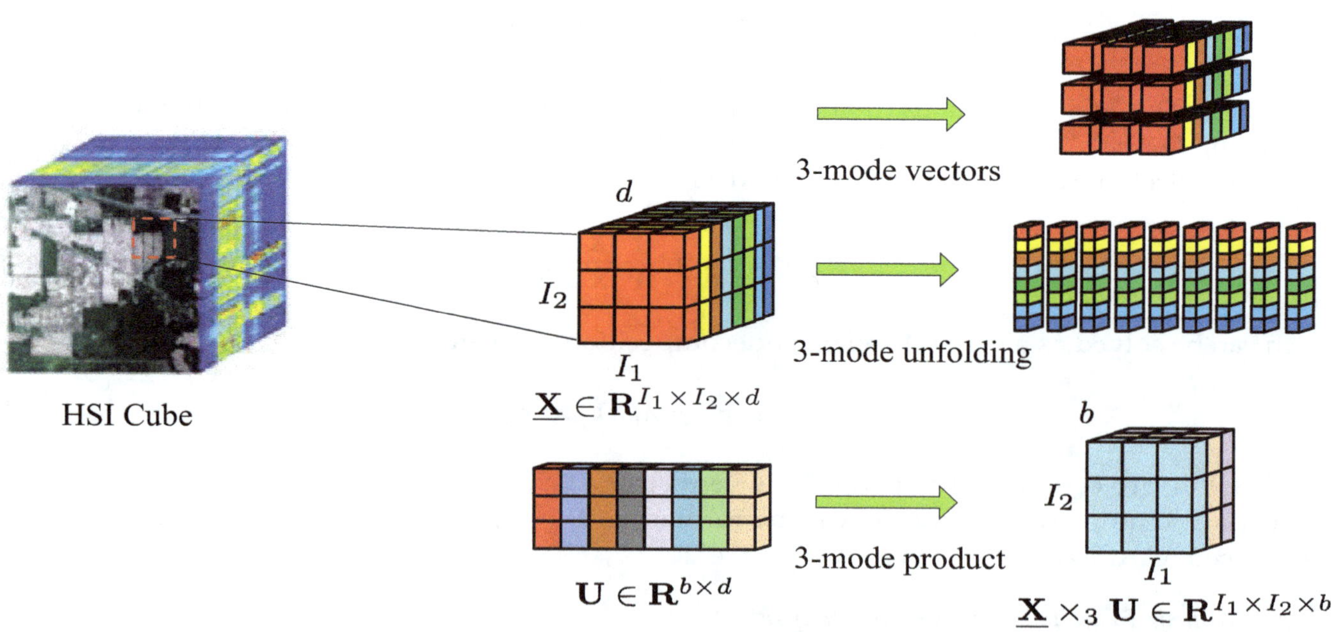

Figure 1. Visual illustration of n-mode vectors, n-mode unfolding, and n-mode product of a third-order tensor from a hyperspectral image.

3.1. Tensor Sparse and Low-Rank Graph

The previous SLGDA framework can capture the local and global structure of hyperspectral data simultaneously by imposing both sparse and low-rank constraints. However, it may lose some important structural information of hyperspectral data, which presents an intrinsic tensor-based data structure. To overcome this drawback, a tensor sparse and low-rank graph is constructed with the objective function

$$\begin{aligned} &\arg\min_{\mathbf{W}^{(l)}} \frac{1}{2}\|\underline{\mathbf{X}}^{(l)} - \underline{\mathbf{X}}^{(l)} \times_4 \mathbf{W}^{(l)}\|_F^2 + \beta\|\mathbf{W}^{(l)}\|_* + \lambda\|\mathbf{W}^{(l)}\|_1, \\ &s.t.\ diag(\mathbf{W}^{(l)}) = 0, \end{aligned} \tag{12}$$

where $\mathbf{W}^{(l)} \in \mathbf{R}^{M_l \times M_l}$ denotes the graph weigh matrix using labeled patches from the lth class only. As such, with the help of class-specific labeled training patches, the global graph weigh matrix $\mathbf{W}$ can be designed as a block-diagonal structure

$$\mathbf{W} = \begin{bmatrix} \mathbf{W}^{(1)} & & \mathbf{0} \\ & \ddots & \\ \mathbf{0} & & \mathbf{W}^{(c)} \end{bmatrix} \tag{13}$$

To obtain the lth class graph weight matrix $\mathbf{W}^{(l)}$, the alternating direction method of multipliers (ADMM) [35] is adopted to solve problem (12). Two auxiliary variables $\mathbf{Z}^{(l)}$ and $\mathbf{J}^{(l)}$ are first introduced to make the objective function separable

$$\begin{aligned} &\arg\min_{\mathbf{Z}^{(l)},\mathbf{J}^{(l)},\mathbf{W}^{(l)}} \frac{1}{2}\|\underline{\mathbf{X}}^{(l)} - \underline{\mathbf{X}}^{(l)} \times_4 \mathbf{W}^{(l)}\|_F^2 + \beta\|\mathbf{Z}^{(l)}\|_* + \lambda\|\mathbf{J}^{(l)}\|_1, \\ &s.t.\ \mathbf{W}^{(l)} = \mathbf{Z}^{(l)},\ \mathbf{W}^{(l)} = \mathbf{J}^{(l)} - diag(\mathbf{J}^{(l)}) \end{aligned} \tag{14}$$

The augmented Lagrangian function of problem (14) is given as

$$\begin{aligned} &L(\mathbf{Z}^{(l)}, \mathbf{J}^{(l)}, \mathbf{W}^{(l)}, \mathbf{D}_1, \mathbf{D}_2) \\ =& \frac{1}{2}\|\underline{\mathbf{X}}^{(l)} - \underline{\mathbf{X}}^{(l)} \times_4 \mathbf{W}^{(l)}\|_F^2 + \beta\|\mathbf{Z}^{(l)}\|_* + \lambda\|\mathbf{J}^{(l)}\|_1 + \langle \mathbf{D}_1, \mathbf{W}^{(l)} - \mathbf{Z}^{(l)}\rangle + \langle \mathbf{D}_2, \mathbf{W}^{(l)} - \mathbf{J}^{(l)} + diag(\mathbf{J}^{(l)})\rangle \\ &+ \frac{\mu}{2}(\|\mathbf{W}^{(l)} - \mathbf{Z}^{(l)}\|_F^2 + \|\mathbf{W}^{(l)} - \mathbf{J}^{(l)} + diag(\mathbf{J}^{(l)})\|_F^2) \end{aligned} \tag{15}$$

where $\mathbf{D}_1$ and $\mathbf{D}_2$ are Lagrangian multipliers, and μ is a penalty parameter.

By minimizing the function $L(\mathbf{Z}^{(l)}, \mathbf{J}^{(l)}, \mathbf{W}^{(l)})$, each variable is alternately updated with other variables being fixed. The updating rules are expressed as

$$\begin{aligned} \mathbf{Z}_{t+1}^{(l)} &= \arg\min_{\mathbf{Z}^{(l)}} \beta\|\mathbf{Z}^{(l)}\|_* + \langle \mathbf{D}_{1,t}, \mathbf{W}_t^{(l)} - \mathbf{Z}^{(l)}\rangle + \frac{\mu_t}{2}\|\mathbf{W}_t^{(l)} - \mathbf{Z}^{(l)}\|_F^2 \\ &= \arg\min_{\mathbf{Z}^{(l)}} \frac{\beta}{\mu_t}\|\mathbf{Z}^{(l)}\|_* + \frac{1}{2}\|\mathbf{Z}^{(l)} - (\mathbf{W}_t^{(l)} + \frac{\mathbf{D}_{1,t}}{\mu_t})\|_F^2 \\ &= \mathbf{\Omega}_{\frac{\beta}{\mu_t}}(\mathbf{W}_t^{(l)} + \frac{\mathbf{D}_{1,t}}{\mu_t}) \end{aligned} \tag{16}$$

$$\begin{aligned} \mathbf{J}_{t+1}^{(l)} &= \arg\min_{\mathbf{J}^{(l)}} \lambda\|\mathbf{J}^{(l)}\|_1 + \langle \mathbf{D}_{2,t}, \mathbf{W}_t^{(l)} - \mathbf{J}\rangle + \frac{\mu_t}{2}\|\mathbf{W}_t^{(l)} - \mathbf{J}^{(l)}\|_F^2 \\ &= \arg\min_{\mathbf{J}^{(l)}} \frac{\lambda}{\mu_t}\|\mathbf{J}^{(l)}\|_1 + \frac{1}{2}\|\mathbf{J}^{(l)} - (\mathbf{W}_t^{(l)} + \frac{\mathbf{D}_{2,t}}{\mu_t})\|_F^2 \\ &= \mathcal{S}_{\frac{\lambda}{\mu_t}}(\mathbf{W}_t^{(l)} + \frac{\mathbf{D}_{2,t}}{\mu_t}), \\ \mathbf{J}_{t+1}^{(l)} &= \mathbf{J}_{t+1}^{(l)} - diag(\mathbf{J}_{t+1}^{(l)}), \end{aligned} \tag{17}$$

where μ_t denotes the learning rate, $\Omega_\tau(\mathbf{\Delta}) = \mathbf{Q}\mathcal{S}_\tau(\sum)\mathbf{V}^T$ is the singular value thresholding operator (SVT), in which $\mathcal{S}_\tau(x) = sgn(x)\max(|x| - \tau, 0)$ is the soft thresholding operator [36]. By fixing $\mathbf{Z}_{t+1}^{(l)}$ and $\mathbf{J}_{t+1}^{(l)}$, the formulation of $\mathbf{W}_{t+1}^{(l)}$ can be written as

$$\begin{aligned}\mathbf{W}_{t+1}^{(l)} &= \arg\min_{\mathbf{W}^{(l)}} \frac{1}{2}\|\underline{\mathbf{X}}^{(l)} - \underline{\mathbf{X}}^{(l)} \times_4 \mathbf{W}^{(l)}\|_F^2 + \langle \mathbf{D}_{1,t}, \mathbf{W}^{(l)} - \mathbf{Z}_{t+1}^{(l)}\rangle + \langle \mathbf{D}_{2,t}, \mathbf{W}^{(l)} - \mathbf{J}_{t+1}^{(l)}\rangle \\ &+ \frac{\mu_t}{2}(\|\mathbf{W}^{(l)} - \mathbf{Z}_{t+1}^{(l)}\|_F^2 + \|\mathbf{W}^{(l)} - \mathbf{J}_{t+1}^{(l)}\|_F^2) \\ &= (\mathbf{H}^{(l)} + 2\mu_t\mathbf{I})^{-1}(\mathbf{H}^{(l)} + \mu_t\mathbf{Z}_{t+1}^{(l)} + \mu_t\mathbf{J}_{t+1}^{(l)} - (\mathbf{D}_{1,t} + \mathbf{D}_{2,t})),\end{aligned} \tag{18}$$

where $\mathbf{H}^{(l)} = [\underline{\mathbf{X}}^{(l)} \otimes \underline{\mathbf{X}}^{(l)}; (\bar{4})(\bar{4})] \in \mathbf{R}^{M_l \times M_l}$, $\mathbf{W}^{(l)} \in \mathbf{R}^{M_l \times M_l}$, and $\mathbf{I} \in \mathbf{R}^{M_l \times M_l}$ is an identity matrix.

The global similarity matrix $\mathbf{W}$ will be obtained depending on Equation (13) when each sub-similarity matrix corresponding to each class is calculated from problem (12). Until now, a tensor sparse and low-rank graph $G = \{\underline{\mathbf{X}}, \mathbf{W}\}$ is completely constructed with vertex set $\underline{\mathbf{X}}$ and similarity matrix $\mathbf{W}$. How to obtain a set of projection matrices $\{\mathbf{U}_n \in \mathbf{R}^{B_n \times I_n}, B_n \leq I_n, n = 1, 2, \ldots, N\}$ is the following task.

3.2. Tensor Locality Preserving Projection

The aim of tensor LPP is to find transformation matrices $\{\mathbf{U}_1, \mathbf{U}_2, \ldots, \mathbf{U}_N\}$ to project high-dimensional data $\underline{\mathbf{X}}_i$ into low-dimensional representation $\widehat{\underline{\mathbf{X}}}_i$, where $\widehat{\underline{\mathbf{X}}}_i = \underline{\mathbf{X}}_i \times_1 \mathbf{U}_1 \times_2 \mathbf{U}_2 \cdots \times_N \mathbf{U}_N$.

The optimization problem for tensor LPP can be expressed as

$$\begin{aligned}&\arg\min J(\mathbf{U}_1, \mathbf{U}_2, \ldots, \mathbf{U}_N) = \sum_{i,j} \|\widehat{\underline{\mathbf{X}}}_i - \widehat{\underline{\mathbf{X}}}_j\|^2 \mathbf{W}_{ij} \\ &= \sum_{i,j} \|\underline{\mathbf{X}}_i \times_1 \mathbf{U}_1 \cdots \times_N \mathbf{U}_N - \underline{\mathbf{X}}_j \times_1 \mathbf{U}_1 \cdots \times_N \mathbf{U}_N\|^2 \mathbf{W}_{ij} \\ &s.t. \sum_i \|\underline{\mathbf{X}}_i \times_1 \mathbf{U}_1 \cdots \times_N \mathbf{U}_N\|^2 \mathbf{C}_{ii} = 1\end{aligned} \tag{19}$$

where $\mathbf{C}_{ii} = \sum_j \mathbf{W}_{ij}$. It can be seen that the corresponding tensors $\widehat{\underline{\mathbf{X}}}_i$ and $\widehat{\underline{\mathbf{X}}}_j$ in the embedded tensor space are expected to be close to each other if original tensors $\underline{\mathbf{X}}_i$ and $\underline{\mathbf{X}}_j$ are greatly similar.

To solve the optimization problem (19), an iterative scheme is employed [33]. First, we assume that $\{\mathbf{U}_1, \ldots, \mathbf{U}_{n-1}, \mathbf{U}_{n+1}, \ldots, \mathbf{U}_N\}$ are known, then, let $\widehat{\underline{\mathbf{X}}}_{i,(n)} = \underline{\mathbf{X}}_i \times_1 \mathbf{U}_1 \ldots \times_{n-1} \mathbf{U}_{n-1} \times_{n+1} \mathbf{U}_{n+1} \ldots \times_N \mathbf{U}_N$. With properties of tensor and trace, the objective function (19) is rewritten as

$$\begin{aligned}\arg\min J_n(\mathbf{U}_n) &= \sum_{i,j} \|\widehat{\underline{\mathbf{X}}}_{i,(n)} \times_n \mathbf{U}_n - \widehat{\underline{\mathbf{X}}}_{j,(n)} \times_n \mathbf{U}_n\|^2 \mathbf{W}_{ij} \\ &= \sum_{i,j} \|\mathbf{U}_n \widehat{\mathbf{X}}_i^n - \mathbf{U}_n \widehat{\mathbf{X}}_j^n\|^2 \mathbf{W}_{ij} \\ &= \sum_{i,j} tr\Big(\mathbf{U}_n((\widehat{\mathbf{X}}_i^n - \widehat{\mathbf{X}}_j^n)(\widehat{\mathbf{X}}_i^n - \widehat{\mathbf{X}}_j^n)^T \mathbf{W}_{ij})\mathbf{U}_n^T\Big) \\ &= tr\Big(\mathbf{U}_n(\sum_{i,j}(\widehat{\mathbf{X}}_i^n - \widehat{\mathbf{X}}_j^n)(\widehat{\mathbf{X}}_i^n - \widehat{\mathbf{X}}_j^n)^T \mathbf{W}_{ij})\mathbf{U}_n^T\Big), \\ &s.t.\ tr(\mathbf{U}_n(\sum_i \widehat{\mathbf{X}}_i^n \widehat{\mathbf{X}}_i^{nT} \mathbf{C}_{ii})\mathbf{U}_n^T) = 1,\end{aligned} \tag{20}$$

where $\widehat{\mathbf{X}}_i^n$ denotes the n-mode unfolding of tensor $\underline{\widehat{\mathbf{X}}}_{i,(n)}$. Finally, the optimal solution of problem (20) is the eigenvectors corresponding to the first B_n smallest nonzero eigenvalues of the following generalized eigenvalue problem

$$\Big(\sum_{i,j}(\widehat{\mathbf{X}}_i^n - \widehat{\mathbf{X}}_j^n)(\widehat{\mathbf{X}}_i^n - \widehat{\mathbf{X}}_j^n)^T\mathbf{W}_{ij}\Big)\mathbf{u} = \lambda\Big(\sum_{i}\widehat{\mathbf{X}}_i^n\widehat{\mathbf{X}}_i^{nT}\mathbf{C}_{ii}\Big)\mathbf{u} \tag{21}$$

Assume $\boldsymbol{\Phi} = \sum_{i,j}(\widehat{\mathbf{X}}_i^n - \widehat{\mathbf{X}}_j^n)(\widehat{\mathbf{X}}_i^n - \widehat{\mathbf{X}}_j^n)^T\mathbf{W}_{ij}$, $\boldsymbol{\Psi} = \sum_i \widehat{\mathbf{X}}_i^n\widehat{\mathbf{X}}_i^{nT}\mathbf{C}_{ii}$, then, problem (21) can be transformed into

$$\boldsymbol{\Phi}\mathbf{u} = \lambda\boldsymbol{\Psi}\mathbf{u} \tag{22}$$

To solve this problem, the function $eig(\cdot)$ embedded in the MATLAB software (R2013a, The MathWorks, Natick, Massachusetts, USA) is adopted, i.e., $[\mathbf{u}, \boldsymbol{\Lambda}] = eig(\boldsymbol{\Phi}, \boldsymbol{\Psi})$, and the eigenvectors in $\mathbf{u}$ corresponding to the first B_n smallest nonzero eigenvalues in $\boldsymbol{\Lambda}$ are chosen to form the projection matrix. The other projection matrices can be obtained in a similar manner. The complete TSLGDA algorithm is outlined in Algorithm 1.

Algorithm 1: Tensor Sparse and Low-Rank Graph-Based Discriminant Analysis for Classification.

Input: Training patches $\underline{\mathbf{X}} = [\underline{\mathbf{X}}^{(1)}, \underline{\mathbf{X}}^{(2)}, \ldots, \underline{\mathbf{X}}^{(c)}]$, testing patches$\underline{\mathbf{Y}}$, regularization parameters β and λ, reduced dimensionality $\{B_1, B_2, B_3\}$.

Initialize: $\mathbf{Z}_0^{(l)} = \mathbf{J}_0^{(l)} = \mathbf{W}_0^{(l)} = \mathbf{0}$, $\mathbf{Y}_{1,0} = \mathbf{Y}_{2,0} = \mathbf{0}$, $\mu_0 = 0.1$, $\mu_{max} = 10^3$, $\rho_0 = 1.1$, $\varepsilon_1 = 10^{-4}$, $\varepsilon_2 = 10^{-3}$, maxIter = 100, $t = 0$.

for $l = 1, 2, \ldots, c$ **do**
  repeat
    Compute $\mathbf{Z}_{t+1}^{(l)}$, $\mathbf{J}_{t+1}^{(l)}$, and $\mathbf{W}_{t+1}^{(l)}$ according to (16)–(18).
    Update the Lagrangian multipliers:
    $\mathbf{Y}_{1,t+1} = \mathbf{Y}_{1,t} + \mu_t(\mathbf{W}_{t+1}^{(l)} - \mathbf{Z}_{t+1}^{(l)})$, $\mathbf{Y}_{2,t+1} = \mathbf{Y}_{2,t} + \mu_t(\mathbf{W}_{t+1}^{(l)} - \mathbf{J}_{t+1}^{(l)})$.
    Update μ: $\mu_{t+1} = \min(\rho\mu_t, \mu_{max})$, where

$$\rho = \begin{cases} \rho_0, & if\ \mu_t \max(\|\mathbf{W}_{t+1}^{(l)} - \mathbf{W}_t^{(l)}\|_F, \|\mathbf{Z}_{t+1}^{(l)} - \mathbf{Z}_t^{(l)}\|_F, \|\mathbf{J}_{t+1}^{(l)} - \mathbf{J}_t^{(l)}\|_F)/\|\underline{\mathbf{X}}^{(l)}\|_F < \varepsilon_2, \\ 1, & otherwise. \end{cases}$$

    Check convergence conditions: $\|\mathbf{W}_{t+1}^{(l)} - \mathbf{Z}_{t+1}^{(l)}\|_\infty < \varepsilon_1$, $\|\mathbf{W}_{t+1}^{(l)} - \mathbf{J}_{t+1}^{(l)}\|_\infty < \varepsilon_1$.
    $t \leftarrow t + 1$.
  until convergence conditions are satisfied or $t >$maxIter.
end for
Construct the block-diagonal weight matrix $\mathbf{W}$ according to (13).
Compute the projection matrices $\{\mathbf{U}_1, \mathbf{U}_2, \mathbf{U}_3\}$ according to (21).
Compute the reduced features:
$\widehat{\underline{\mathbf{X}}} = \underline{\mathbf{X}} \times_1 \mathbf{U}_1 \times_2 \mathbf{U}_2 \times_3 \mathbf{U}_3$, $\widehat{\underline{\mathbf{Y}}} = \underline{\mathbf{Y}} \times_1 \mathbf{U}_1 \times_2 \mathbf{U}_2 \times_3 \mathbf{U}_3$.
Determine the class label of $\widehat{\underline{\mathbf{Y}}}$ by NN classifier.
Output: The class labels of test patches.

---

4. Experiments and Discussions

In this section, three hyperspectral datasets are used to verify the performance of the proposed method. The proposed TSLGDA algorithm is compared with some state-of-the-art approaches, including unsupervised methods (e.g., PCA [8], MPCA [25]) and supervised methods (e.g., LDA [9], LFDA [10], SGDA [19], GDA-SS [15], SLGDA [23], G-LTDA (local tensor discriminant analysis with Gabor filters) [29]). SGDA is implemented using the SPAMS (SPArse Modeling Software) toolbox [38]. The nearest neighbor classifier (NN classifier) is exploited to classify the projected features obtained by these DR methods. The class-specific accuracy , overall accuracy (OA), average accuracy (AA), and kappa coefficient (κ) are reported for quantitative assessment after ten runs. All experiments are implemented on an Inter Core i5-4590 CPU personal computer (Santa Clara, CA, USA).

4.1. Experimental Datasets

The first dataset [39] was acquired by Airborne Visible/Infrared Imaging Spectrometer (AVIRIS) sensor over northwest Indiana's Indian Pine test site in June 1992. The AVIRIS sensor generates the wavelength range of 0.4–2.45-μm covered 220 spectral bands. After removing 20 water-absorption bands (bands 104–108, 150–163, and 220), a total of 200 bands is used in experiments. The image with 145 × 145 pixels represents a rural scenario having 16 different land-cover classes. The numbers of training and testing samples in each class are listed in Table 1.

Table 1. Number of training and testing samples for the Indian Pines and University of Pavia datasets.

	Indian Pines			University of Pavia		
Class	**Name**	**Training**	**Testing**	**Name**	**Training**	**Testing**
1	Alfalfa	5	41	Asphalt	40	6591
2	Corn-notill	143	1285	Meadows	40	18,609
3	Corn-mintill	83	747	Gravel	40	2059
4	Corn	24	213	Tree	40	3024
5	Grass-pasture	48	435	Painted metal sheets	40	1305
6	Grass-trees	73	657	Bare Soil	40	4989
7	Grass-pasture-mowed	3	25	Bitumen	40	1290
8	Hay-windrowed	48	430	Self-blocking bricks	40	3642
9	Oats	2	18	Shadows	40	907
10	Soybean-notill	97	875			
11	Soybean-mintill	246	2209			
12	Soybean-clean	59	534			
13	Wheat	21	184			
14	Woods	127	1138			
15	Buildings-Grass-Trees-Drive	39	347			
16	Stone-Steel-Towers	9	84			
Total		1027	9222		360	42,416

The second dataset [39] is the University of Pavia collected by the Reflective Optics System Imaging Spectrometer (ROSIS) sensor in Italy. The image has 103 bands after removing 12 noisy bands with a spectral coverage from 0.43 to 0.86 μm, covering a region of 610 × 340 pixels. There are nine ground-truth classes, from which we randomly select training and testing samples as shown in Table 1.

The third dataset [39] was also collected by the AVIRIS sensor over the Valley of Salinas, Central Coast of California, in 1998. The image comprises 512 × 217 pixels with a spatial resolution of 3.7 m, and only preserves 204 bands after 20 water-absorption bands removed. Table 2 lists 16 land-cover classes and the number of training and testing samples.

Table 2. Number of training and testing samples for the Salinas dataset.

	Salinas		
Class	**Name**	**Training**	**Testing**
1	Brocoli-green-weeds-1	40	1969
2	Brocoli-green-weeds-2	75	3651
3	Fallow	40	1936
4	Fallow-rough-plow	28	1366
5	Fallow-smooth	54	2624
6	Stubble	79	3880
7	Celery	72	3507
8	Grapes-untrained	225	11,046
9	Soil-vinyard-develop	124	6079
10	Corn-senesced-green-weeds	66	3212
11	Lettuce-romaine-4wk	21	1047
12	Lettuce-romaine-5wk	39	1888
13	Lettuce-romaine-6wk	18	898
14	Lettuce-romaine-7wk	21	1049
15	Vinyard-untrained	145	7123
16	Vinyard-vertical-trellis	36	1771
Total		1083	53,046

4.2. Parameters Tuning

For the proposed method, four important parameters (i.e., regularization parameters β and λ, window size, and the number of spectral dimension) that can be divided into three groups need to be determined before proceeding to the following experiments. β and λ control the effect of sparse term and low-rank term in the objective function, respectively, which can be tuned together, while window size and the number of spectral dimension are another two groups that can be determined separately. When analyzing one group specific parameter, the other group parameters are fixed on their corresponding chosen values. According to many existing DR methods [22–24] and tensor-based research [26,28], window size is the first set as 9 for the Indian Pines and Salinas datasets, and 7 for the University of Pavia dataset; the initial value for the number of spectral dimension is given as 30 for all three datasets, and the performance basically reaches steady state with this dimension.

4.2.1. Regularization Parameters for TSLGDA

With the initial values of window size and the number of spectral dimension fixed, β and λ are first tuned to achieve better classification performance. Figure 2 shows the overall classification accuracy with respect to different β and λ by fivefold cross validation for three experimental datasets. It can be clearly seen that the OA values can reach the maximum values for some β and λ. Accordingly, for the Indian Pines dataset, the optimal values of β and λ can be set as $(0.01, 0.1)$, which is also an appropriate choice for the University of Pavia dataset, while $(0.001, 0.1)$ is chosen for the Salinas data.

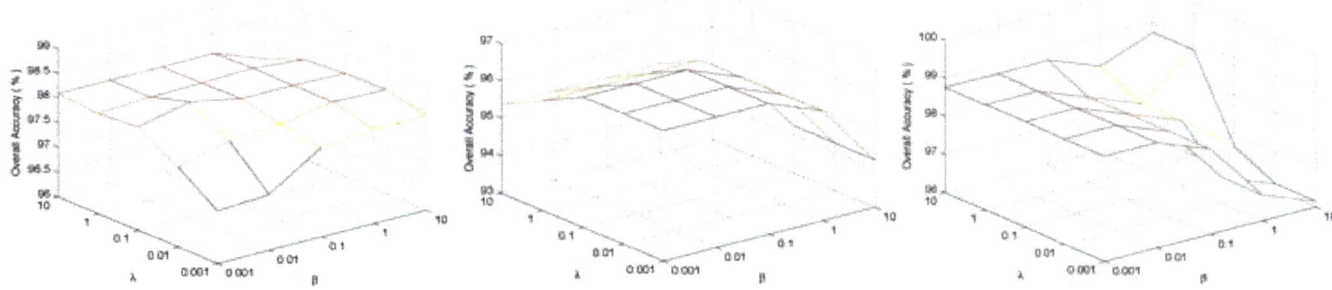

(a) (b) (c)

Figure 2. Parameter tuning of β and λ for the proposed TSLGDA algorithm using three datasets: (**a**) Indian Pines; (**b**) University of Pavia; (**c**) Salinas.

4.2.2. Window Size for Tensor Representation

For tensor-based DR methods, i.e., MPCA and TSLGDA, window size (or patch size) is another important parameter. Note that small windows may fail to cover enough spatial information, whereas large windows may contain multiple classes, resulting in complicated analysis and heavy computational burden. Therefore, the window size is searched in the range of $\{3 \times 3, 5 \times 5, 7 \times 7, 9 \times 9\}$. β and λ are fixed on the tuned values, while the numbers of spectral dimension are still set as initial values for three datasets, respectively. Figure 3 presents the variation of classification performances of MPCA and TSLGDA with different window sizes for experimental datasets. It can be seen that the window sizes for MPCA and TSLGDA can be both chosen as 9×9 for the Indian Pines and Salinas datasets, while the optimal values are 5×5 and 7×7, respectively, for the University of Pavia dataset. This may be because the formers represent a rural scenario containing large spatial homogeneity while the Pavia University data is obtained from an urban area with small homogeneous regions. To evaluate the classification performance using the low-dimensional data, 1NN classifier is adopted in this paper.

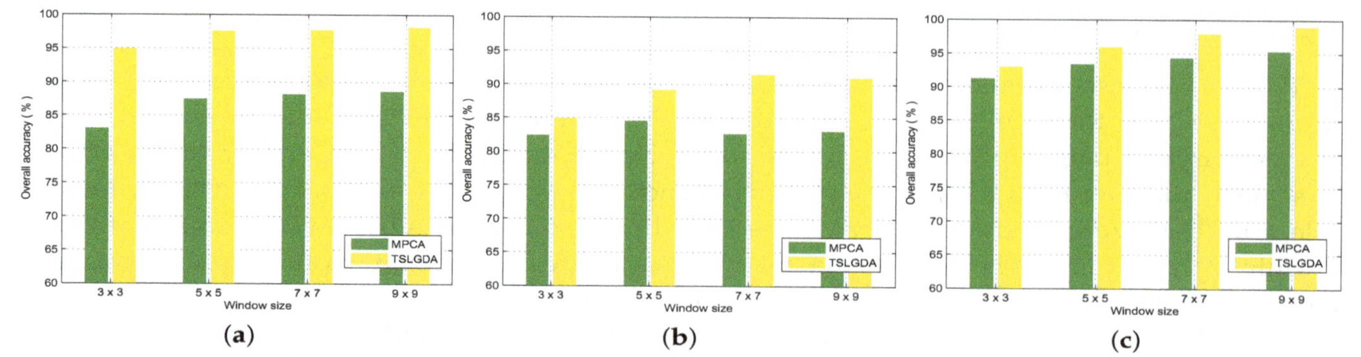

Figure 3. Parameter tuning of window size for MPCA and TSLGDA using three datasets: (**a**) Indian Pines; (**b**) University of Pavia; (**c**) Salinas.

4.2.3. The Number of Spectral Dimension for TSLGDA

According to [28], $\{1,1\}$ is set as the reduced dimensionality of the first two dimensions (i.e., two spatial dimensions). The third dimension (i.e., spectral dimension) is considered carefully by keeping the tuned values of β, λ, and window size is fixed. Figure 4 shows the overall classification accuracy with respect to spectral dimension for three hyperspectral datasets. Obviously, due to the spatial information contained in tensor structure, tensor-based DR methods (i.e., MPCA, TSLGDA) outperform vector-based DR methods (i.e., PCA, SGDA, GDA-SS, SLGDA). According to [29,37], G-LTDA can automatically obtain the optimal reduced dimensions during the optimization procedure; therefore, the number of spectral dimension for G-LTDA is not discussed here. For the Indian Pines dataset, the performances of all considered methods increase when the spectral dimension increases, and then keep stable at the maximum values. The similar results can also be observed from the University of Pavia and Salinas datasets. In any case, TSLGDA outperforms other DR methods even when the spectral dimension is as low as 5. In the following assessment, $\{1,1,30\}$ and $\{1,1,20\}$ dimensions are used to conduct classification for two AVIRIS datasets and one ROSIS dataset, respectively.

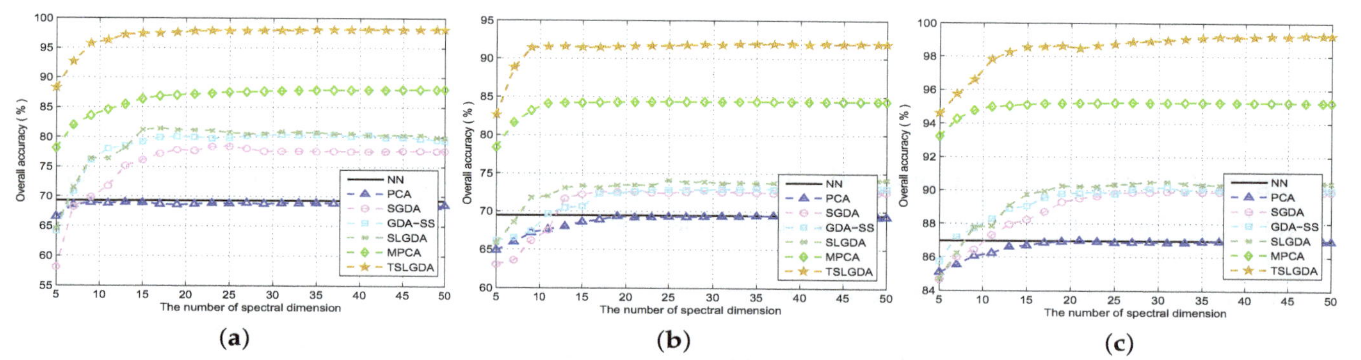

Figure 4. Overall accuracy versus the reduced spectral dimension for different methods using three datasets: (**a**) Indian Pines; (**b**) University of Pavia; (**c**) Salinas.

4.3. Classification Results

4.3.1. Classification Accuracy

Tables 3–5 present the classification accuracy of individual class, OA, AA, and kappa coefficient for three experimental datasets, respectively. Obviously, the proposed method provides the best results than other compared methods on almost all of classes; meanwhile, OA, AA, and kappa coefficient are also better than those of other methods. Specifically, by comparing to all considered methods, TSLGDA yields about 2% to 30%, 5% to 20%, and 2% to 12% gain in OA with limited training sets for three datasets, respectively. Even for classes with few labeled training samples, such as class 1, class 7, and class 9 in the Indian Pines data, the proposed TSLGDA algorithm offers great improvement in performance as well. Besides TSLGDA, MPCA and G-LTDA also obtain much higher accuracies than other vector-based methods, which effectively demonstrates the advantage of tensor-based techniques. In addition, SLGDA yields better results than SGDA (about 3%, 1%, and 0.6% gain) by simultaneously exploiting the properties of sparsity and low-rankness, while GDA-SS is superior to SGDA by considering the spectral similarity measurement based on spectral characteristics when constructing the graph.

Table 3. Classification accuracy (%) and standard deviation of different methods for the Indian Pines data when the reduced dimension is 30.

No.	Origin	PCA	LDA	LFDA	SGDA	GDA-SS	SLGDA	MPCA	G-LTDA	TSLGDA
1	39.02	54.15	33.66	44.88	65.04	49.59	48.78	71.34	**92.20**	91.71
	±8.27	±11.1	±17.8	±15.5	±7.45	±12.2	±6.90	±9.63	±4.69	±8.02
2	55.92	52.96	57.28	67.78	69.31	74.24	73.04	81.09	96.47	**97.32**
	±2.68	±1.53	±2.13	±3.56	±2.37	±3.95	±1.93	±2.74	±1.01	±0.68
3	49.83	50.15	58.34	66.75	62.65	69.57	67.00	82.26	93.98	**97.51**
	±2.68	±2.34	±2.57	±2.82	±1.76	±5.56	±0.28	±2.74	±2.34	±0.91
4	42.07	40.19	38.12	54.93	49.14	58.06	62.68	87.91	96.53	**97.37**
	±7.75	±4.56	±4.00	±7.69	±5.40	±8.24	±12.3	±4.65	±3.93	±1.90
5	82.95	84.47	81.20	88.25	89.55	92.03	93.32	91.13	93.15	**97.00**
	±2.93	±4.58	±3.87	±2.41	±1.74	±1.34	±0.98	±2.00	±1.44	±2.50
6	90.75	93.06	93.36	94.64	95.38	96.91	96.27	97.53	94.76	**99.27**
	±1.00	±2.95	±1.47	±1.59	±0.61	±0.89	±0.11	±1.14	±2.94	±0.46
7	81.60	72.00	76.00	79.20	88.00	88.00	88.00	94.00	95.20	**96.80**
	±8.29	±13.6	±12.3	±22.5	±4.00	±8.00	±5.66	±7.66	±7.15	±3.35
8	96.28	93.02	95.26	99.12	99.53	97.91	99.19	98.37	97.81	**99.86**
	±1.78	±1.52	±2.71	±1.47	±0.40	±2.02	±0.49	±1.66	±0.67	±0.31
9	26.67	34.44	25.56	43.33	50.00	37.04	25.00	54.17	78.89	**93.33**
	±4.65	±12.0	±16.5	±9.94	±33.8	±16.9	±11.8	±19.4	±15.4	±7.24
10	66.06	63.91	65.40	69.04	69.64	73.64	74.03	84.12	95.93	**96.52**
	±2.04	±3.49	±3.61	±3.05	±5.81	±3.02	±0.32	±1.32	±1.35	±1.56
11	71.75	71.41	73.65	72.43	78.18	79.45	79.52	90.30	96.32	**98.53**
	±3.00	±2.00	±1.81	±1.83	±1.42	±1.23	±2.08	±0.78	±1.41	±0.59
12	43.41	41.46	48.63	67.20	67.29	74.78	76.83	73.73	93.60	**96.17**
	±6.34	±2.55	±3.25	±1.56	±2.19	±4.59	±1.99	±2.38	±1.70	±1.75
13	91.41	94.02	93.59	98.70	96.01	97.83	98.64	98.23	91.85	**99.46**
	±2.44	±2.40	±1.11	±0.62	±0.63	±1.63	±1.15	±1.12	±4.21	±0.67
14	90.04	89.65	89.44	93.83	94.58	94.00	96.05	95.78	97.72	**99.67**
	±1.96	±2.10	±2.16	±1.56	±0.89	±1.18	±0.87	±0.40	±0.66	±0.43
15	37.98	36.54	41.15	61.04	48.90	56.20	56.48	88.26	95.91	**98.67**
	±2.18	±2.30	±3.73	±2.89	±1.92	±3.20	±2.85	±4.69	±1.62	±1.16
16	88.43	88.67	91.08	89.64	92.37	91.27	93.98	93.07	84.29	**97.35**
	±6.30	±3.02	±3.47	±5.56	±3.03	±2.99	±1.70	±4.33	±8.68	±1.32
OA	69.25	68.52	70.86	76.60	77.65	80.51	80.76	88.34	95.67	**98.08**
	±1.16	±0.88	±0.76	±0.82	±1.44	±0.31	±0.08	±0.51	±0.49	±0.30
AA	65.89	66.26	66.36	74.42	75.97	76.91	76.80	86.33	93.41	**97.28**
	±1.19	±1.62	±2.30	±1.79	±2.37	±2.38	±1.98	±1.17	±0.56	±0.85
κ	64.90	64.04	66.73	73.32	74.40	77.70	78.01	86.70	95.07	**97.81**
	±1.30	±0.98	±0.92	±0.93	±1.68	±0.38	±0.14	±0.59	±0.56	±0.34

Table 4. Classification accuracy (%) and standard deviation of different methods for the University of Pavia data when the reduced dimension is 20.

No.	Origin	PCA	LDA	LFDA	SGDA	GDA-SS	SLGDA	MPCA	G-LTDA	TSLGDA
1	56.13	55.98	64.77	60.56	47.44	52.88	52.84	84.20	72.41	**91.15**
	±1.99	±2.90	±2.11	±5.24	±2.00	±6.58	±1.98	±1.49	±2.03	±1.46
2	69.68	70.30	68.75	77.05	82.15	78.88	80.92	84.60	89.24	**92.59**
	±5.59	±3.27	±3.44	±4.42	±2.71	±2.80	±3.74	±3.31	±0.93	±2.68
3	68.02	67.34	69.90	66.47	63.83	64.27	61.17	80.24	**89.48**	86.83
	±3.95	±1.49	±3.10	±3.94	±10.5	±3.28	±3.26	±3.01	±5.68	±2.44
4	90.21	86.98	88.92	91.33	90.73	91.26	92.54	92.20	71.28	**96.04**
	±4.43	±3.70	±2.23	±2.01	±2.25	±2.10	±0.07	±1.85	±4.90	±2.23
5	99.39	99.49	99.51	99.88	99.73	99.79	99.66	99.72	98.41	**100**
	±0.38	±0.23	±0.25	±0.10	±0.18	±0.08	±0.27	±0.26	±1.10	±0.00

Table 4. *Cont.*

No.	Origin	PCA	LDA	LFDA	SGDA	GDA-SS	SLGDA	MPCA	G-LTDA	TSLGDA
6	59.11 ±2.25	61.68 ±6.60	66.35 ±6.62	65.36 ±7.09	59.47 ±5.18	65.07 ±2.72	63.97 ±0.50	77.99 ±4.68	**95.04** ±2.35	93.06 ±3.12
7	83.36 ±4.59	83.22 ±3.57	86.34 ±2.25	75.78 ±1.97	82.25 ±5.40	79.04 ±3.64	81.71 ±1.75	89.22 ±2.09	**98.26** ±1.37	97.50 ±0.90
8	68.06 ±2.72	66.89 ±4.34	68.24 ±3.24	60.81 ±4.18	61.16 ±8.92	64.67 ±4.21	65.46 ±2.87	76.30 ±3.07	**93.31** ±1.32	86.07 ±3.27
9	95.94 ±1.52	95.90 ±1.36	97.00 ±1.82	83.95 ±4.64	84.04 ±6.01	87.81 ±2.20	85.17 ±1.01	**99.49** ±0.32	88.00 ±2.23	98.39 ±1.03
OA	69.47 ±2.16	69.65 ±0.88	71.38 ±1.10	73.04 ±0.70	72.59 ±0.68	73.01 ±1.47	73.80 ±1.91	84.30 ±1.05	86.92 ±0.42	**92.33** ±0.93
AA	76.66 ±0.52	76.42 ±0.70	78.86 ±0.92	75.69 ±1.55	74.53 ±1.82	75.96 ±0.74	75.94 ±0.25	87.11 ±0.71	88.38 ±0.43	**93.52** ±0.53
κ	61.22 ±2.30	61.43 ±0.88	63.79 ±1.19	65.31 ±0.83	64.39 ±0.89	65.22 ±1.74	66.10 ±2.14	79.57 ±1.24	82.88 ±0.50	**89.93** ±1.17

Table 5. Classification accuracy (%) and standard deviation of different methods for the Salinas data when the reduced dimension is 30.

No.	Origin	PCA	LDA	LFDA	SGDA	GDA-SS	SLGDA	MPCA	G-LTDA	TSLGDA
1	98.07 ±0.44	98.73 ±0.80	98.98 ±0.81	99.44 ±0.10	99.49 ±0.13	99.39 ±0.14	99.61 ±0.23	98.00 ±0.98	96.94 ±1.63	**99.92** ±0.15
2	98.68 ±0.38	98.90 ±0.25	98.88 ±0.29	99.23 ±0.17	99.54 ±0.28	99.25 ±0.21	99.50 ±0.37	99.47 ±0.55	98.73 ±0.81	99.98 ±0.03
3	96.20 ±0.25	96.85 ±0.61	95.13 ±1.05	99.16 ±0.25	99.28 ±0.05	99.59 ±0.15	99.57 ±0.17	98.17 ±0.19	93.65 ±1.88	**99.97** ±0.06
4	99.24 ±0.08	99.39 ±0.35	99.51 ±0.18	99.12 ±0.46	99.41 ±0.13	99.12 ±0.41	99.15 ±0.30	**99.71** ±0.87	93.92 ±3.27	98.41 ±0.68
5	94.55 ±0.66	93.45 ±1.85	95.63 ±0.81	98.79 ±0.09	98.64 ±0.87	98.42 ±0.62	**99.03** ±0.12	97.95 ±1.28	96.50 ±1.76	98.87 ±1.33
6	99.67 ±0.16	99.63 ±0.25	99.56 ±0.11	99.79 ±0.21	99.77 ±0.05	99.70 ±0.13	99.87 ±0.13	99.24 ±1.27	98.74 ±0.52	**100** ±0.00
7	98.87 ±0.53	99.40 ±0.11	99.34 ±0.24	99.43 ±0.24	99.44 ±0.09	99.64 ±0.30	99.64 ±0.08	98.18 ±0.35	96.21 ±2.39	**99.99** ±0.02
8	72.41 ±2.03	73.59 ±2.33	74.13 ±0.49	73.01 ±3.40	76.25 ±4.74	78.11 ±0.42	78.86 ±1.50	90.80 ±0.19	**97.93** ±0.60	97.73 ±0.22
9	97.82 ±0.01	97.91 ±0.88	98.79 ±0.50	98.92 ±0.18	99.10 ±0.19	98.78 ±1.46	99.65 ±0.12	99.54 ±0.07	98.71 ±1.07	**100** ±0.00
10	87.70 ±4.21	89.62 ±0.33	91.68 ±1.05	95.24 ±0.44	96.07 ±1.28	94.88 ±1.65	95.42 ±1.12	94.77 ±0.67	94.96 ±2.25	**99.77** ±0.37
11	93.82 ±1.38	96.85 ±1.92	93.47 ±4.81	95.03 ±2.28	96.49 ±3.75	95.61 ±2.83	97.29 ±3.54	94.58 ±1.72	90.58 ±4.90	**100** ±0.00
12	99.75 ±0.16	99.93 ±0.12	99.45 ±0.46	99.95 ±0.09	99.91 ±0.06	99.95 ±0.07	99.82 ±0.17	99.44 ±0.98	97.17 ±1.53	**100** ±0.00
13	97.29 ±0.17	96.14 ±1.56	97.14 ±0.17	98.36 ±0.73	97.84 ±0.89	97.94 ±0.08	98.59 ±0.84	99.74 ±0.28	95.01 ±2.11	**100** ±0.00
14	92.49 ±1.53	93.89 ±0.87	95.00 ±0.98	94.91 ±1.63	96.91 ±1.39	95.23 ±2.02	97.23 ±0.25	94.97 ±2.23	93.16 ±5.57	**99.87** ±0.15
15	62.04 ±1.48	58.38 ±2.25	64.37 ±1.98	69.36 ±4.08	67.05 ±5.23	67.51 ±1.65	66.31 ±1.88	88.63 ±0.62	96.22 ±1.10	**96.77** ±1.47
16	94.75 ±1.41	94.44 ±0.85	98.00 ±0.58	98.78 ±0.40	98.57 ±0.31	98.76 ±0.16	99.30 ±0.46	96.95 ±1.68	91.91 ±7.30	**100** ±0.00
OA	86.97 ±0.63	86.96 ±0.49	88.23 ±0.27	89.34 ±0.79	89.86 ±0.45	90.13 ±0.42	90.43 ±0.07	95.27 ±0.04	96.73 ±0.89	**98.98** ±0.15
AA	92.71 ±0.58	92.94 ±0.23	93.69 ±0.40	94.91 ±0.43	95.24 ±0.38	95.12 ±0.24	95.55 ±0.18	96.70 ±0.06	95.65 ±1.41	**99.46** ±0.08
κ	85.50 ±0.70	85.48 ±0.53	86.90 ±0.30	88.15 ±0.88	89.02 ±0.49	88.33 ±0.46	89.34 ±0.08	94.74 ±0.05	96.35 ±0.99	**98.86** ±0.16

4.3.2. Classification Maps

In order to show the classification results more directly, classification maps of all considered methods are provided in Figures 5–7 for three experimental datasets, respectively. From Figure 5, it can be clearly seen that the proposed method can obtain much smoother classification regions than other methods, especially for class 1 (Alfalfa), class 2 (Corn-notill), class 3 (Corn-mintill), and class 12

(Soybean-clean) whose spectral characteristics are highly correlated with other classes. The similar results can also be observed from Figures 6 and 7, where class 1 (Asphalt), class 6 (Bare Soil), and class 8 (Self-blocking bricks) in the second dataset, and class 8 (Grapes untrained), class 15 (Vineyard untrained) in the third dataset are labeled more precisely. These observations are consistent with the quantitative results listed in Tables 3–5.

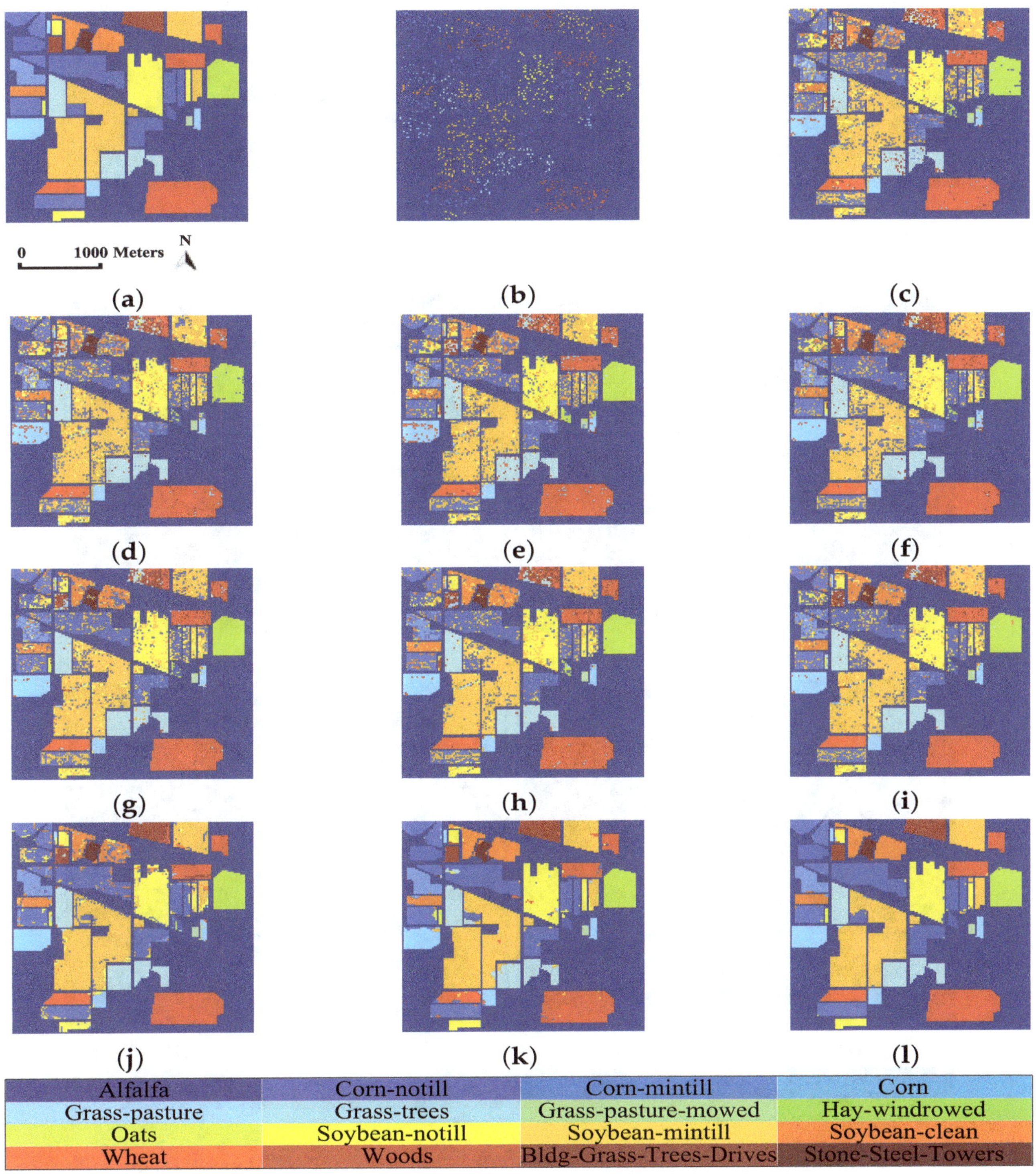

Figure 5. Classification maps of different methods for the Indian Pines dataset: (**a**) ground truth; (**b**) training set; (**c**) origin; (**d**) PCA; (**e**) LDA; (**f**) LFDA; (**g**) SGDA; (**h**) GDA-SS; (**i**) SLGDA; (**j**) MPCA; (**k**) G-LTDA; and (**l**) TSLGDA.

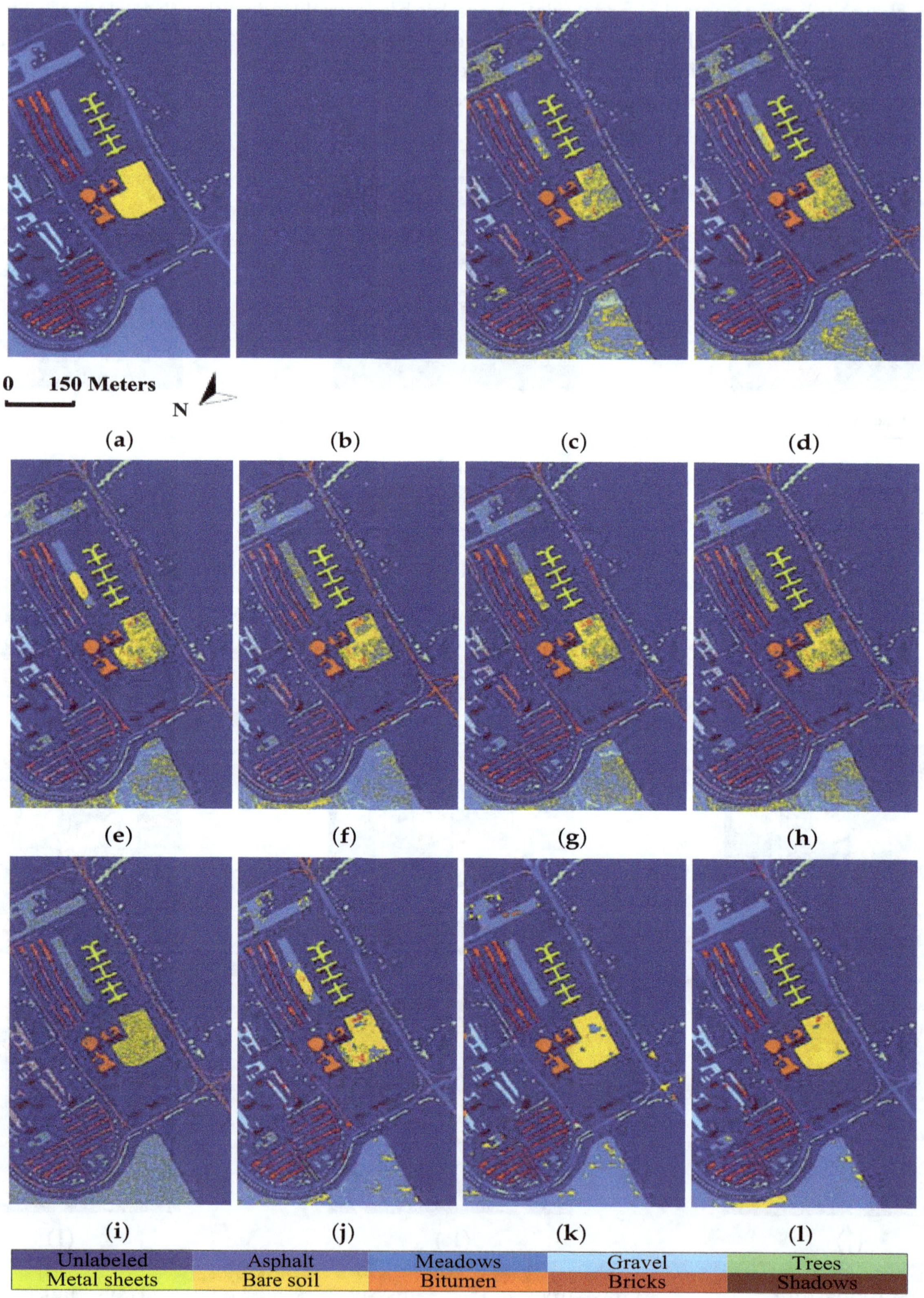

Figure 6. Classification maps of different methods for the University of Pavia dataset: (**a**) ground truth; (**b**) training set; (**c**) origin; (**d**) PCA; (**e**) LDA; (**f**) LFDA; (**g**) SGDA; (**h**) GDA-SS; (**i**) SLGDA; (**j**) MPCA; (**k**) G-LTDA; and (**l**) TSLGDA.

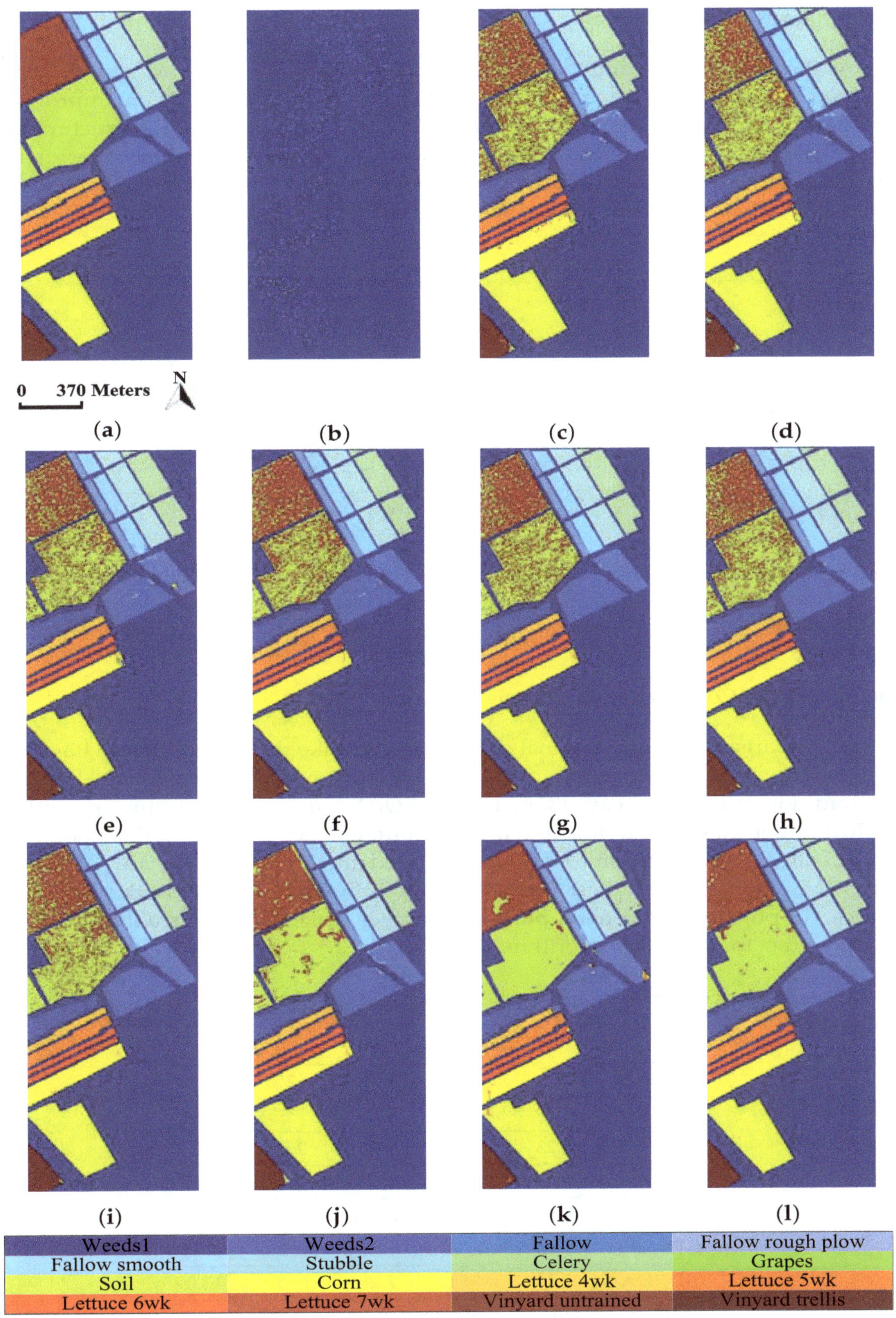

Figure 7. Classification maps of different methods for the Salinas dataset: (**a**) ground truth; (**b**) training set; (**c**) origin; (**d**) PCA; (**e**) LDA; (**f**) LFDA; (**g**) SGDA; (**h**) GDA-SS; (**i**) SLGDA; (**j**) MPCA; (**k**) G-LTDA; and (**l**) TSLGDA.

4.3.3. The Influence of Training Size

To show the influence of training size, some considered DR methods are tested. The results are given in Figure 8, from which we can see that the OA values of all methods are improved when the number of training samples increases for three datasets. Due to the spatial structure information contained in the tensor, the proposed method always performs better than other methods in all cases. In addition, with the label information, the supervised DR methods (i.e., SGDA, GDA-SS, SLGDA, G-LTDA, TSLGDA) achieve better results than the corresponding unsupervised DR methods (i.e., PCA, MPCA).

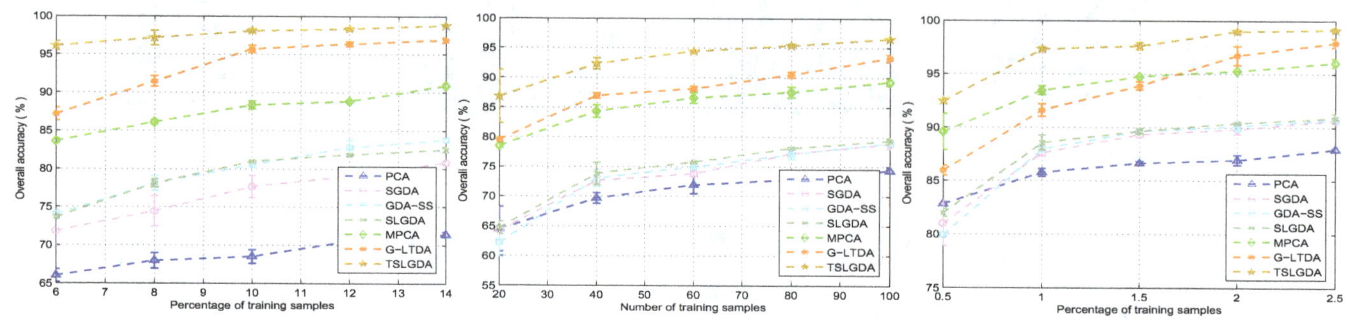

(a) (b) (c)

Figure 8. Overall classification accuracy and standard deviation versus different numbers of training samples per class for all methods using three datasets: (**a**) Indian Pines; (**b**) University of Pavia; (**c**) Salinas.

4.3.4. The Analysis of Computational Complexity

For the comparison of computational complexity, we take the Indian Pines data as an example. Table 6 shows the time requirements of all considered methods, from which it can be clearly seen that traditional methods (e.g., PCA, LDA, LFDA) run faster than other recently proposed methods. In addition, due to complicated tensor computation, tensor-based DR methods (e.g., MPCA, G-LTDA, TSLGDA) cost more time than vector-based methods (e.g., SGDA, GDA-SS, SLGDA). Although TSLGDA has the highest computational complexity, it yields the best classification performance. In practice, the general-purpose graphics processing units (GPUs) can be adopted to greatly accelerate the TSLGDA algorithm.

Table 6. Execution time (in seconds) of different methods for the Indian Pines data with different training size.

Methods	6%	8%	10%	12%	14%
PCA	1.23	1.49	1.86	2.35	2.54
LDA	1.23	1.51	1.88	2.34	2.54
LFDA	1.24	1.57	1.93	2.40	2.62
SGDA	10.60	14.11	18.53	23.90	29.30
GDA-SS	1.13	1.36	1.67	2.15	2.45
SLGDA	3.24	4.81	7.20	10.19	13.09
MPCA	115.94	150.00	161.06	182.37	203.94
G-LTDA	30.96	40.24	49.86	62.41	74.83
TSLGDA	183.91	225.06	281.19	349.44	456.84

5. Conclusions

In this paper, we have proposed a tensor sparse and low-rank graph-based discriminant analysis method (i.e., TSLGDA) for dimensionality reduction of hyperspectral imagery. The hyperspectral data cube is taken as a third-order tensor, from which sub-tensors (local patches) centered at the training samples are extracted to construct the sparse and low-rank graph. On the one hand, by imposing both the sparse and low-rank constraints on the objective function, the proposed method is capable of capturing the local and global structure simultaneously. On the other hand, due to the spatial structure information introduced by tensor data, the proposed method can improve the graph structure and enhance the discriminative ability of reduced features. Experiments conducted on three hyperspectral datasets have consistently confirmed the effectiveness of our proposed TSLGDA algorithm, even for small training size. Compared to some state-of-the-art methods, the overall classification accuracy of TSLGDA in the low-dimensional space improves about 2% to 30%, 5% to 20%, and 2% to 12% for three experimental datasets, respectively, with increased computational complexity.

Acknowledgments: This work was supported by the National Natural Science Foundation of China under Grant 61371165 and Grant 61501018, and by the Frontier Intersection Basic Research Project for the Central Universities under Grant A0920502051714-5. The authors would like to thank Prof. David A. Landgreve from Purdue University for providing the AVIRIS image of Indian Pines and Prof. Paolo Gamba from University of Pavia for providing the ROSIS dataset. The authors would like to thank Dr. Zisha Zhong for sharing the code of Gabor filters and giving some useful suggestions. Last but not least, we would like to thank the editors and the anonymous reviewers for their detailed comments and suggestions, which greatly helped us to improve the clarity and presentation of our manuscript.

Author Contributions: All of the authors made significant contributions to the work. Lei Pan and Heng-Chao Li designed the research model, analyzed the results and wrote the paper. Yang-Jun Deng provided codes about tensor processing. Fan Zhang and Xiang-Dong Chen reviewed the manuscript. Qian Du contributed to the editing and review of the manuscript.

References

1. He, Z.; Li, J.; Liu, L. Tensor block-sparsity based representation for spectral-spatial hyperspectral image classification. *Remote Sens.* **2016**, *8*, 636.
2. Peng, B.; Li, W.; Xie, X.M.; Du, Q.; Liu, K. Weighted-fusion-based representation classifiers for hyperspectral imagery. *Remote Sens.* **2015**, *7*, 14806–14826.
3. Yu, S.Q.; Jia, S.; Xu, C.Y. Convolutional neural networks for hyperspectral image classification. *Neurocomputing* **2017**, *219*, 88–98.
4. Jimenez, L.O.; Landgrebe, D.A. Supervised classification in high-dimensional space: Geometrical, statistical, and asymptoticla properitier of multivariate data. *IEEE Trans. Syst. Man Cybern. Part C Appl. Rev.* **1998**, *28*, 39–54.
5. Chang, C.-I.; Safavi, H. Progressive dimensionality reduction by transform for hyperspectral imagery. *Pattern Recognit.* **2011**, *44*, 2760–2773.
6. Yuan, Y.; Lin, J.Z.; Wang, Q. Dual-clustering-based hyperspectral band selection by contextual analysis. *IEEE Trans. Geosci. Remote Sens.* **2016**, *54*, 1431–1445.
7. Wang, Q.; Lin, J.Z.; Yuan, Y. Salient band selection for hyperspectral image classification via manifold ranking. *IEEE Trans. Neural Netw. Learn. Syst.* **2016**, *27*, 1279–1289.
8. Jolliffe, I.T. *Principal Component Analysis*; Springer-Verlag: New York, NY, USA, 2002.
9. Bandos, T.V.; Bruzzone, L.; Camps-Valls, G. Classification of hyperspectral images with regularized linear discriminant analysis. *IEEE Trans. Geosci. Remote Sens.* **2009**, *47*, 862–873.
10. Li, W.; Prasad, S.; Fowler, J.E.; Bruce, L.M. Locality-preserving dimensionality reduction and classification for hyperspectral image analysis. *IEEE Trans. Geosci. Remote Sens.* **2012**, *50*, 1185–1198.
11. Tan, K.; Zhou, S.Y.; Du, Q. Semisupervised discriminant analysis for hyperspectral imagery with block-sparse graph. *IEEE Geosci. Remote Sens. Lett.* **2015**, *12*, 1765–1769.

12. Chatpatanasiri, R.; Kijsirikul, B. A unifier semi-supervised dimensionality reduction framework for manifold learning. *Neurocomputing* **2010**, *73*, 1631–1640.
13. Yan, S.C.; Xu, D.; Zhang, B.Y.; Zhang, H.-J.; Yang, Q.; Lin, S. Graph embedding and extensions: A general framework for dimensionality reduction. *IEEE Trans. Pattern Anal. Mach. Intell.* **2007**, *29*, 40–51.
14. He, X.F.; Cai D.; Yan, S.C.; Zhang, H.-J. Neighborhood preserving embedding. In Proceedings of the 2005 IEEE Conference on Computer Vision (ICCV), Beijing, China, 17–20 October 2005; pp. 1208–1213.
15. Feng, F.B.; Li, W.; Du, Q.; Zhang, B. Dimensionality reduction of hyperspectral image with graph-based discriminant analysis considering spectral similarity. *Remote Sens.* **2017**, *9*, 323.
16. Wright, J.; Yang, A.Y.; Ganesh, A.; Sastry, S.S.; Ma, Y. Robust face recognition via sparse representation. *IEEE Trans. Pattern Anal. Mach. Intell.* **2009**, *31*, 210–227.
17. Wright, J.; Ma, Y.; Mairal, J.; Sapiro, G.; Huang, T.S.; Yan, S.C. Sparse representation for computer vision and pattern recognition. *Proc. IEEE* **2010**, *98*, 1031–1044.
18. Cheng, B.; Yang, J.C.; Yan, S.C.; Fu, Y.; Huang, T.S. Learning with l_1-graph for image analysis. *IEEE Trans. Image Process.* **2015**, *23*, 2241–2253.
19. Ly, N.H.; Du, Q.; Fowler, J.E. Sparse graph-based discriminant analysis for hyperspectral imagery. *IEEE Trans. Geosci. Remote Sens.* **2014**, *52*, 3872–3884.
20. He, W.; Zhang, H.Y.; Zhang, L.P.; Philips, W.; Liao, W.Z. Weighted sparse graph based dimensionality reduction for hyperspectral images. *IEEE Geosci. Remote Sens. Lett.* **2016**, *13*, 686–690.
21. Ly, N.H.; Du, Q.; Fowler, J.E. Collaborative graph-based discriminant analysis for hyperspectral imagery. *IEEE J. Sel. Top. Appl. Earth Observ. Remote Sens.* **2014**, *7*, 2688–2696.
22. Li, W.; Du, Q. Laplacian regularized collaborative graph for discriminant analysis of hyperspectral imagery. *IEEE Trans. Geosci. Remote Sens.* **2016**, *54*, 7066–7076.
23. Li, W.; Liu, J.B.; Du, Q. Sparse and low-rank graph for discriminant analysis of hyperspectral imagery. *IEEE Trans. Geosci. Remote Sens.* **2016**, *54*, 4094–4105.
24. Xue, Z.H.; Du, P.J.; Li, J.; Su, H.J. Simultaneous sparse graph embedding for hyperspectral image classification. *IEEE Trans. Geosci. Remote Sens.* **2015**, *53*, 6114–6132.
25. Lu, H.P.; Plataniotis, K.N.; Venetsanopoulos, A.N. MPCA: Multilinear principal component analysis of tensor objects. *IEEE Trans. Neural Netw.* **2008**, *19*, 18–39.
26. Guo, X.; Huang, X.; Zhang, L.F.; Zhang, L.P.; Plaza, A.; Benediktssom, J.A. Support tensor machines for classification of hyperspectral remote sensing imagery. *IEEE Trans. Geosci. Remote Sens.* **2016**, *54*, 3248–3264.
27. An, J.L.; Zhang, X.R.; Jiao, L.C. Dimensionality reduction based on group-based tensor model for hyperspectral image classification. *IEEE Geosci. Remote Sens. Lett.* **2016**, *13*, 1497–1501.
28. Zhang, L.P.; Zhang, L.F.; Tao, D.C.; Huang, X. Tensor discriminative locality alignment for hyperspectral image spectral-spatial feature extraction. *IEEE Trans. Geosci. Remote Sens.* **2013**, *51*, 242–256.
29. Zhong, Z.S.; Fan, B.; Duan, J.Y.; Wang, L.F.; Ding, K.; Xiang, S.M.; Pan, C.H. Discriminant tensor spectral-spatial feature extraction for hyperspectral image classification. *IEEE Geosci. Remote Sens. Lett.* **2015**, *12*, 1028–1032.
30. Benediktsson, J.A.; Palmason, J.A.; Sveinsson, J.R. Classification of hyperspectral data from urban areas based on extended morphological profiles. *IEEE Trans. Geosci. Remote Sens.* **2005**, *43*, 480–491.
31. Dallal Mura, M.; Benediktsson, J.A.; Waske, B.; Bruzzone, L. Morphological attribute profiles for the analysis of very high resolution images. *IEEE Trans. Geosci. Remote Sens.* **2010**, *48*, 3747–3762.
32. Rajadell, O.; Garcia-Sevilla, P.; Pla, F. Spectral-spatial pixel characterization using Gabor filters for hyperspectral image classification. *IEEE Geosci. Remote Sens. Lett.* **2013**, *10*, 860–864.
33. Zhao, H.T.; Sun, S.Y. Sparse tensor embedding based multispectral face recognition. *Neurocomputing* **2014**, *133*, 427–436.
34. Lai, Z.H.; Xu, Y.; Chen, Q.C.; Yang, J.; Zhang, D. Multilinear sparse principal component analysis. *IEEE Trans. Neural Netw. Learn. Syst.* **2014**, *25*, 1942–1950.

35. Boyd, S.; Parikh, N.; Chu, E.; Peleato, B.; Eckstein, J. Distributed optimization and statistical learning via the alternating direction method of multipliers. *Found. Trends Mach. Learn.* **2011**, *3*, 1–122.
36. Cai, J.-F.; Candés, E.J.; Shen, Z. A singular value thresholding algorithm for matrix completion. *SIAM J. Optim.* **2010**, *24*, 1956–1982.
37. Nie, F.P.; Xiang, S.M.; Song, Y.Q.; Zhang, C.S. Extracting the optimal dimensionality for local tensor discriminant analysis. *Pattern Recognit.* **2009**, *42*, 105–114.
38. SPArse Modeling Software. Available online: http://spams-devel.gforge.inria.fr/index.html (accessed on 20 January 2017).
39. Hyperspectral Remote Sensing Scenes. Available online: http://www.ehu.eus/ccwintco/index.php?title=Hyperspectral_Remote_Sensing_Scenes (accessed on 20 January 2017).

8

Hyperspectral Target Detection via Adaptive Joint Sparse Representation and Multi-Task Learning with Locality Information

Yuxiang Zhang [1], Ke Wu [1], Bo Du [2,*], Liangpei Zhang [3] and Xiangyun Hu [1]

[1] Hubei Subsurface Multi-Scale Imaging Key Laboratory, Institute of Geophysics and Geomatics, China University of Geosciences, Wuhan 430074, China; zyx_070504@163.com (Y.Z.); tingke2000@126.com (K.W.); xyhu@cug.edu.cn (X.H.)
[2] School of Computer, Wuhan University, Wuhan 430079, China
[3] State Key Laboratory of Information Engineering in Surveying, Mapping, and Remote Sensing, Wuhan University, Wuhan 430079, China; zlp62@whu.edu.cn
* Correspondence: gunspace@163.com.

Academic Editors: Qi Wang, Nicolas H. Younan, Carlos López-Martínez, Xiaofeng Li and Prasad S. Thenkabail

Abstract: Target detection from hyperspectral images is an important problem but encounters a critical challenge of simultaneously reducing spectral redundancy and preserving the discriminative information. Recently, the joint sparse representation and multi-task learning (JSR-MTL) approach was proposed to address the challenge. However, it does not fully explore the prior class label information of the training samples and the difference between the target dictionary and background dictionary when constructing the model. Besides, there may exist estimation bias for the unknown coefficient matrix with the use of ℓ_1/ℓ_2 minimization which is usually inconsistent in variable selection. To address these problems, this paper proposes an adaptive joint sparse representation and multi-task learning detector with locality information (JSRMTL-ALI). The proposed method has the following capabilities: (1) it takes full advantage of the prior class label information to construct an adaptive joint sparse representation and multi-task learning model; (2) it explores the great difference between the target dictionary and background dictionary with different regularization strategies in order to better encode the task relatedness; (3) it applies locality information by imposing an iterative weight on the coefficient matrix in order to reduce the estimation bias. Extensive experiments were carried out on three hyperspectral images, and it was found that JSRMTL-ALI generally shows a better detection performance than the other target detection methods.

Keywords: hyperspectral image; target detection; multi-task learning; sparse representation; locality information

---

1. Introduction

Target detection is essentially a binary classification problem, which aims to separate specific target pixels from various backgrounds with prior knowledge of the targets [1,2]. With the characteristic of high spectral resolution [3], hyperspectral images (HSIs) with hundreds or even thousands of spectral bands can distinguish subtle spectral differences, even between very similar materials, providing a unique advantage for target detection [4,5]. Target detection has therefore attracted much attention in many HSI applications, and it has been successfully used in real-world applications such as detecting rare minerals in geology, oil pollution in environmental research, landmines in the public safety and defense domain, and man-made objects in reconnaissance and surveillance applications [6–9].

The current target detection methods mainly utilize the detailed spectral information from the HSI data and use different techniques to distinguish the targets and the background, such as the statistical hypothesis testing theory [10–12], filtering or projection technique [13–15], and sparse representation technique [16–19]. These existing target detection methods, using a uniform vector of test pixel's spectrum as input, usually employ all the original bands to both construct the model and perform the detection. In other words, these methods fully and uniformly utilize the discriminative information within all single-band images, without considering the inherent similarity between the adjacent single-band images of HSI. In fact, the spectral resolution of HSIs is so high that the adjacent single-band images present a great spectral similarity or redundancy, and this spectral redundancy provides an obstacle for effective target detection. Many methods via dimension reduction for hyperspectral target detection have been proposed in order to relieve this problem [20–24]. However, none of them can guarantee that all the valuable discriminative spectral information underlying the HSI data is preserved, since the HSI data dimension is greatly reduced after the dimension reduction process. To summarize, there exists a dilemma to simultaneously reduce spectral redundancy and preserve discriminative information for Hyperspectral target detection.

In recent years, the multi-task learning (MTL) technique has attracted much interest [25–28] and has been employed to address the above dilemma for hyperspectral target detection in [29], labeled as the joint sparse representation and multi-task learning (JSR-MTL) approach. The approach explores the spectral similarity between the adjacent single-band images to construct multiple sub-HSIs with a band cross-grouping strategy, which leads to multiple related detection tasks. The approach further explores the similarity between the sub-HSIs to analyze the latent sparse representation of each task. Then multiple sparse representation models via the union target and background dictionary are integrated via a unified multitask learning technique. In this way, the redundancy in each detection task can be effectively avoided; and the spectral information behind the high dimension original HSI dataset fully used, so that the discriminative information is not lost [29].

However, there still exist several problems with the JSR-MTL approach. Firstly, it does not fully incorporate the class label (prior) information of the training samples, which only utilizes the class label information in post-processing when calculating the residuals for each class and ignores the class label information when constructing the sparse representation models. Secondly, it encourages shared sparsity among the columns of the coefficient matrix corresponding to the union dictionary, which lead to the same sparsity constraint among the tasks corresponding to both the target dictionary and background dictionary. However, as the size and the spectral variability of the target dictionary are much different from the background dictionary, it is therefore not appropriate to impose the same sparsity constraint for both the coefficient matrices corresponding to the target dictionary and background dictionary. Finally, it does not take the locality information between the test pixel and all the neighboring background training samples into consideration, which may make a contribution for better signal reconstruction, due to the fact that the samples similar to the test pixel are more likely to be selected for signal reconstruction.

To address the above problems, this paper proposes an adaptive joint sparse representation and multi-task learning detector with locality information (JSRMTL-ALI). The proposed method explores the prior class label information of the training samples to construct two joint sparse representation and multi-task learning models, where the test pixel is separately modeled via the target dictionary or background dictionary. Considering also the great difference between the target dictionary and background dictionary, different regularization strategies encoding the task relatedness are employed for the two joint sparse representation and multi-task learning models based on the target dictionary or background dictionary. Besides, a locality information descriptor is introduced to indicate the difference between the central test pixel and the neighboring background training samples. Additionally, inspired by the idea that the coefficient matrix may have estimation bias in [30,31], since the ℓ_1/ℓ_2 minimization used in the regularization strategy is usually inconsistent in variable selection, a locality information descriptor-based weight is employed to iteratively constrain the regularization term to reduce the estimation bias.

The rest of this paper is organized as follows. Section 2 briefly introduces the original JSR-MTL method. The proposed JSRMTL-ALI method is then presented in Section 3. The experimental results of the proposed method with several HSIs are presented in Section 4. Finally, the discussion and conclusions are drawn in Sections 5 and 6.

2. Brief Introduction to the JSR-MTL Method

For the hyperspectral imagery (HSI), as discussed in [29], the adjacent single band images are similar to each other and MTL technology is introduced to utilize the spectral similarity for hyperspectral target detection.

The MTL methodology was proposed by Caruana [28]. It is an inductive transfer method that uses the domain-specific information contained in the training signals of related tasks, which can guarantee that the related tasks can learn from each other and make the inductive transfer method work. There are two key techniques of MTL. One is the construction of multiple tasks with commonality. The other key technique is the relevance analysis of multiple tasks. The multiple tasks can be constructed in various ways, which may depend on the specific application [25,26]. Tasks can be related in various ways. There are two commonly used approaches: (1) tasks may be related by assuming that all the learned functions are close to each other in some norm, such as the linear regression function [25]; and (2) tasks may also be related in that they all share a common underlying representation [32], such as sparsity, a manifold constraint, or a graphical model structure.

In JSR-MTL [29], multiple related detection tasks are constructed through band cross-grouping strategy. In accordance with the band order of the original HSI, the multiple adjacent single-band images are cross-grouped into different groups. Each group then forms a sub-HSI, as shown in Figure 1. Based on the spectral similarity between the adjacent single-band images, multiple sub-HSIs from the original HSI are related with each other. Therefore, these multiple related sub-HSIs naturally correspond to multiple related detection tasks [29].

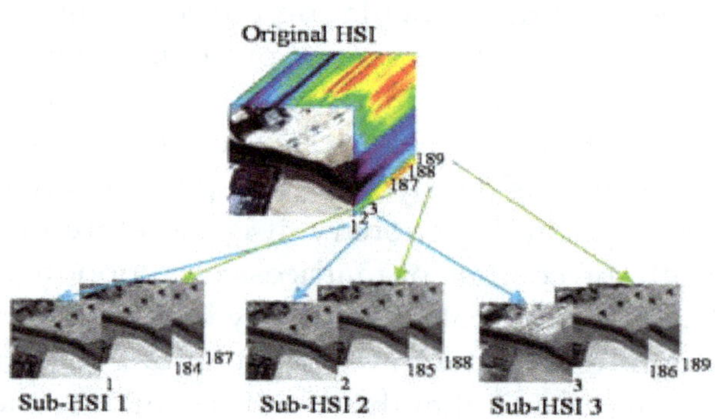

Figure 1. Illustration of the band cross-grouping strategy for the multiple detection tasks. HSI = hyperspectral image.

For the relevance analysis of the multiple tasks, the spectral similarity of the multiple sub-HSIs naturally guarantees the relevance of the multiple detection tasks. Therefore the multiple detection tasks are likely to share a common sparse representation [29], which has shown effectiveness in hyperspectral target detection [16–19].

Considering hyperspectral data $\mathbf{X} \in R^{h \times w \times B}$ with training samples $\mathbf{D} = \left[\mathbf{D}^t, \mathbf{D}^b\right]$, where $\mathbf{D}^t$ is the target dictionary generated via the target training samples $\{\mathbf{d}_i{}^t\}_{i=1}^{N_t} \in R^B$, and $\mathbf{D}^b$ is the background dictionary generated via the background training samples $\{\mathbf{d}_i{}^b\}_{i=1}^{N_b} \in R^B$. N_b and N_t are the number of background and target training samples, respectively. Let $\mathbf{x}$ be a test pixel in the original HSI, and $\left\{\mathbf{x}^k \in R^{B^k}\right\}_{k=1}^{K}$ represents the partial test pixel in each sub-HSI.

For the k-th sub-HSI, the partial test pixel $\left\{\mathbf{x}^k\right\}_{k=1}^{K} \in R^{B^k}$ can be modeled to lie in the union of the background and target subspaces respectively spanned by the background training samples $\{\mathbf{d}_i^{kb} \in R^{B^k \times N_b}\}_{i=1}^{N_b}$ and the target training samples $\{\mathbf{d}_i^{kt} \in R^{B^k \times N_t}\}_{i=1}^{N_t}$. Therefore, $\mathbf{x}^k$ can be represented by a sparse linear combination of the training samples

$$\begin{aligned}\mathbf{x}^k &= \left(w_1^{kb}\mathbf{d}_1^{kb} + w_2^{kb}\mathbf{d}_2^{kb} + \cdots + w_{N_b}^{kb}\mathbf{d}_{N_b}^{kb}\right) + \left(w_1^{kt}\mathbf{d}_1^{kt} + w_2^{kt}\mathbf{d}_2^{kt} + \cdots + w_{N_t}^{kt}\mathbf{d}_{N_t}^{kt}\right) + \varsigma^k \\ &= \mathbf{D}^{kb}\mathbf{w}^{kb} + \mathbf{D}^{kt}\mathbf{w}^{kt} + \varsigma^k \\ &= \mathbf{D}^k\mathbf{w}^k + \varsigma^k\end{aligned} \tag{1}$$

where ς^k is the random noise. $\mathbf{D}^{kb}$ and $\mathbf{D}^{kt}$ are the $B^k \times N_b$ background sub-dictionary and $B^k \times N_t$ target sub-dictionary, respectively. $\mathbf{w}^k \in R^{N_b+N_t}$ is a concatenation of $\mathbf{w}^{kb}$ and $\mathbf{w}^{kt}$, which are the coefficient sub-vectors over the k-th sub-dictionary $\mathbf{D}^{kb}$ and $\mathbf{D}^{kt}$.

Since the K groups of partial test pixels are highly related to each other, the sparse representation for a single-task case can be generalized to a multiple-task case. Thus, for the multiple detection tasks, the original pixel $\mathbf{x} \in R^B$ decomposed into K sub-vectors can be represented as

$$\begin{aligned}\mathbf{x}^1 &= \mathbf{D}^{1b}\mathbf{w}^{1b} + \mathbf{D}^{1t}\mathbf{w}^{1t} + \varsigma^1 = \mathbf{D}^1\mathbf{w}^1 + \varsigma^1 \\ &\vdots \qquad\qquad\qquad\qquad \vdots \\ \mathbf{x}^K &= \mathbf{D}^{Kb}\mathbf{w}^{Kb} + \mathbf{D}^{Kt}\mathbf{w}^{Kt} + \varsigma^K = \mathbf{D}^K\mathbf{w}^K + \varsigma^K\end{aligned} \tag{2}$$

These can be incorporated into the joint sparse representation and multi-task learning model

$$\widehat{\mathbf{W}} = \arg\min_{\mathbf{w}^k} \sum_{k=1}^{K} ||\mathbf{x}^k - \mathbf{D}^k\mathbf{w}^k||_2^2 + \rho||\mathbf{W}||_{2,1} \tag{3}$$

where $\mathbf{W} \in R^{(N_b+N_t)\times K}$ is the coefficient matrix formed by stacking the vectors $\mathbf{w}^k \in R^{N_b+N_t}$. ρ is the regularization parameter to trade off the data fidelity term and the regularization term, which penalizes the $\ell_{2,1}$-norm of the coefficient matrix $\mathbf{W}$. The $\ell_{2,1}$-norm of $\mathbf{W}$ is obtained by first computing the ℓ_2-norm of the rows $\{\mathbf{w}_i\}_{i=1}^{N_b+N_t}$ (across the tasks) of the matrix $\mathbf{W}$, and then computing the ℓ_1-norm of the vector $b(\mathbf{W}) = (||\mathbf{w}_1||_2, \cdots, ||\mathbf{w}_{N_b+N_t}||_2)^T$. This norm encourages the sparsity of each column of the matrix $\mathbf{W}$, and simultaneously encourages shared sparsity among the columns of the matrix $\mathbf{W}$.

3. Adaptive JSR-MTL with Locality Information Detector

3.1. Adaptive JSR-MTL Model

Some sparse representation classifiers employ the sparsity within a class for the classification, and show that a few background samples are adequate to reconstruct a test background sample in HSI [18]. Thus, If $\mathbf{x}$ is a background pixel, for the k-th sub-HSI, the partial test pixel $\left\{\mathbf{x}^k\right\}_{k=1}^{K} \in R^{B^k}$ can be approximately represented as a linear combination of the background training samples $\{\mathbf{d}_i^{kb}\}_{i=1}^{N_b} \in R^{B^k}$ as follows:

$$\begin{aligned}\mathbf{x}^k &= \left(w_1^{kb}\mathbf{d}_1^{kb} + w_2^{kb}\mathbf{d}_2^{kb} + \cdots + w_{N_b}^{kb}\mathbf{d}_{N_b}^{kb}\right) + \varsigma^{\mathrm{kb}} \\ &= \mathbf{D}^{kb}\mathbf{w}^{kb} + \varsigma^{kb}\end{aligned} \tag{6}$$

where ς^{kb} is the random noise. $\mathbf{D}^{kb}$ is the $B^k \times N_b$ background sub-dictionary. $\mathbf{w}^{kb} \in R^{N_b}$ is the coefficient sub-vector over the sub-dictionary $\mathbf{D}^{kb}$.

For the multiple detection tasks, the original background pixel $\mathbf{x} \in R^B$ decomposed into K sub-vectors can be represented as

$$\begin{aligned} \mathbf{x}^1 &= \mathbf{D}^{1b}\mathbf{w}^{1b} + \varsigma^{1b} \\ &\vdots \qquad\quad \vdots \\ \mathbf{x}^K &= \mathbf{D}^{Kb}\mathbf{w}^{Kb} + \varsigma^{Kb} \end{aligned} \tag{7}$$

These models can be incorporated into the following joint sparse representation and multi-task learning model

$$\widehat{\mathbf{W}}^b = \arg\min_{\mathbf{w}^{kb}} \sum_{k=1}^{K} ||\mathbf{x}^k - \mathbf{D}^{kb}\mathbf{w}^{kb}||_2^2 + \rho^b \Omega(\mathbf{W}^b) \tag{8}$$

where $\mathbf{W}^b \in R^{N_b \times K}$ is the matrix formed by stacking the vectors $\mathbf{w}^{kb} \in R^{N_b}$. $\Omega(\mathbf{W}^b)$ is the regularization term to further encode the task relatedness. ρ^b is the regularization parameter to trade off the data fidelity term and the regularization term.

Similarly, a target pixel $\mathbf{x} \in R^B$ decomposed into K sub-vectors can be represented as

$$\begin{aligned} \mathbf{x}^1 &= \mathbf{D}^{1t}\mathbf{w}^{1t} + \varsigma^{1t} \\ &\vdots \qquad\quad \vdots \\ \mathbf{x}^K &= \mathbf{D}^{Kt}\mathbf{w}^{Kt} + \varsigma^{Kt} \end{aligned} \tag{9}$$

where ς^{kt} is the random noise. $\mathbf{D}^{kt}$ is the $B^k \times N_t$ target sub-dictionary. $\mathbf{w}^{kt} \in R^{N_t}$ is the coefficient sub-vector over the k-th sub-dictionary $\mathbf{D}^{kt}$.

These models can also be incorporated into the following joint sparse representation and multi-task learning model

$$\widehat{\mathbf{W}}^t = \arg\min_{\mathbf{w}^{kt}} \sum_{k=1}^{K} ||\mathbf{x}^k - \mathbf{D}^{kt}\mathbf{w}^{kt}||_2^2 + \rho^t \Omega(\mathbf{W}^t) \tag{10}$$

where $\mathbf{W}^t \in R^{N_t \times K}$ is the matrix formed by stacking the vectors $\mathbf{w}^{kt} \in R^{N_t}$.

In the detection problems, we are given a set of training samples with corresponding labels. The above two JSR-MTL models in Equations (8) and (10) make the assumption that a test sample should be represented by atoms from the same classes that the test sample belongs to, which means that the test sample is modeled separately for target and background pixel. Therefore, the above two JSR-MTL models in Equations (8) and (10) are more complete and realistic than the basic JSR-MTL model in [29]. In the above two JSR-MTL models in Equations (8) and (10), the test samples are modeled separately with more reasonable dictionaries, with only the background training samples for the null hypothesis, and the target training samples for the alternative hypothesis. In the case of the basic JSR-MTL model in [29], either the target test samples or the background test samples are represented by both the background and target training samples. In other words, the basic JSR-MTL model in [29] does not fully incorporate the class label (prior) information of the data set; it only utilizes the class label (background and target) information in post-processing when calculating the residuals for each class and ignores it when constructing models and calculating sub-vectors.

What is more, as noted, the regularization terms $\Omega(\mathbf{W})$ in Equations (8) and (10) are employed to further encode the task relatedness for the background pixel and target pixel, respectively. It can be seen that different assumptions on the task relatedness lead to different regularization terms. Whether the same regularization terms should be used for both the target and background pixel is an interesting problem, which needs further discussion. As we know, in the basic JSR-MTL model [29], multiple partial test pixels $\mathbf{x}^k$ in each sub-HSI are sparsely represented via the union target and background dictionary $\mathbf{D}^k = [\mathbf{D}^{kt}, \mathbf{D}^{kb}]$, and the $\ell_{2,1}$-norm of $\mathbf{W}$ encourages shared sparsity among the columns of the matrix $\mathbf{W}$which is formed by stacking the vectors $\mathbf{w}^k \in R^{N_b+N_t}$. This will lead to the same sparsity constraint among the columns of the matrix $\mathbf{W}^b$ and $\mathbf{W}^t$ corresponding to the target dictionary and background dictionary. This is inappropriate when considering the construction of target and background dictionary. In target detection applications, the number of target pixels is usually small. The target dictionary is therefore constructed from some of the target pixels in the global image scene [17,18]. The background dictionary is generated locally for each test pixel through a dual concentric window which separates the local area around each pixel into two regions, a small inner window region (IWR) centered within a larger outer window region (OWR), which can better represent and capture the spectral signature of the test sample [17,18]. The background dictionary consists of many locally neighboring background training samples whose spectra are likely to be similar to each other. Thus, for the background pixel, multiple columns of the coefficient matrix $\mathbf{W}^b$ corresponding to multiple background sub-dictionaries are likely to share consistent sparsity among different tasks. However, the case for the target pixel is much different from the background pixel. The size of the target dictionary is smaller than the background dictionary, and the target training samples selected from the whole image are likely to show spectral variability [11]. Therefore it is inappropriate to assume consistent sparsity among multiple columns of the coefficient matrix $\mathbf{W}^t$ corresponding to multiple target sub-dictionaries. In brief, different regularization terms should be used for the target pixel and background pixel.

For the background pixel in (8), the $\ell_{2,1}$-norm can be enforced to the matrix $\mathbf{W}^b$ as is done in [29]. For the target pixel in Equation (10), the ℓ_1-norm is applied for the matrix $\mathbf{W}^t$, which is obtained by the sum of absolute values in the matrix. The difference between ℓ_1-norm and $\ell_{2,1}$-norm of the matrix $\mathbf{W}$ is that, ℓ_1-norm imposes element wise sparsity and does not require consistent feature selection among columns (tasks), while $\ell_{2,1}$-norm by grouping rows together can achieve consistent sparsity among different columns (tasks).

Therefore, Equations (8) and (10) can be rewritten as the following adaptive JSR-MTL model, which can be labeled as the JSRMTL-A model:

$$\widehat{\mathbf{W}}^b = \arg\min_{\mathbf{w}^{kb}} \sum_{k=1}^{K} ||\mathbf{x}^k - \mathbf{D}^{kb}\mathbf{w}^{kb}||_2^2 + \rho^b ||\mathbf{W}^b||_{2,1} \tag{11}$$

$$\widehat{\mathbf{W}}^t = \arg\min_{\mathbf{w}^{kt}} \sum_{k=1}^{K} ||\mathbf{x}^k - \mathbf{D}^{kt}\mathbf{w}^{kt}||_2^2 + \rho^t ||\mathbf{W}^t||_1 \tag{12}$$

3.2. Locality Information Descriptor-Based Weight

The background dictionary is further discussed in this section. It can be seen from Equation (11) that, all the training samples (atoms) in the background dictionary are treated equally for signal representation, which ignores locality information, such as differences between the neighboring pixels and the central test pixel. However, some surrounding pixels may be quite similar to the center pixel and are likely to be selected for signal representation; some are quite different from the center pixel, such as the pixel which has a different kind of material from the central pixel, which should be limited or even prohibited for signal representation. The differences between the test pixel and the target atoms are not discussed here due to the small size of the target atoms and the global target atoms selection method.

To preserve the locality difference between the central test pixel and the neighboring background atoms, a distance based locality information descriptor is introduced, which can be expressed as

$$\alpha_i^k = \exp\left(\frac{||\mathbf{x}^k - \mathbf{d}_i^{kb}||_2^2}{2}\right) \tag{13}$$

where α_i^k is the sample-specific descriptor for training sample $i(i = 1, 2, \cdots, N_b)$ in the k-th background sub-dictionary $\mathbf{D}^{kb}$. It is clear that a smaller α_i^k indicates $\mathbf{x}^k$ is more similar to the atom $\mathbf{d}_i^{kb}$, and vice versa.

Once the above descriptor is included, all the atoms in the background dictionary will be adaptively treated for signal representation via the $\ell_{2,1}$-norm. However, there may still exist estimation bias for the signal representation. As stated in [30,31] the estimation bias can be large due to the fact that the ℓ_1/ℓ_2 minimization is generally inconsistent in variable selection. Many efforts have been made to reduce the estimation bias, such as adaptive Lasso method [30] and the reweighted ℓ_1

minimization [31]. Inspired by the reweighted ℓ_1 minimization, a weight strategy on the $\ell_{2,1}$-based regularization term is introduced to reduce the estimation bias as follows.

$$\widehat{\mathbf{W}}^b = \arg\min_{\mathbf{w}^{kb}} \sum_{k=1}^{K} ||\mathbf{x}^k - \mathbf{D}^{kb}\mathbf{w}^{kb}||_2^2 + \rho^b ||\mathbf{\Psi} \odot \mathbf{W}^b||_{2,1} \tag{14}$$

where $\mathbf{\Psi} = \left\{\Psi_i^k\right\}$ is the weighting matrix, and Ψ_i^k is the weight for atom $i(i = 1,2,\cdots,N_b)$ and column (task) $k(k = 1,2,\cdots,K)$, $\odot$ denotes modifying the element in the coefficients matrix $\mathbf{W}^b$ by iteratively multiplying a weight during the coefficient optimization.

In order to impose a relatively higher penalty for smaller coefficients and a lower penalty for larger coefficients, the weight can be computed as inversely proportional to the sparse coefficient

$$\varphi_i^k = \frac{1}{|w_i^{kb}|} \tag{15}$$

Combining the above locality information descriptor and weight strategy, we obtain the locality information descriptor-based weight defined as

$$\Psi_i^k = \frac{\varphi_i^k \alpha_i^k}{\max\limits_{i,k} \varphi_i^k \alpha_i^k} \tag{16}$$

3.3. Model Optimization

For the model optimization, we use the popular accelerated proximal gradient (APG) algorithm [33,34] to efficiently solve the problem in Equations (12) and (14). The APG algorithm alternately updates a matrix sequence $\widehat{\mathbf{W}}^t = \left[w_i^{k,t}\right]$ and an aggregation matrix sequence $\widehat{\mathbf{V}}^t = \left[v_i^{k,t}\right]$.

Given the current matrix aggregation matrix $\widehat{\mathbf{V}}^t$, a generalized gradient mapping step is employed to update matrix $\widehat{\mathbf{W}}^{t+1}$ as follows

$$\begin{aligned} \widehat{\mathbf{w}}^{k,t+1} &= \widehat{\mathbf{v}}^{k,t} - \eta^t \nabla^{k,t}, \quad t \geq 1, \\ \widehat{\mathbf{W}}^{t+1} &= f(\widehat{\mathbf{W}}^{t+1}), k = 1,2,\cdots K \end{aligned} \tag{17}$$

where $\nabla^{k,t} = -(\mathbf{D}^k)^T \mathbf{x}^k + (\mathbf{D}^k)^T \mathbf{D}^k \widehat{\mathbf{v}}^{k,t}$, $\eta^t = 1/2^t$ is the step size. $f(\cdot)$ is a function of $\widehat{\mathbf{W}}^{t+1} = \left[\widehat{w}_i^{k,t+1}\right]$, which has a different format for (12) and (14).

For (12), the matrix $\widehat{\mathbf{W}}^{t+1}$ can be updated as follows

$$\widehat{\mathbf{W}}^{t+1} = \begin{cases} P_{\Omega_1}(\widehat{\mathbf{W}}^{t+1} - \frac{\rho}{2^t}), & \Omega_1 : (\widehat{\mathbf{W}}^{t+1})_{i,k\in\Omega_1} > \frac{\rho}{2^t} \\ P_{\Omega_2}(\widehat{\mathbf{W}}^{t+1} + \frac{\rho}{2^t}), & \Omega_2 : (\widehat{\mathbf{W}}^{t+1})_{i,k\in\Omega_2} < -\frac{\rho}{2^t} \\ P_{\Omega_3}(\mathbf{0} \in R^{N_t \times K}), & \Omega_3 : (\Omega_1 \cup \Omega_2)^{\perp}, \quad i = 1,2,\cdots,N_b \end{cases} \tag{18}$$

where P_Ω is the projection of a matrix onto an entry set, and Ω is the index of the entry set.

For (14), the matrix $\widehat{\mathbf{W}}^{t+1}$ can be updated as follows

$$\begin{aligned} \widehat{\mathbf{w}}_i^{t+1} &= \left[1 - \frac{\rho}{2^t ||\widehat{\mathbf{w}}_i^{t+1}||_2}\right]_+ \widehat{\mathbf{w}}_i^{t+1}, \quad i = 1,2,\cdots,N_b \\ \widehat{w}_i^{k,t+1} &= \widehat{w}_i^{k,t+1} \times \frac{\alpha_i^k / |\widehat{w}_i^{k,t+1}|}{\max\limits_{i=1,2,\cdots N_b, k=1,2,\cdots K} \alpha_i^k / |\widehat{w}_i^{k,t+1}|} \end{aligned} \tag{19}$$

where $[\cdot]_+ = \max(\cdot, 0)$.

An aggregation forward step is then employed to update $\widehat{\mathbf{V}}^{t+1}$ by linearly combining $\widehat{\mathbf{W}}^{t+1}$ and $\widehat{\mathbf{W}}^{t}$ as follows

$$\widehat{\mathbf{V}}^{t+1} = (1+\tau^t)\widehat{\mathbf{W}}^{t+1} - \tau^t \widehat{\mathbf{W}}^{t} \tag{20}$$

where the sequence τ^t is conventionally set to $\tau^t = 2(t-1)/(1+\sqrt{1+4t^2})$, as applied in our implementation.

The optimization methods for the problem in Equations (12) and (14) can be summarized as Algorithms 1 and 2, respectively.

Algorithm 1. The Coefficients over Target Dictionary Optimization Algorithm.

Input: Data $\left\{\mathbf{D}^k, \mathbf{x}^k\right\}_{k=1}^{K}$, regularization parameter ρ

Output: Coefficient vectors $\left\{\widehat{\mathbf{w}}^{k}\right\}_{k=1}^{K}$

Step (1): Initialization: $\widehat{\mathbf{w}}^{k,0} = (\mathbf{D}^k)^T \mathbf{x}^k$, $\widehat{\mathbf{v}}^{k,0} = \widehat{\mathbf{w}}^{k,0}$, $\tau^0 = -1$, $t := 0$

Step (2): Repeat {Main loop}

a) $\widehat{\mathbf{w}}^{k,t+1} = \widehat{\mathbf{v}}^{k,t} - \frac{1}{2^t}\left[-(\mathbf{D}^k)^T\mathbf{x}^k + (\mathbf{D}^k)^T\mathbf{D}^k\widehat{\mathbf{v}}^{k,t}\right], \quad k = 1, \cdots K$

b) $\widehat{w}_i^{k,t+1} = \begin{cases} \widehat{w}_i^{k,t+1} - \rho/2^t, & \Omega_1 \\ \widehat{w}_i^{k,t+1} + \rho/2^t, & \Omega_2 \\ 0, & \Omega_3 \end{cases}$

c) $\tau^t = \frac{2(t-1)}{1+\sqrt{1+4t^2}}$, $\widehat{\mathbf{v}}^{k,t+1} = (1+\tau^t)\widehat{\mathbf{w}}^{k,t+1} - \tau^t\widehat{\mathbf{w}}^{k,t}$

d) $t := t+1$

Until: convergence is attained

Algorithm 2. The Coefficients over Background Dictionary Optimization Algorithm.

Input: Data $\left\{\mathbf{D}^k, \mathbf{x}^k\right\}_{k=1}^{K}$, regularization parameter ρ, locality information descriptor $\left\{\alpha_i^k\right\}$

Output: Coefficient vectors $\left\{\widehat{\mathbf{w}}^{k}\right\}_{k=1}^{K}$

Step (1): Initialization: $\widehat{\mathbf{w}}^{k,0} = (\mathbf{D}^k)^T \mathbf{x}^k$, $\widehat{\mathbf{v}}^{k,0} = \widehat{\mathbf{w}}^{k,0}$, $\tau^0 = -1$, $t := 0$

Step (2): Repeat {Main loop}

a) $\widehat{\mathbf{w}}^{k,t+1} = \widehat{\mathbf{v}}^{k,t} - \frac{1}{2^t}\left[-(\mathbf{D}^k)^T\mathbf{x}^k + (\mathbf{D}^k)^T\mathbf{D}^k\widehat{\mathbf{v}}^{k,t}\right], \quad k = 1, \cdots K$

b) $\widehat{\mathbf{w}}_i^{t+1} = \left[1 - \frac{\rho}{2^t \|\widehat{\mathbf{w}}_i^{t+1}\|_2}\right]_+ \widehat{\mathbf{w}}_i^{t+1}, \quad i = 1, 2, \cdots, N_b$

$\widehat{w}_i^{k,t+1} = \widehat{w}_i^{k,t+1} \times \frac{\alpha_i^k / |\widehat{w}_i^{k,t+1}|}{\max_{i,k} \alpha_i^k / |\widehat{w}_i^{k,t+1}|}$

c) $\tau^t = \frac{2(t-1)}{1+\sqrt{1+4t^2}}$, $\widehat{\mathbf{v}}^{k,t+1} = (1+\tau^t)\widehat{\mathbf{w}}^{k,t+1} - \tau^t\widehat{\mathbf{w}}^{k,t}$

d) $t := t+1$

Until: convergence is attained

3.4. Final Sketch of the JSRMTL-ALI Detector

Once given the recovery of the coefficient vectors $\widehat{\mathbf{w}}^{kb}$ and $\widehat{\mathbf{w}}^{kt}$ corresponding to the background dictionary $\mathbf{D}^{kb}$ and target dictionary $\mathbf{D}^{kt}$ for each task, we can then calculate the residual errors for the background and target between the multiple signals in the sub-HSIs $\left\{\mathbf{x}^k\right\}_{k=1}^{K}$ and the approximations recovered via their corresponding sub-dictionaries $\left\{\mathbf{D}^{kb}\right\}_{k=1}^{K}$ and $\left\{\mathbf{D}^{kt}\right\}_{k=1}^{K}$ as follows.

$$r^b = \sum_{k=1}^{K} ||\mathbf{x}^k - \mathbf{D}^{kb}\widehat{\mathbf{w}}^{kb}||_2$$
$$r^t = \sum_{k=1}^{K} ||\mathbf{x}^k - \mathbf{D}^{kt}\widehat{\mathbf{w}}^{kt}||_2 \quad (21)$$

where $\widehat{\mathbf{w}}^{kb}$ and $\widehat{\mathbf{w}}^{kt}$ are the subsets of the coefficient vector $\widehat{\mathbf{w}}^{k}$ associated with the background and target. The output of the test pixel $\mathbf{x}$ is then calculated by

$$D(\mathbf{x}) = r^b - r^t \quad (22)$$

Finally, a visual illustration of the proposed LWAJSR-MTL algorithm for HSIs is shown in Figure 2. Given a hyperspectral image, multiple sub-HSIs are extracted via the band cross-grouping strategy. We construct the multiple-signals for each pixel $\left\{\mathbf{x}^k\right\}_{k=1}^{K}$, multiple background dictionary $\left\{\mathbf{D}^{kb}\right\}_{k=1}^{K}$ with the local dual window, and multiple target dictionary $\left\{\mathbf{D}^{kt}\right\}_{k=1}^{K}$ via the target training samples. Each pixel is represented by the multi-task sparse representation model via the target dictionary and background dictionary, respectively. The coefficient matrices corresponding to the target dictionary and dictionary are recovered via the Algorithms 1 and 2, respectively. Finally, the detection decision rules in favor of the target class or the background class with the lowest total reconstruction error difference accumulated over all the tasks.

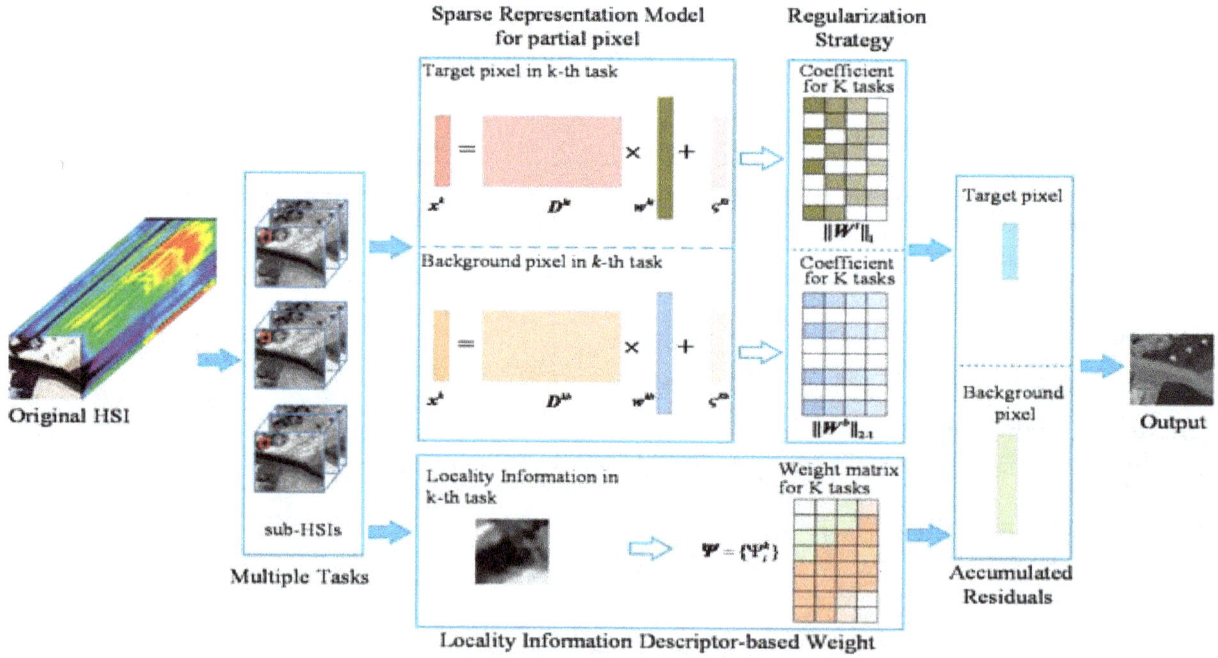

Figure 2. Schematic illustration of the adaptive joint sparse representation and multi-task learning detector with locality information (JSRMTL-ALI) algorithm.

4. Experiments and Analysis

4.1. Dataset Description

Three hyperspectral datasets were used in this study to evaluate the effectiveness of the proposed detector introduced in Section 3.

The first dataset was collected by the Airborne Visible/Infrared Imaging Spectrometer (AVIRIS) sensor from San Diego, CA, USA. The spatial resolution of this image is 3.5 m per pixel. The image has 224 spectral channels in wavelengths ranging from 370 to 2510 nm. After removing the bands that correspond to the water absorption regions, low-SNR, and bad bands (1–6, 33–35, 97, 107–113, 153–166, and 221–224), 189 bands were retained in the experiments. An area of 100 × 100 pixels was used for the experiments. The image scene is shown in Figure 3a. There are three planes in the image, which consist of 58 pixels, as shown in Figure 3b. We selected one pixel from each plane as the target atoms, $N_t = 3$.

The second dataset was gathered by the Airborne Visible/Infrared Imaging Spectrometer (AVIRIS) sensor over the Indian Pines test site in Northwest Indiana and consists of 145 × 145 pixels and 224 spectral reflectance bands in the wavelength range 0.4–2.5 μm. The false color image of the Indian Pines image is shown in Figure 4a. We also reduced the number of bands to 200 by removing bands covering the regions of water absorption: 104–108, 150–163, and 220, as referred to in [35]. This image contains 16 ground-truth classes via a ground truth labels, and the stone-steel-towers was selected as the target of interest to be detected, which has 93 pixels, as shown in Figure 4b. We selected three pixels from the target as the target atoms, $N_t = 3$.

The third data set was acquired by the Nuance Cri hyperspectral sensor. This sensor can acquire imagery with a spectral resolution of 10 nm. The image scene covers an area of 400 × 400 pixels, as shown in Figure 5a, with 46 spectral bands in wavelengths ranging from 650 to 1100 nm. There are ten rocks located in the grassy scene, which consist of 1254 pixels, as shown in Figure 5b. We selected one pixel from each rock as the target atoms, $N_t = 10$.

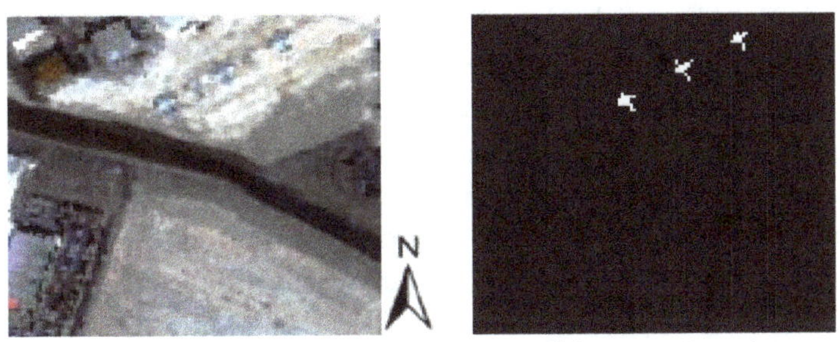

(**a**) Image scene (**b**) Ground truth

Figure 3. The AVIRIS dataset.

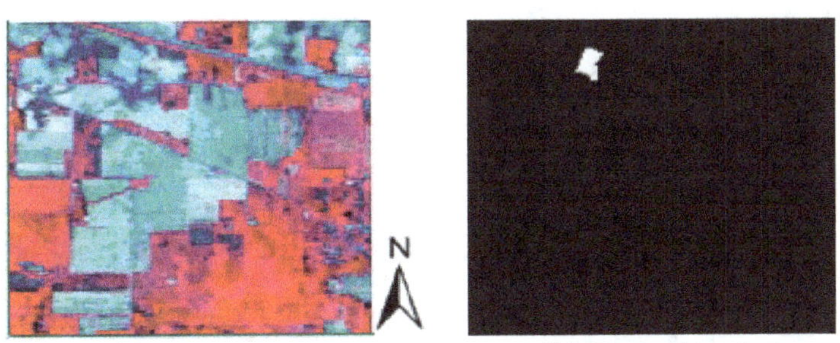

(**a**) Image scene (**b**) Ground truth

Figure 4. The Indian dataset.

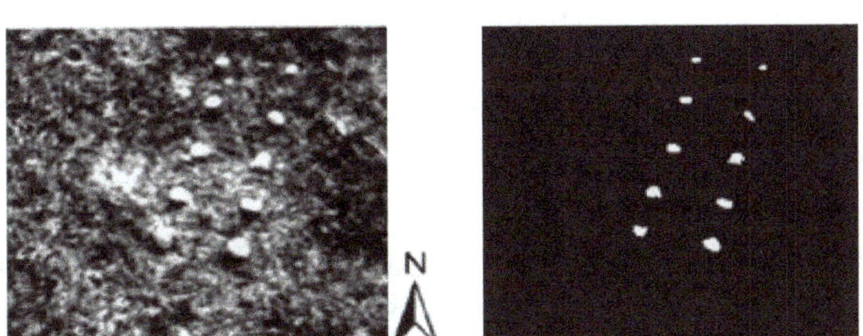

(**a**) Image scene (**b**) Ground truth

Figure 5. The Cri dataset.

4.2. Evaluation of JSRMTL-ALI Model

Firstly, the effectiveness of the JSRMTL-ALI model was investigated and compared with the original JSR-MTL model in Equation (3) and the adaptive JSR-MTL (labeled as JSRMTL-A) model in Equations (11) and (12). We took three detection tasks ($K = 3$) as an example, and the JSR-MTL with $K = 1$ which indicates the detection performance without the multi-task learning technique. For simplicity, all regularization parameters used in the four models are set with the same value ($\rho = 0.1$). The sizes of the OWR for the three datasets were respectively set as 17×17, 23×23, and 23×23. The sizes of the IWR are related to the size of the target, and were set as 7×7, 15×15, and 15×15 for the AVIRIS, Indian, and Cri datasets, respectively. The numbers of the background training samples for the three datasets were therefore $N_b = 240$, $N_b = 304$, and $N_b = 304$, respectively. The detection performance for the four models with three datasets are provided by the area under the receiver operation characteristics (ROC) curves, as shown in Figure 6.

For the AVIRIS dataset, as shown Figure 6a, the ROC curve of JSRMTL-ALI is not above that of JSRMTL-A, however, it is above that of JSR-MTL with $K = 3$ and $K = 1$. For the Indian dataset, as shown in Figure 6b, the ROC curve of JSRMTL-ALI is always above those of the other models, and the ROC curve of JSRMTL-A is always above that of JSR-MTL with $K = 3$ and $K = 1$. For the Cri dataset, as shown Figure 6c, the ROC curve of JSRMTL-ALI is successively above that of JSRMTL-A, JSR-MTL with $K = 1$, and JSR-MTL with $K = 3$.

Overall, the results show that the performance of the JSR-MTL model is generally better than that without the multi-task learning technique, especially for the AVIRIS and Indian datasets. The JSRMTL-A model can also obtain a better detection performance compared to the JSR-MTL model for all three datasets, which shows the effectiveness of the adaptive JSR-MTL (JSRMTL-A) model. This result demonstrates that it is useful to explore the prior class label information of the training samples and the difference between the target dictionary and background dictionary for hyperspectral target detection. What is more, the JSRMTL-ALI model can further improve the detection performance of the JSRMTL-A model, especially for the Indian and Cri datasets. This result confirms that the locality information descriptor-based weight can improve the detection performance, which can remain as the locality information between the central test pixel and neighboring background training samples, and also reduce the estimation bias caused by the ℓ_1/ℓ_2 minimization. In addition, we can further adjust the number of detection tasks, the regularization parameter, and the window size to obtain an even better performance.

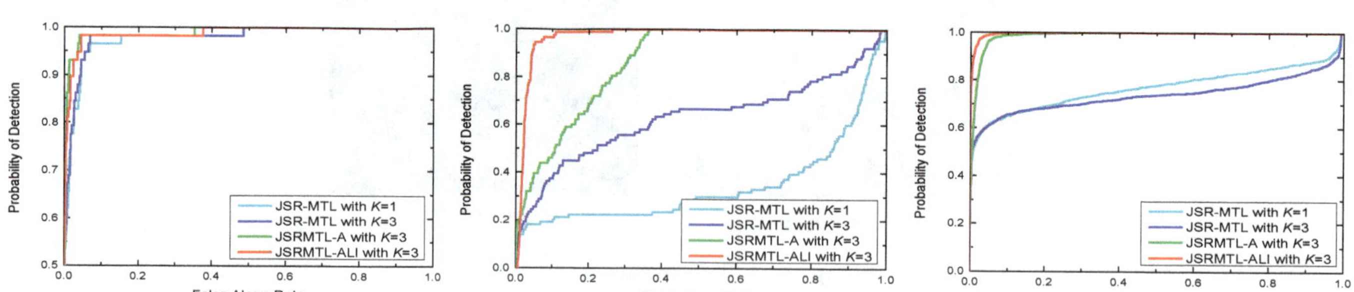

(**a**) AVIRIS dataset (**b**) Indian dataset (**c**) Cri dataset

Figure 6. Receiver operation characteristic (ROC) curves for the effectiveness investigation of JSRMTL-ALI model.

4.3. Parameter Analysis for the JSRMTL-ALI Algorithm

In this section, we examine the effect of the parameters on the detection performance of the JSRMTL-ALI algorithm with the three datasets. We fixed the other parameters and focused on one specific parameter at a time. There are three key parameters in the JSRMTL-ALI algorithm: the detection task number parameter K, the regularization parameter ρ, and the size of the dual window. As is done in [29], the range of K was set as [1, 2, 3, 4, 5, 6, 7, 8, 9] and the range of ρ was set as [1, 0.5, 10−1, 10−2, 10−3, 10−4, 10−5]. For the size of the dual window, the size of the IWR is related to the size of the target. When the size of the IWR is set too large, the background training samples in the OWR will not effectively represent the local background characteristic. Thus, the sizes of the IWR were fixed as above-mentioned 7×7, 15×15, and 15×15 for the AVIRIS, Indian, and Cri datasets, respectively. The range of the size of the OWR for AVIRIS dataset was set as [17, 19, 21, 23, 25], and it was set as [23, 25, 27, 29, 31] for the Indian and Cri datasets. The experimental results are provided through the AUC values, as shown in Figures 7–9. The X-axes and the Y-axes respectively represent the value range of the corresponding parameter and the AUC values.

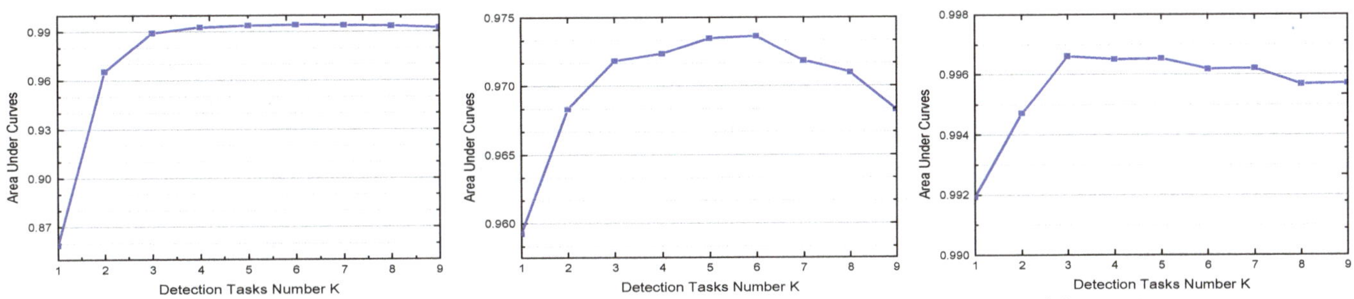

(**a**) AVIRIS dataset (**b**) Indian dataset (**c**) Cri dataset

Figure 7. Detection performance of JSRMTL-ALI versus the detection task number K.

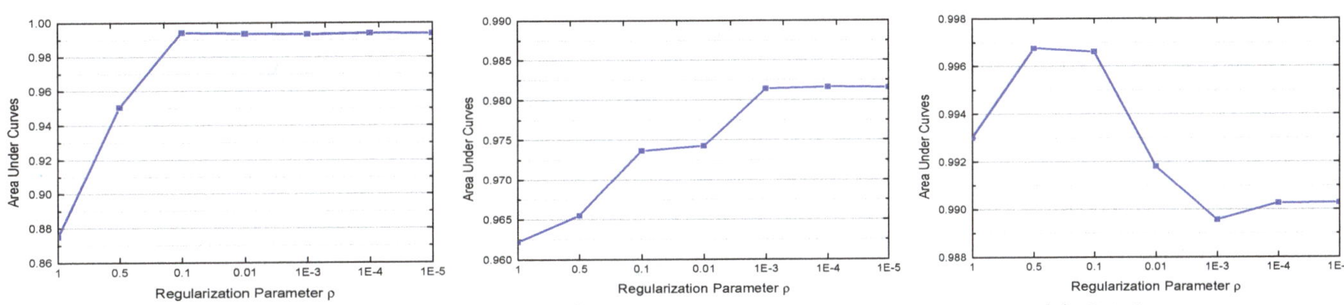

(**a**) AVIRIS dataset (**b**) Indian dataset (**c**) Cri dataset

Figure 8. Detection performance of JSRMTL-ALI versus the detection task number ρ.

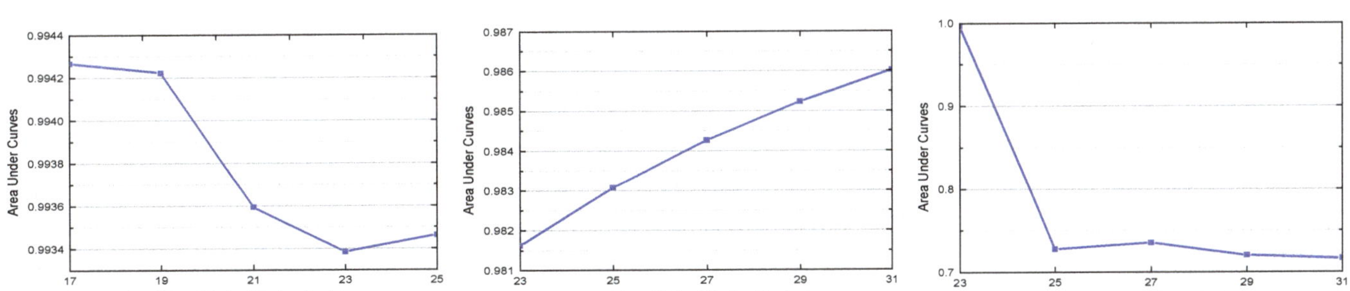

(**a**) AVIRIS dataset (**b**) Indian dataset (**c**) Cri dataset

Figure 9. Detection performance of JSRMTL-ALI versus the size of the outer window region (OWR).

For the AVIRIS dataset in Figure 7a, the AUC value of the JSRMTL-ALI algorithm improves as the detection task number parameter K increases to 6. After that, the detection performance slowly decreases as K increases to 9. For the Indian dataset in Figure 7b, the AUC value generally improves as K increases to 6 and then decreases as K increases to 9. For the Cri dataset in Figure 7c, the AUC value improves as K increases to 3, gently decreases as K increases to 9. Based on these results, it can be generally concluded that the performance of the JSRMTL-ALI algorithm improves as the detection task number parameter K increases and then begins to decrease after the maximum value. The reason for this may be as follows. As discussed in [29], a large detection task number K results in too many detection tasks which leads to too many unknown coefficients; however, the rows of the dictionary for the multiple representation models will be significantly decreased. This can lead to a weakened estimation for the unknown coefficient matrix, which will affect the detection performance. Besides, the advantage of the multi-task learning technique for hyperspectral image lies in the fact that it can explore the relatedness within the corresponding single-band images in the same position in each sub-HIS. However, a large detection task number is highly likely to reduce the relatedness within multiple sub-HSIs, and the effectiveness of MTL will decrease in return.

For the AVIRIS dataset in Figure 8a, the AUC value of the JSRMTL-ALI algorithm improves when the regularization parameter ρ decreases from 1 to 10^{-1}, and the AUC values gradually decrease as ρ decreases from 10^{-2} to 10^{-5}. For the Indian dataset in Figure 8b, the AUC value improves as ρ decreases from 1 to 10^{-4}, and decreases as ρ decreases to 10^{-5}. For the Cri dataset in Figure 8c, the AUC value improves as ρ decreases from 1 to 0.5, and generally decreases as ρ decreases to 10^{-5}. Based on these results, it can be generally concluded that a too small or too large regularization parameter ρ can decrease the detection performance of JSRMTL-ALI. The reasons may be listed as follows. A too small regularization parameter makes the dominant part of Equations (11) and (14) become the first term $||\mathbf{x}^k - \mathbf{D}^k\mathbf{w}^k||_2^2$, which will weaken the effect of the multiple detection task combination, and will affect the final detection performance of JSRMTL-ALI. A too large regularization parameter makes the dominant part of Equations (11) and (14) become the second term, which will weaken the effect of the data representation, and again affect the final detection performance of JSRMTL-ALI.

For the AVIRIS dataset in Figure 9a, the AUC value of the JSRMTL-ALI algorithm decreases as the size of the OWR increases to 23, and then slightly increases as the size of the OWR increases to 25. For the Indian dataset in Figure 9b, the AUC value improves as the size of the OWR increases to 31. For the Cri dataset in Figure 9c, the AUC value of the JSRMTL-ALI algorithm generally decreases as the size of the OWR increases to 31. Based on these results, it can be seen that the detection performance decreases as the size of Outer Window Region (OWR) increases for the AVIRIS dataset and Cri dataset, while the case is totally different for the Indian dataset. Although the regular patter of the size of the OWR for all datasets is not obvious; it can still generally be concluded that a too large or too small size of OWR can affect the detection performance of JSRMTL-ALI. The reason for this may be as follows. For a too large size of OWR, the background training samples in the OWR will not effectively represent the local background characteristic, which may include some other background materials. For a too small size of OWR, the background training samples in the OWR are not sufficient to represent the local background characteristic. Both cases will lead to a weakened detection performance. Therefore, it is not easy to select a proper value for the size of OWR in a practical application.

4.4. Detection Performance

In this section, the detection performance of the proposed JSRMTL-ALI algorithm was further analyzed and compared with traditional detectors of local adaptive coherence/cosine estimator (LACE), local constrained energy minimization (LCEM), reweighted adaptive coherence/cosine estimator (rACE) [10], hierarchical constrained energy minimization (hCEM) [14], STD [17], RBBHD [18], and JSR-MTL [29]. The parameters of the JSRMTL-ALI algorithm were set as the optimal parameter values for the three datasets. The detection task number parameter K was respectively set as 6, 6, and 3 for the three datasets. The regularization parameter ρ was respectively set as 10^{-1}, 10^{-4}, 0.5 for the three datasets. The size of the OWR was set as 17, 31, and 23 for the three datasets. For the comparison methods, the parameters were also tested, such as the sparsity level for the sparsity-based detectors (STD, SRBBHD), and so on. The optimal parameter values were experimentally set for the comparison methods. For all the detectors, we used the same given target spectra as a priori target spectra. In the case of hCEM and LCEM, the mean of the target atoms was used as the target signature. We adopted the pixels falling in the OWR to estimate the background covariance matrix for LACE, to estimate the background correlation matrix for LCEM, and to construct the background dictionary for STD, SRBBHD, JSR-MTL, and JSRMTL-ALI. The detection performance of the eight detectors are provided through the receiver operation characteristics (ROC) curves, as shown in Figure 10.

For the AVIRIS dataset, as shown in Figure 10a, the ROC curve of JSRMTL-ALI is above that of the other detectors, except for rACE. For the Indian dataset, as shown in Figure 10b, the ROC curve of JSRMTL-ALI is always above those of the other detectors. For the Cri dataset, as shown in Figure 10c, rACE and hCEM obtain the best result, and the ROC curve of JSRMTL-ALI is above those of the rest of the detectors.

Overall, the results generally show that the JSRMTL-ALI algorithm obtains a better detection performance than the other detectors, especially for the Indian dataset. For the AVIRIS and Cri dataset, JSRMTL-ALI does not perform as well as rACE or hCEM. However, the detection performances of rACE and hCEM are much different for the three datasets and a robust detection performance is not shown. For example, rACE obtains a good performance for AVIRIS and Cri datasets, while obtains a weak performance for the Indian dataset. hCEM obtains a good performance for Indian and Cri datasets, while obtaining a weak performance for the AVIRIS dataset.

The separability between target and background was evaluated via separability maps, as shown in Figure 11. After statistical calculation of the detection values of each pixel, boxes were drawn to enclose the main parts of the pixels, excluding the biggest 10% and the smallest 10%. There are target and background columns for each detector. The lines at the top and bottom of each column are the extreme values, which are normalized to [1]. The orange boxes illustrate the distribution of the target pixel values, and the line in the middle of the box is the mean of the pixels. In a similar way, the green boxes enclose the middle 80% of the pixels of the background pixels. The position of the boxes reflects the tendency and compactness of the distribution of the pixels. In other words, the position reflects the separability between target and background.

For the AVIRIS dataset, as shown in Figure 11a, STD and rACE can effectively suppress the background information; and LACE, LCEM, STD, SRBBHD, rACE and hCEM can effectively suppress the middle 80% of the background pixels. Compared to these detectors, the gaps between the target box and the background box for rACE, JSR-MTL, and JSRMTL-ALI are very obvious, and the gap for JSRMTL-ALI is larger than JSR-MTL. The target box and the background box for rACE, JSR-MTL, and JSRMTL-ALI are overlapping, but the overlapped region for JSR-MTL is slightly less. For the Indian dataset, as shown in Figure 11b, rACE can specially, effectively suppress the middle 80% of the background pixels. Compared to these detectors, the gap between the two boxes for JSRMTL-ALI is very obvious, and the two boxes for the other detectors are overlapping. For the Cri dataset, as shown in Figure 11c, JSRMTL-ALI, hCEM, and rACE can gradually and successively increase the gap between the target box and the background box. Based on these results, it can be seen that, the proposed JSRMTL-ALI algorithm can perform well at the distinguishing target from the background.

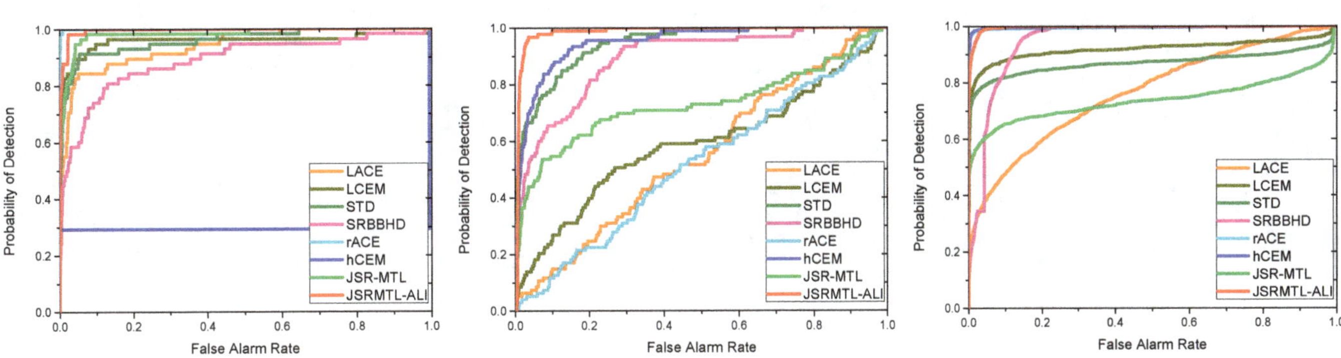

(**a**) AVIRIS dataset (**b**) Indian dataset (**c**) Cri dataset

Figure 10. Detection performance of eight detectors for three datasets.

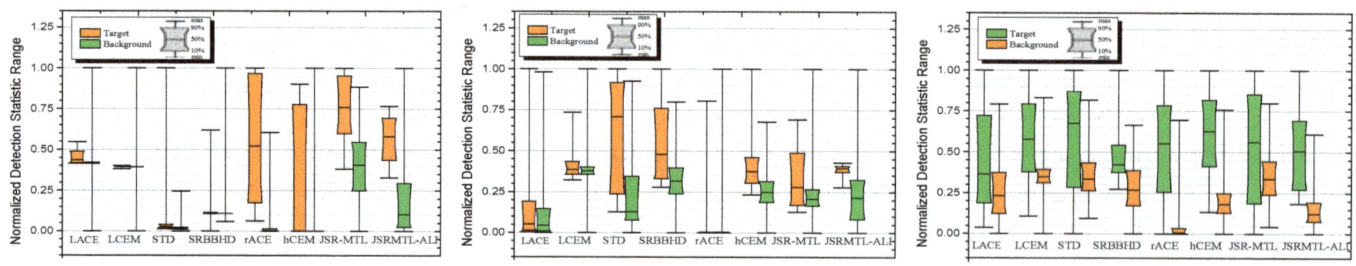

(**a**) AVIRIS dataset (**b**) Indian dataset (**c**) Cri dataset

Figure 11. The separability maps of eight detectors for three datasets.

Finally, 2-D plots of the detection map of all the comparison algorithms with the three data sets are shown in Figures 12–14. For the AVIRIS dataset, as shown in Figure 12, we can see that the proposed JSRMLT-ALI shows high statistical values for the target pixels as well as STD, rACE, and JSR-MTL. However, compared with JSRMLT-ALI, STD and JSR-MTL also show high values for some tree or grass pixels, particularly in the bottom/right left corner in the image. Also rACE shows a good performance for suppressing background. For the Indian dataset, as shown in Figure 13, none of these detectors show a clearly distinguishable statistic map, but JSRMLT-ALI generally shows relatively higher statistical values for the target pixels compared with all the other detectors. For the Cri dataset, as shown in Figure 14, the proposed JSRMLT-ALI shows low statistical values for the background pixels as well as rACE and hCEM. However, compared with rACE, JSRMTL-ALI does not show a clearly distinguishable statistic map between target and background.

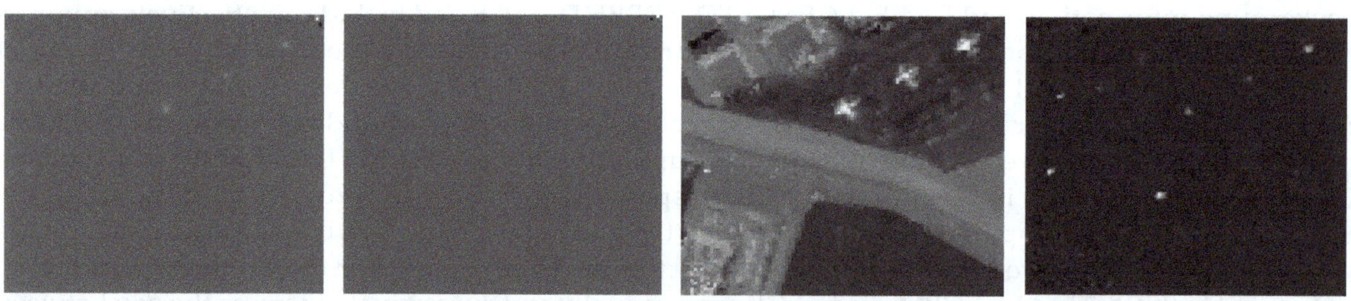

(**a**) LACE (**b**) LCEM (**c**) STD (**d**) SRBBHD

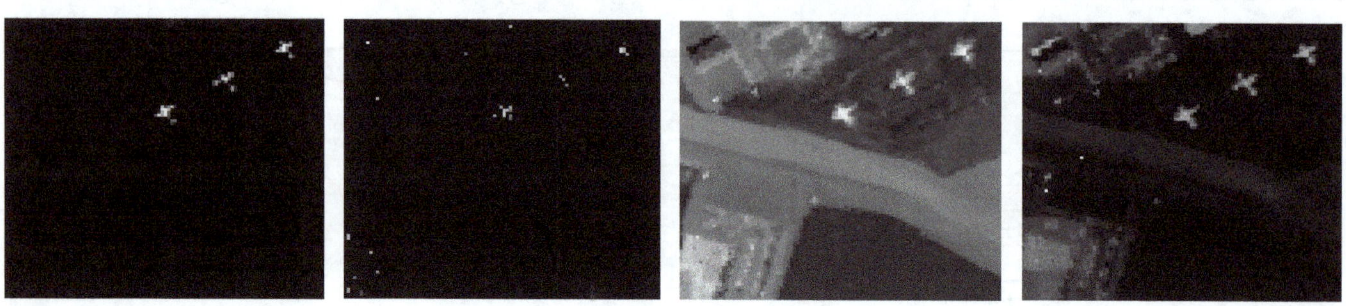

(**e**) rACE (**f**) hCEM (**g**) JSR-MTL (**h**) JSRMTL-ALI

Figure 12. Two-dimensional plots of the detection map for the AVIRIS dataset.

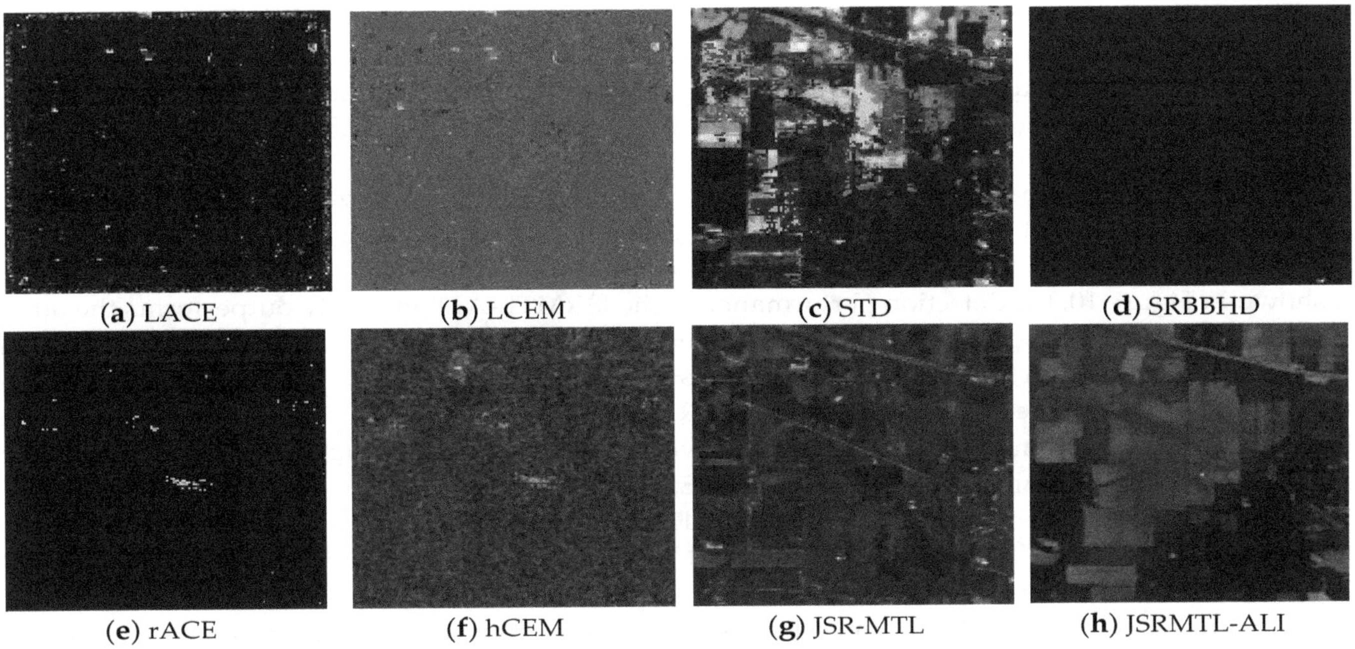

Figure 13. Two-dimensional plots of the detection map for the Indian dataset.

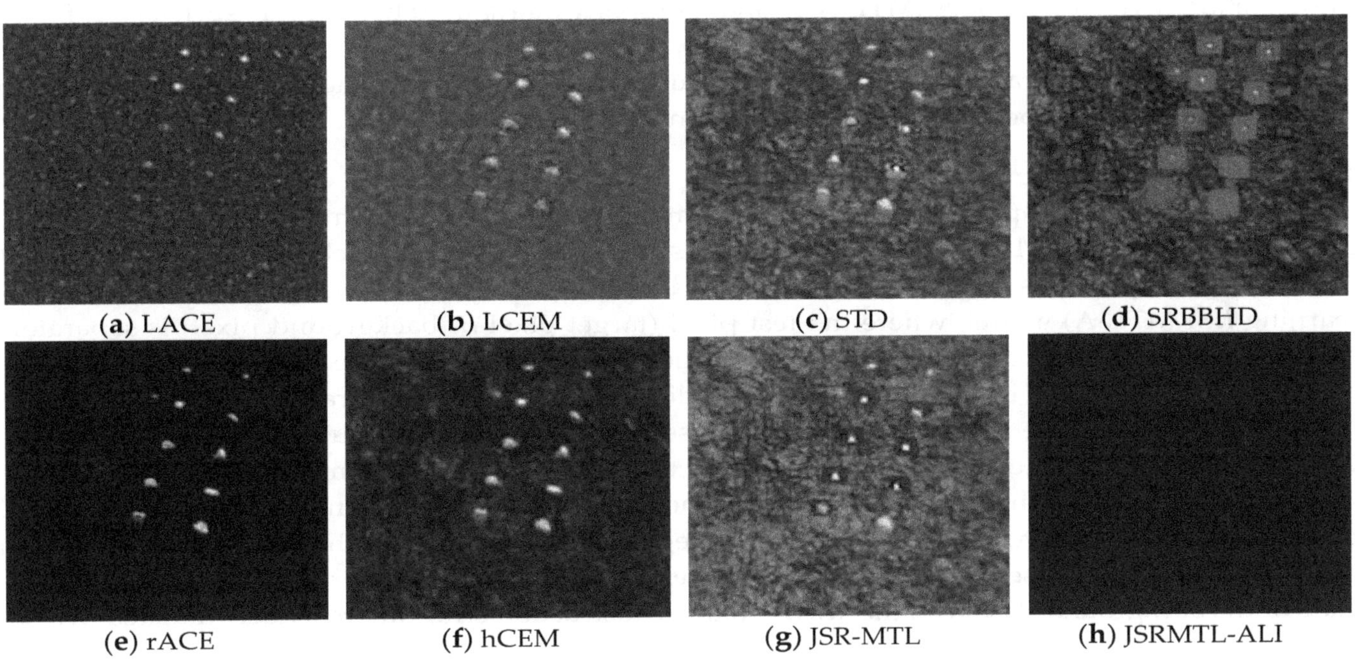

Figure 14. Two-dimensional plots of the detection map for the Cri dataset.

5. Discussion

An adaptive joint sparse representation and multi-task learning detector with locality information (JSRMTL-ALI) is proposed in this paper. In order to fully explore the prior class label information of the training samples, JSRMTL-ALI constructs two joint sparse representation and multi-task learning models corresponding to the target and background classes. In order to consider the difference between the target dictionary and background dictionary, JSRMTL-ALI then employs different regularization strategies encoding the task relatedness for the two models, where the $\ell_{2,1}$-norm is enforced to the coefficient matrix $\mathbf{W}^b$ corresponding to the background dictionary, while the ℓ_1-norm is applied for the coefficient matrix $\mathbf{W}^t$ corresponding to the target dictionary. These two contributions lead to the so-called JSRMTL-A model. What is more, in order to keep the locality information between the central test pixel and neighboring background training samples, JSRMTL-ALI employs the locality information descriptor-based weight to the joint sparse representation and multi-task learning model corresponding to the background class, which can also reduce the estimation bias caused by the ℓ_1/ℓ_2 minimization.

From the above experimental results, it can be seen that Figure 6 shows the superiority of the JSRMTL-A model to the traditional JSR-MTL. The JSRMTL-ALI model can generally further improve the detection performance of the JSRMTL-A model. In the detection performance analysis section, as shown in Figure 10, the detection performance of the JSRMTL-ALI generally outperforms the other detectors for all the datasets, especially the robustness of the JSRMTL-ALI compared to the rACE and hCEM algorithms. From the separability maps as shown in Figure 11 and detection map as shown in Figure 12, it can be seen that, the proposed JSRMTL-ALI algorithm generally performs well at a distinguishing target from the background. However, the performance of suppressing background for the JSRMTL-ALI algorithm is not as good as rACE, which needs further consideration in the future.

There are three key parameters of the JSRMTL-ALI algorithm, which have been analyzed as depicted in Figures 7–9. As shown in Figure 7, it can be seen that a large detection task number K can affect the detection performance of JSRMTL-ALI. A larger value for detection task number K is recommended for the dataset with more bands, such as 6 for the AVIRIS and Indian datasets and a lower value is recommended for the dataset with fewer bands, such as 3 for the Cri dataset. Then as shown in Figure 8, it can be seen that a too small or too large regularization parameter ρ can decrease the detection performance of JSRMTL-ALI, and a proper value should be set for ρ, such as 0.1. Based on the results as shown in Figure 9, it can be seen that, it is not easy to recommend a regular value for the size of OWR in practical application. Our future research will investigate the construction of a global background dictionary in order to avoid tuning the size of the OWR.

6. Conclusions

In this paper, the adaptive joint sparse representation and multi-task learning detector with locality information (JSRMTL-ALI) algorithm was proposed. Based on the prior class label information of the training samples, this algorithm constructs an adaptive joint sparse representation and multi-task learning (JSRMTL-A) model, where the test pixel (target pixel or background pixel) is separately modeled via the target dictionary or background dictionary. Considering the great difference between the target dictionary and background dictionary, different regularization strategies encoding the task relatedness are employed for the two joint sparse representation and multi-task learning models based on the target dictionary or background dictionary. A locality information descriptor is then introduced to indicate the difference between the central test pixel and neighboring background training samples. A descriptor based weight strategy is applied to reduce the estimation bias caused by ℓ_1/ℓ_2 minimization used in the JSRMTL-A model. The detection decision rules in favor of the target class or the background class with the lowest total reconstruction error difference accumulated over all the tasks.

Experiments in hyperspectral target detection with three datasets confirmed the superior performance of the multiple detection task combination in the proposed JSRMTL-ALI algorithm. With the integration of the JSRMTL-A model and local information descriptor based weight strategy, the JSRMTL-ALI shows its superiority to the traditional JSR-MTL for hyperspectral target detection. In general, the JSR-MTL presents a better detection performance and better separability than the other common detectors.

Acknowledgments: The authors would like to thank the handling editor and anonymous reviewers for their careful reading and helpful remarks. This work was supported in part by the Fundamental Research Funds for the Central Universities, China University of Geosciences (Wuhan) under Grant CUG170617, Grant CUGL140410 and Grant 26420160125; in part by the China Postdoctoral Science Found under Grant 2017M612533; in part by the Natural Science Foundation of Hubei Province, China under Grant 2014CFA052; in part by the National Natural Science Foundation of China under Grant 61471274, Grant 61372153 and Grant 41630317.

Author Contributions: All the authors made significant contributions to the work. Yuxiang Zhang and Bo Du conceived, designed and performed the experiments; Ke Wu and Xiangyun Hu analyzed the data; Liangpei Zhang provided advice for the preparation and revision of the paper.

References

1. Manolakis, D.; Truslow, E.; Pieper, M.; Cooley, T.; Brueggeman, M. Detection Algorithms in Hyperspectral Imaging Systems: An Overview of Practical Algorithms. *IEEE Signal Process. Mag.* **2014**, *31*, 24–33. [CrossRef]
2. Nasrabadi, N.M. Hyperspectral Target Detection: An Overview of Current and Future Challenges. *IEEE Signal Process. Mag.* **2014**, *31*, 34–44. [CrossRef]
3. Kang, X.; Li, S.; Benediktsson, J.A. Feature Extraction of Hyperspectral Images with Image Fusion and Recursive Filtering. *IEEE Trans. Geosci. Remote Sens.* **2014**, *52*, 3742–3752. [CrossRef]
4. Landgrebe, D. Hyperspectral Image Data Analysis. *IEEE Signal Process. Mag.* **2002**, *19*, 17–28. [CrossRef]
5. Yuan, Y.; Ma, D.; Wang, Q. Hyperspectral Anomaly Detection by Graph Pixel Selection. *IEEE Trans. Cybern.* **2016**, *46*, 3123–3134. [CrossRef] [PubMed]
6. Stefanou, M.S.; Kerekes, J.P. Image-derived prediction of spectral image utility for target detection applications. *IEEE Trans. Geosci. Remote Sens.* **2010**, *48*, 1827–1833. [CrossRef]
7. Datt, B.; McVicar, T.R.; Niel, T.G.V.; Jupp, D.L.B.; Pearlman, J.S. Preprocessing EO-1 Hyperion hyperspectral data to support the application of agricultural indexes. *IEEE Trans. Geosci. Remote Sens.* **2003**, *41*, 1246–1259. [CrossRef]
8. Eismann, M.T.; Stocker, A.D.; Nasrabadi, N.M. Automated hyperspectral cueing for civilian search and rescue. *Proc. IEEE* **2009**, *97*, 1031–1055. [CrossRef]
9. Manolakis, D.; Marden, D.; Shaw, G.A. Hyperspectral Image Processing for Automatic Target Detection Applications. *Linc. Lab. J.* **2003**, *14*, 79–116.
10. Wang, T.; Du, B.; Zhang, L. An automatic robust iteratively reweighted unstructured detector for hyperspectral imagery. *IEEE J. Sel. Top. Appl. Earth Obs. Remote Sens.* **2014**, *7*, 2367–2382. [CrossRef]
11. Gao, L.; Yang, B.; Du, Q.; Zhang, B. Adjusted Spectral Matched Filter for Target Detection in Hyperspectral Imagery. *Remote Sens.* **2015**, *7*, 6611–6634. [CrossRef]
12. Liu, Y.; Gao, G.; Gu, Y. Tensor Matched Subspace Detector for Hyperspectral Target Detection. *IEEE Trans. Geosci. Remote Sens.* **2017**, *54*, 1967–1974. [CrossRef]
13. Geng, X.; Ji, L.; Sun, K.; Zhao, Y. CEM: More Bands, Better Performance. *IEEE Geosci. Remote Sens. Lett.* **2014**, *11*, 1876–1880. [CrossRef]
14. Zou, Z.; Shi, Z. Hierarchical Suppression Method for Hyperspectral Target Detection. *IEEE Trans. Geosci. Remote Sens.* **2016**, *54*, 330–342. [CrossRef]
15. Harsanyi, J.C.; Chang, C.I. Hyperspectral Image Classification and Dimensionality Reduction: An Orthogonal Subspace Projection Approach. *IEEE Trans. Geosci. Remote Sens.* **1994**, *32*, 779–785. [CrossRef]
16. Huang, Z.; Shi, Z.; Yang, S. Nonlocal Similarity Regularized Sparsity Model for Hyperspectral Target Detection. *IEEE Geosci. Remote Sens. Lett.* **2013**, *10*, 1532–1536. [CrossRef]
17. Chen, Y.; Nasrabadi, N.M.; Tran, T.D. Sparse representation for target detection in hyperspectral imagery. *IEEE J. Sel. Top. Appl. Earth Obs. Remote Sens.* **2011**, *5*, 629–640. [CrossRef]
18. Zhang, Y.; Du, B.; Zhang, L. A Sparse Representation Based Binary Hypothesis Model for Target Detection in Hyperspectral Imagery. *IEEE Trans. Geosci. Remote Sens.* **2015**, *53*, 1346–1354. [CrossRef]
19. Niu, Y.; Wang, B. Extracting Target Spectrum for Hyperspectral Target Detection: An Adaptive Weighted Learning Method Using a Self-Completed Background Dictionary. *IEEE Trans. Geosci. Remote Sens.* **2017**, *55*, 1604–1617. [CrossRef]
20. Farrell, M.D.; Mersereau, R.M. On the impact of PCA dimension reduction for hyperspectral detection of difficult targets. *IEEE Geosci. Remote Sens. Lett.* **2005**, *2*, 192–195. [CrossRef]
21. Fowler, J.E.; Du, Q. Anomaly Detection and Reconstruction from Random Projections. *IEEE Trans. Image Process.* **2012**, *21*, 184–195. [CrossRef] [PubMed]
22. Wang, Q.; Lin, J.; Yuan, Y. Salient Band Selection for Hyperspectral Image Classification via Manifold Ranking. *IEEE Trans. Neural Netw. Learn. Syst.* **2016**, *27*, 1279–1289. [CrossRef] [PubMed]

23. Binol, H.; Ochilov, S.; Alam, M.S.; Bal, A. Target oriented dimensionality reduction of hyperspectral data by Kernel Fukunaga—Koontz Transform. *Opt. Laser Eng.* **2016**, *89*, 123–130. [CrossRef]
24. Sun, K.; Geng, X.; Ji, L. A New Sparsity-Based Band Selection Method for Target Detection of Hyperspectral Image. *IEEE Geosci. Remote Sens. Lett.* **2015**, *12*, 329–333. [CrossRef]
25. Jalali, A.; Ravikumar, P.; Sanghavi, S.; Ruan, C. A Dirty Model for Multi-task Learning. In Proceedings of the Neural Information Processing Systems Conference, Hyatt Regency, Vancouver, BC, Canada, 6–11 December 2010.
26. Yuan, X.; Liu, X.; Yan, S. Visual classification with multi-task joint sparse representation. *IEEE Trans. Image Process.* **2012**, *21*, 4349–4360. [CrossRef] [PubMed]
27. Yuan, Y.; Lin, J.; Wang, Q. Hyperspectral Image Classification via Multi-Task Joint Sparse Representation and Stepwise MRF Optimization. *IEEE Trans. Cybern.* **2016**, *46*, 2966–2977. [CrossRef] [PubMed]
28. Caruana, R. Multitask learning. *Mach. Learn.* **1997**, *28*, 41–75. [CrossRef]
29. Zhang, Y.; Du, B.; Zhang, L. Joint Sparse Representation with Multitask Learning for Hyperspectral Target Detection. *IEEE Trans. Geosci. Remote Sens.* **2017**, *55*, 894–906. [CrossRef]
30. Zou, H. The adaptive lasso and its oracle properties. *J. Am. Stat. Assoc.* **2006**, *101*, 1418–1429. [CrossRef]
31. Candes, E.J.; Wakin, M.B.; Boyd, S.P. Enhancing sparsity by reweighted ℓ minimization. *J. Fourier Anal. Appl.* **2008**, *14*, 877–905. [CrossRef]
32. Ben-David, S.; Schuller, R. Exploiting task relatedness for multiple task learning. In Proceedings of the Conference on Computational Learning Theory, Washington, DC, USA, 24–27 August 2003.
33. Chen, X.; Pan, W.; Kwok, J.; Garbonell, J. Accelerated gradient method for multi-task sparse learning problem. In Proceedings of the IEEE International Conference on Data Mining, Miami, FL, USA, 6–9 December 2009; pp. 746–751.
34. Tseng, P. On accelerated proximal gradient methods for convex-concave optimization. *SIAM J. Optim.* **2008**, submitted.
35. Kang, X.; Li, S.; Fang, L.; Benediktsson, J.A. Intrinsic Image Decomposition for Feature Extraction of Hyperspectral Images. *IEEE Trans. Geosci. Remote Sens.* **2015**, *53*, 2241–2253. [CrossRef]

9

One-Dimensional Convolutional Neural Network Land-Cover Classification of Multi-Seasonal Hyperspectral Imagery in the San Francisco Bay Area, California

Daniel Guidici [1] and Matthew L. Clark [2,*]

1 Department of Engineering Science, Sonoma State University, 1801 E Cotati Ave, Rohnert Park, CA 94928, USA; daniel.giudici@gmail.com
2 Center for Interdisciplinary Geospatial Analysis (CIGA), Department of Geography, Environment and Planning, Sonoma State University, 1801 E Cotati Ave, Rohnert Park, CA 94928, USA
* Correspondence: mateolclark@gmail.com.

Academic Editors: Qi Wang, Nicolas H. Younan, Carlos López-Martínez and Prasad S. Thenkabail

Abstract: In this study, a 1-D Convolutional Neural Network (CNN) architecture was developed, trained and utilized to classify single (summer) and three seasons (spring, summer, fall) of hyperspectral imagery over the San Francisco Bay Area, California for the year 2015. For comparison, the Random Forests (RF) and Support Vector Machine (SVM) classifiers were trained and tested with the same data. In order to support space-based hyperspectral applications, all analyses were performed with simulated Hyperspectral Infrared Imager (HyspIRI) imagery. Three-season data improved classifier overall accuracy by 2.0% (SVM), 1.9% (CNN) to 3.5% (RF) over single-season data. The three-season CNN provided an overall classification accuracy of 89.9%, which was comparable to overall accuracy of 89.5% for SVM. Both three-season CNN and SVM outperformed RF by over 7% overall accuracy. Analysis and visualization of the inner products for the CNN provided insight to distinctive features within the spectral-temporal domain. A method for CNN kernel tuning was presented to assess the importance of learned features. We concluded that CNN is a promising candidate for hyperspectral remote sensing applications because of the high classification accuracy and interpretability of its inner products.

Keywords: hyperspectral imagery; 1-dimensional (1-D); Convolutional Neural Network (CNN); Support Vector Machine (SVM); Random Forests (RF); machine learning; deep learning; TensorFlow; multi-seasonal; regional land cover

1. Introduction

Land-cover maps provide information for natural resource and ecosystem service management, conservation planning, urban planning, agricultural monitoring, and the assessment of long-term land change. The automated classification of land cover from satellite imagery is a challenging task due to spectral mixing, intra-class spectral variability, and low spectral contrast among classes. Hyperspectral, or imaging spectroscopy, data consist of hundreds of spectral bands, and capture more spectral detail and variability relative to conventional multispectral sensors used for mapping land cover. Terrestrial hyperspectral applications have shown success in mapping composition, physiology, and biochemistry of vegetation, ecosystem disturbance, and built-up environments [1]. The analysis of hyperspectral data presents issues in classification, due to large data volumes, and increased spectral variability as recorded by hundreds of correlated bands [2]. Additionally, the classification task becomes more difficult when presented with the larger spatial extents and temporally detailed data collected by spaceborne hyperspectral sensors with repeat measurements.

A traditional classification method for hyperspectral imagery involves Multiple Endmember Spectral Mixture Analysis (MESMA), which consists of unmixing image spectra with pure spectral profiles (endmembers), and assigning the class through endmembers selected in the unmixing solution [3]. This family of methods typically requires regionally specific libraries of pure spectral profiles that are from field spectra, synthetically generated, or selected from large libraries of image spectra [4]. This type of classification is the most closed form solution available currently, analytically processing an exhaustive combination of endmembers that most closely match the data to be classified [3]. Further, MESMA assumes linear mixing, which is often violated by the interaction of photons with components within an individual pixel and from nearby pixels.

Machine learning is an alternative domain of classification techniques that can accurately distinguish land cover in hyperspectral and multi-seasonal imagery [5]. In contrast to MESMA, these classifiers can learn non-linear decision spaces and do not require training data optimization steps to select spectrally pure endmembers. There are many different varieties of machine learning algorithms implemented on different platforms. Picking the classifier to analyze a dataset at times falls to the user's familiarity with the computational platform and/or algorithm for a specific field. Random Forests (RF) and Support Vector Machines (SVM) are widely adopted machine learning classifiers in the remote sensing community [6,7]. These algorithms have provided robust results across many platforms and datasets, surpassing many other families and implementations of classifiers [8].

Convolutional Neural Network (CNN) is a leading machine learning classifier for image recognition tasks that use 2-dimensional (2-D) image data [9,10], such as identifying faces in photographs of people. Applications of CNN have extended into the classification of other contiguous data types, like speech recognition utilizing 1-dimensional (1-D) data [11]. In remote sensing, there have been several recent applications of CNNs and other similar "deep" network topologies [12–18]. The form in which CNNs are applied to remote sensing data can vary significantly depending on the data available, as there is no universal deep network classification architecture. The application of a CNN to classify land cover from 2-D visual images has been performed with good results in their respective applications. For example, Kussel et al. [15] achieved a 95% overall accuracy with land-cover classification and Li et al. [19] had a 96% correct detection of plants within a scene. In some applications, segmentation and previously learned features can be transferred and leveraged as part of the CNN classification task [12,20–22]. The method within [18,23,24] applies down-selected spectra to a 2-D CNN architecture, thereby mainly exploiting the spatial extent of the data. The spectral dimension is reduced because large number of features present with hyperspectral imagery typically poses a problem for some classifiers, and many studies use dimensionality reduction (DR) to aid in the classification process [13,18,23–26]. In [23–25] the spectral dimension was reduced, for example by applying principal component analysis [23,24], comparing salient band vectors in a manifold ranking space to consider hyperspectral data structure [25], or clustering similar bands and extracting features [26]. All these methods reduce the number of spectral bands before subjecting the data to the classifier. The method provided in [6] pre-calculates features based on physical and chemical composition. All these DR methods have shown an increase in classification accuracy as compared to using the full hyperspectral data. In contrast, some studies have shown increased classification accuracy when utilizing the full spectra, invalidating the need for complex DR preprocessing. For example, the work done in [13] shows the comparison between a 3-dimensional (3-D) CNN, which utilizes the full spectral dimension, and other methods that do not utilize the full spectral dimension. The 3-D CNN spanning the full spectral dimension increased accuracy by up to 3%. In this case, the CNN classifier determines what are the distinctive features from the initial data, without a DR pre-processing step. As another example, Hu et al. [17] utilized single-season hyperspectral data and 1-D CNN across the full spectral dimension to classify land cover with 90 to 93% overall accuracy, and CNN outperformed SVM by 1 to 3%. A distinction between the extraction of key spectra by using DR and spectral feature generation from the CNN is that the CNN approach extracts features based on spectral characteristics directly driven from reducing classification error. This makes CNN a promising method for the exploitation of the distinctive aspects of the spectral dimension of the data and warrants further investigation with hyperspectral land-cover mapping applications at various spatial and temporal scales.

A broad goal of this study is to assess the accuracy of a 1-D CNN for classifying land cover from multi-seasonal hyperspectral imagery. Accuracy from this CNN is compared to those from the two leading machine learning classification methods in remote sensing, RF and SVM. These methods were utilized as a control group due to their high accuracy rates, high prevalence within the field and the robust libraries available for their implementation. In order to support applications based on spaceborne hyperspectral imagery, our analyses were performed with simulated Hyperspectral Infrared Imager (HyspIRI) imagery, a satellite mission currently being considered by NASA. Our analyses are regional in scale, covering the San Francisco Bay Area, California, and land-cover classes followed the global Land-Cover Classification System (LCCS) [27].

A CNN architecture requires the data to be in a contiguous format as convolutional layers of the network distinguish, or filter, local features or patterns from neighboring regions throughout the data. The neural network then performs the subsequent classification based on these learned features. A potentially useful feature of CNN architectures is that the inner resulting data products, such as properties of convolutional filters, can provide some insight into what the classifier has learned to make its classification. The inner data products of CNN architectures as applied to 1-D hyperspectral data have not been discussed in the literature. Thus, another goal of this study is to show how the inner data products of CNN can provide insight into the classification task and the features extracted from the spectra through the training of the network. With the analysis of feature maps that the convolutional layer of this network creates, local regions within the spectral dimension of the data that are excited by the convolutional layer of the network can be shown to have an impact on the classification accuracy. The importance of these learned features can be then traced back to how important they are to the classification task, by zeroing them from the network and re-computing the accuracy. Additional processing of the feature maps extracted from the CNN enables an illustrative visual that reveals which spectral areas assist in separating the classes. By standardizing, scaling and capturing only the magnitude of the convolutional kernel feature maps, the spectral-temporal band importance can be visually explored.

2. Materials and Methods

2.1. Study Area

The study area was the San Francisco Bay Area in northern California, USA (Figure 1) and is described in Clark and Kilhman [6]. Natural vegetation in the study area includes evergreen needleleaf forests (conifer), evergreen broadleaf forests, deciduous broadleaf forests, mixed forests, shrublands and grasslands. Anthropomorphic land cover includes dense urban areas around the San Francisco Bay and perennial crops (e.g., vineyards, fruit orchards) and annual crops (e.g., strawberries, cotton, rice).

2.2. Simulated HyspIRI Imagery

Hyperspectral imagery for the ~30,000 km^2 study area was from NASA's Airborne Visible Infrared Imaging Spectrometer (AVIRIS) "Classic" sensor, flown in spring, summer and fall of year 2015 [1]. The AVIRIS-C sensor images spectral radiance in 224 bands from 370 nm (visible) to 2500 nm (shortwave infrared, SWIR) with 10 nm sampling and SNR of >1000:1 at 600 nm and >400:1 at 2200 nm [28,29]. There were twelve, ~12 km swath flight runs per season that provided 20% image overlap among runs. Complete spring and fall runs were used while different dates were used to make a complete, cloud-free summer dataset (Table 1, Figure 1).

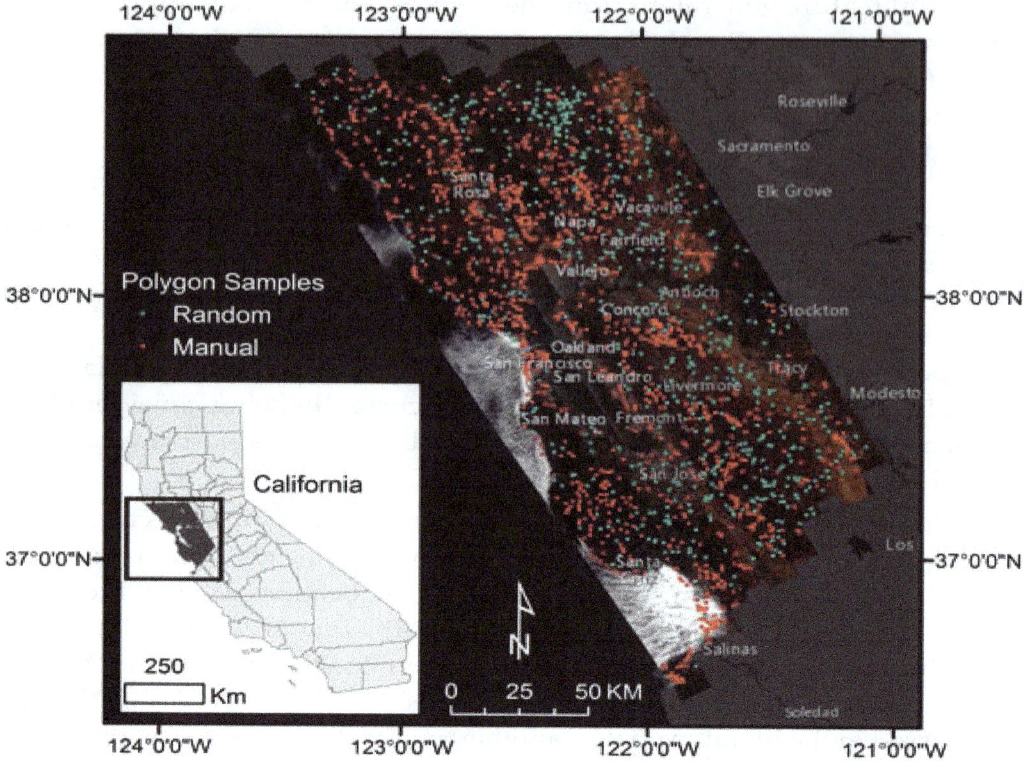

Figure 1. Study area overview with an AVIRIS-C, 11 June 2015 RGB mosaic of 12 individual flight runs. Reference data are red and cyan points. Inset shows the multi-seasonal image extent with water and cloud mask applied.

Table 1. Summary of San Francisco Bay Area AVIRIS-C images for the year 2015.

Season	Image Collection Dates (UTC)	Runs Used	Average Solar Zenith	Average Solar Azimuth
Spring	30 April (16:48–21:25)	Runs 06–17	26.7°	147.0°
Summer	11 June (18:30–23:42)	Runs 09–20	20.1°	227.2°
	15 June (21:32–22:51)	Runs 12, 15	33.3°	254.5°
Fall	2 October (17:19–22:32)	Runs 06–13; Runs 16–20	42.2°	191.9°

Hyperspectral images from AVIRIC-C were used to simulate HyspIRI as part of a preparatory science campaign [1,6]. The current configuration of HyspIRI is a satellite sensor with 30-m spatial resolution, 185 km swath width, and 16-day repeat global coverage [1]. The measurements would cover 380 to 2510 nm in $\leq$10-nm contiguous bands (~214 bands). The AVIRIS-C radiance data were processed by NASA's Jet Propulsion Laboratory into HyspIRI Precursor Data Products (Figure 1) and can be downloaded at http://aviris.jpl.nasa.gov. The 30-m simulated HyspIRI products include, in order: orthrorectification and 90 $\times$ 90-m Gaussian-weighted resampling of at-sensor radiance to 30 m pixels, addition of noise approximating a HyspIRI visible-SWIR Noise Equivalent Delta Radiance (NEdL) function, and ATREM-based per-pixel atmospheric correction and apparent surface reflectance retrieval [6,30]. Bands in the shortwave infrared with strong atmospheric water vapor absorption and poor signal-to-noise were removed, leaving 186 bands for analysis per season. Clouds and water in simulated images were masked prior to image classification using a Random Forests classifier developed in Clark and Kilham [6]. The analysis of multi-season data was performed on an image cube of seasonal spectral data stacked in temporal sequence (spring, summer, fall), resulting in an input vector of 558 bands; $N_d = 186\ (bands) \times 3\ (seasons)$.

2.3. Land-Cover Reference Data

Reference data for training and testing were collected using visual interpretation of high-resolution imagery in Google Earth. Detailed methods are found in Clark and Kilham [6] and summarized below. Data were percent cover of the following twelve "land-cover components" within a polygon: evergreen needleleaf trees (ENT); evergreen broadleaf trees (EBT); deciduous broadleaf trees (DBT); shrubs; herbaceous; dune vegetation; tidal marsh; annual crops; perennial crops; impervious surfaces; urban landscape; and, bare non-vegetated (beaches, dunes, rocks, bare soil). Percent cover data were visually estimated in 10% intervals using high-resolution Google Earth imagery in 100-, 250- or 500-m square polygons. Different polygon sizes were chosen depending on patch size in order to maximize pixels collected in large patches (e.g., 500-m square chosen), while minimizing mixed pixels in relatively small patches (e.g., 100-m square chosen). Polygons were located over areas of well-mixed land-cover components, with most centers further than 1000 m to a neighboring sample. Initially, a simple random method was used to locate samples, which led to under-represented samples of some classes. This necessitated manual placement to ensure an adequate sample of class types (Figure 1). Each polygon was classified into one of twelve discrete Land-Cover Classification System [27] classes using a decision-tree of rules applied to a polygon's percent cover data (Table A1;[6]). Open-canopy trees (woodlands) and shrubs (shrublands) have >10–65% tree or shrub cover, respectively; closed-canopy trees (forests) and shrubs (thickets) have >65% tree or shrub cover, respectively. In this study, we focus on mapping closed-canopy tree and shrub classes.

There were 1495 total reference polygons for the twelve LCCS classes. Reference reflectance spectra were extracted from polygons overlaid on the three-season image cube for each run (Figure 2). To exclude potentially mixed pixels at edges and minimize geolocation error, polygons were buffered inward using a half-pixel buffer, and only pixels that were 100% contained by the buffered polygon were selected. Due to scene overlap, there were some polygons that were located in two images. In these cases, all available pixels were extracted. Reference data were split at the polygon level into training and testing sets using the same polygon designations found in Clark & Kilham [6], except polygons in areas of fires between years 2013 and 2015 were removed from the analysis. Training spectra were then filtered at the polygon level to remove outliers following methods in Clark & Kilham [6]. Single- and three-season data were processed for outliers separately for each group of variables. There were a total of 71,362 training pixels and 52,510 testing pixels (Figure 3).

2.4. Classifier Architectures and Tuning

Three machine classifiers were assessed in this work: Random Forest (RF), Support Vector Machine (SVM) and Convolutional Neural Network (CNN). We used Python's Scikit-learn machine learning library for training and classification. This library provides access to established data mining and data analysis tools and algorithms for implementing RF and SVM. The adaptations of the respective CNN architectures require a relatively flexible platform for development. A CNN is a deep network topology that can be formed as a computational graph and Google's open-source computational platform, TensorFlow™, is directly suited for this application. In this study, we used TensorFlow™ for the CNN definition, training and classification portion of this work. This platform had the additional advantage of using the GPU to speed data processing, whereas Scikit-learn relied solely on the CPU.

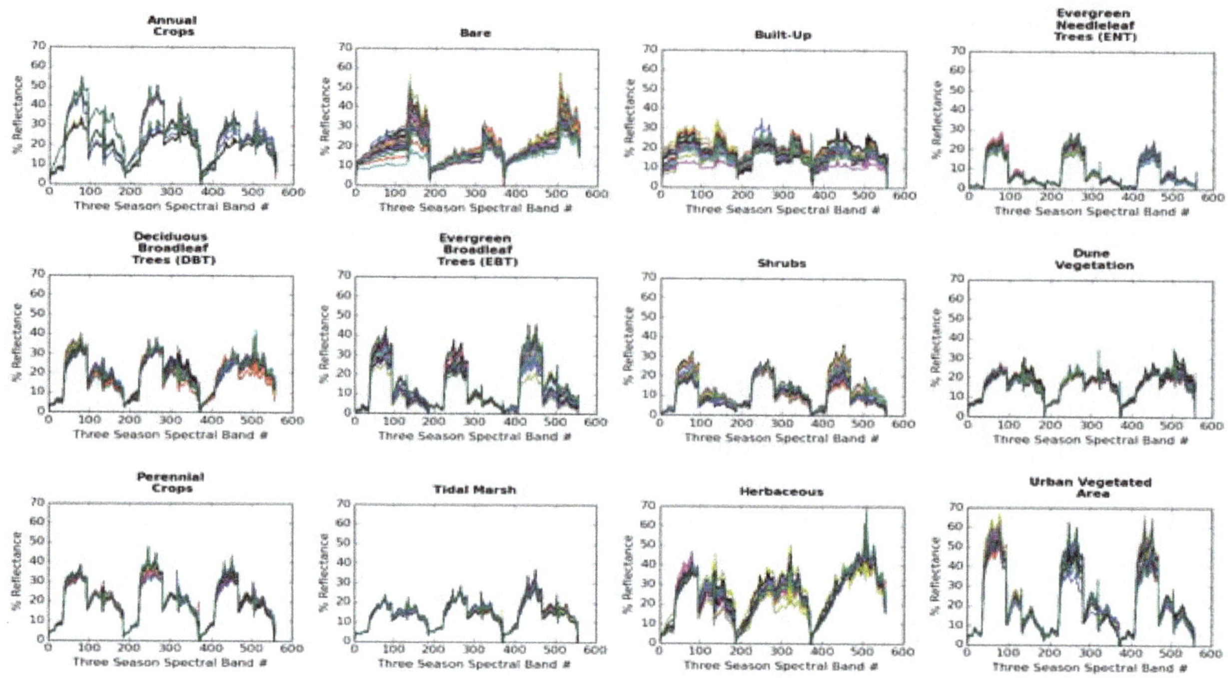

Figure 2. Twenty-five randomly selected three-season reflectance spectra from each of the twelve LCCS classes. Note that the x-axis is the band number (1–558) with seasons in spring- summer-fall sequence. Bad bands have been removed within each season (186 bands per season).

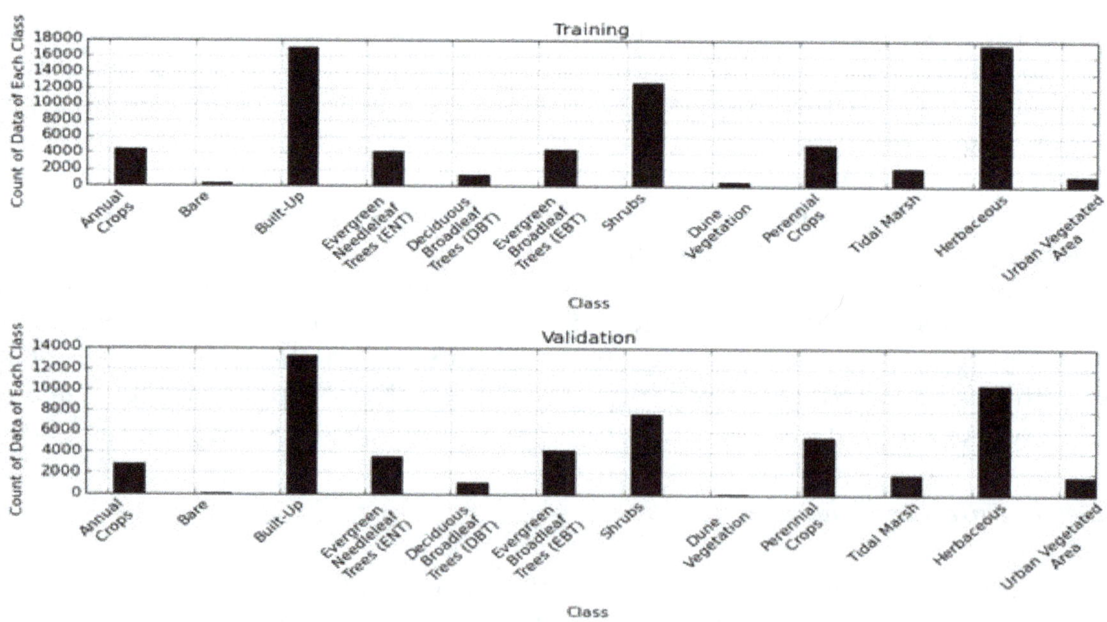

Figure 3. Training and testing reference data distributions.

Below we provide a brief background and explanation of the RF, SVM and CNN classifiers and their respective hyper-parameters. An overview of the method is shown in Figure 4. Training and testing reference data included duplicate pixels from areas of scene overlap, thus providing spectral variation for differences in sensor view and sun angle in the classification process. We did not have a validation dataset due to the limited number of samples of some classes within the data. Training data were thus used to train the classifier, while hyper-parameter tuning was performed by observing training and test data accuracies. The final accuracy assessment was performed by extracting the most trusted classification from each classifier in the areas of scene overlap. This consists of utilizing data classifications that result from classifications that have the highest confidence for RF or highest probability for SVM and CNN (Section 2.6; Figure 4, LCCS-labeled Land-Cover Samples).

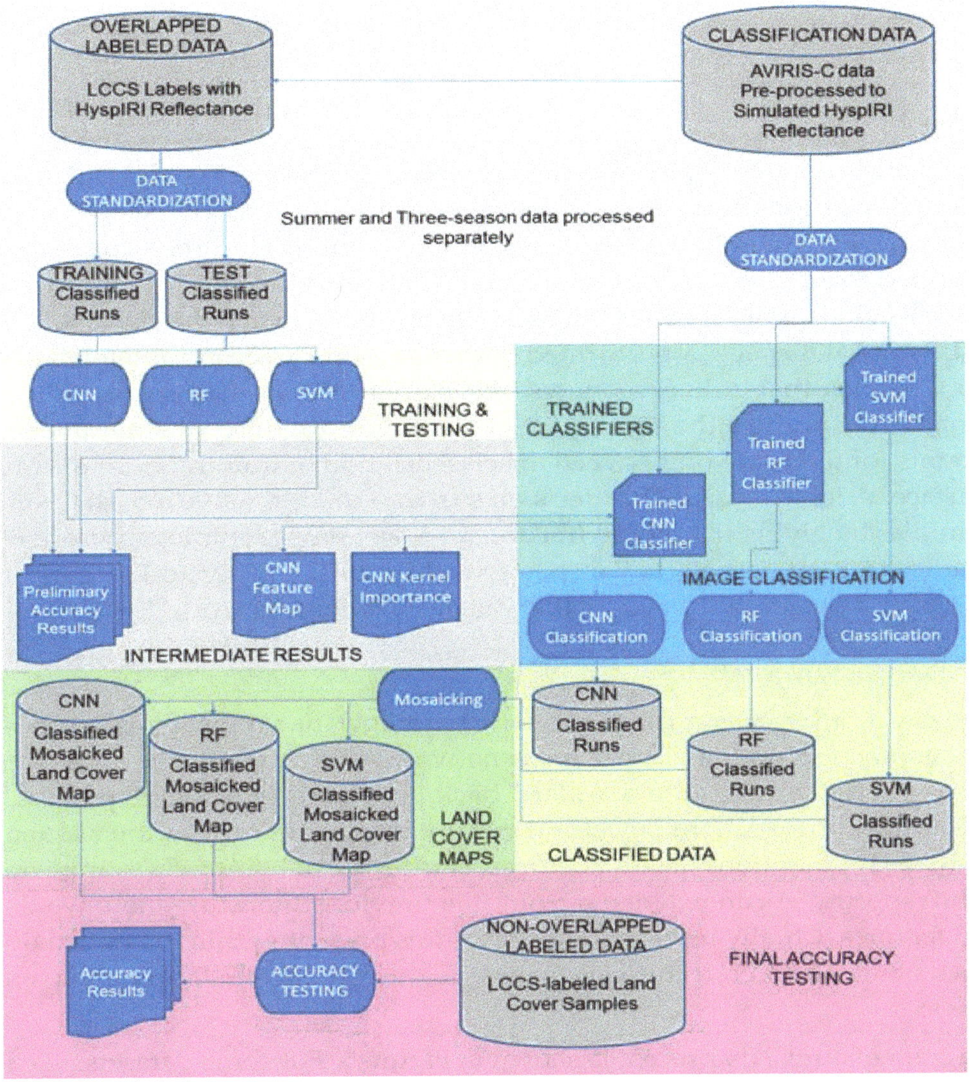

Figure 4. Overview of machine learning classification and accuracy assessment methodology.

2.4.1. Random Forest

The RF classifier is a technique of training an ensemble of decision trees through randomized draws of training data. Once the ensemble of trees is created, data are applied to the classifier and the prediction across the individual trees is captured [31]. The plurality of predictions is the resulting classification of this algorithm.

There are three main parameters when tuning a Random Forest classifier: the number of estimators (or trees), the maximum number of features utilized at each decision point within a tree and the minimum samples at each leaf. As a general rule, the number of features utilized in RF classifiers is $\sqrt{N_f}$, where N_f is the number of features. In hyperspectral imagery, each feature is an individual reflectance band. The number of trees included in the ensemble typically improves classification accuracy until a critical point is reached, beyond which accuracy does not increase further. For the minimum leaf size, smaller values result in RF classifiers that are more prone to capturing noise in the training data. Optimization of these parameters should be considered when training a RF classifier on a particular dataset [32].

For this work the standard $Max\ Number\ of\ Features = \sqrt{N_f}$ was utilized. The number of trees was increased until there was not a significant change in prediction accuracy, resulting in a parameter of 1000 trees. A search across the minimum leaf sizes was performed, resulting in a parameter of 1 sample in leaves.

2.4.2. Support Vector Machine

The SVM classifier separates training data by defining a hyperplane(s) through the data to segment the classes. In two dimensions, this can be visualized as a line drawn between two classes that defines the maximum separation in spectral space. For non-linear and high dimensional classification tasks, the data are mapped to an even higher dimensional space and the "line" that is drawn between the classes is, in effect, a plane which creates the most separation between the classes. How this mapping to the higher-dimensional space is performed is dictated by hyper-parameters. A standard linear Support Vector Classifier that utilizes a one-vs.-rest classification scheme was utilized for this work. This implementation simplifies the tuning of the classifier down to one parameter, cost (C). This parameter accounts for the trade-off between misclassification of training examples and simplicity of the decision surface. A low C makes this decision surface smooth, while a high C aims at classifying all training examples by giving the model freedom to select more samples as support vectors [33]. A linear search for this tuning parameter was performed from 1.0×10^{-2} to 1.0×10^{10} and the highest accuracy classifier was chosen; a C value of 0.1 was utilized for this work.

2.4.3. Convolutional Neural Network

At a broad level, a CNN is a deep-network topology that typically combines convolutional filter layers in conjunction with a classification network, which for this work is a fully connected Neural Network (NN). Through the standard back-propagation training process, convolutional filters are trained to capture salient structural feature information from the sequential input data. Hu and colleagues [17] mention these structural features as the "intraclass appearance and shape variation" within spectra. As an extension from this previous literature, here we demonstrate that these structural features actually represent those features present within the data that distinguish the classes from each other. The architecture feeds these features, or filtered input data, into a subsequent classification process.

A flow diagram of our CNN process is shown in Figure 5. For some datasets, like those consisting of 2-D imagery, larger networks are required.

CNN Architecture

The feature generation and feature classification nature of CNN can become complex with multiple layers that extract different levels of features from the data. A few examples of more complex CNN architectures as implemented for image classification tasks can be reviewed in [9,10]. These complex networks required the learning of a large number of features and a high level of neural network complexity. These traditional CNN architectures are implemented in two dimensions (i.e., 2-D, width and a height). The convolutional operation for these networks occurs with multi-dimensional kernels and in various configurations, with the goal to classify the full or subsets of the image. In contrast, hyperspectral data classification can be a pixel-based operation spanning only the spectral dimension (i.e., 1-D), or in the spatial and spectral dimensions (i.e., 3-D). To simplify implementation and compare results to RF and SVM, we chose to apply the CNN across the spectral domain of hyperspectral pixels; and thus, the convolutional operation of the network only needs to operate on this single dimension. This in turn greatly simplifies the architecture of the hyperspectral CNN relative to 2-D or 3-D implementations [9,10,13,15].

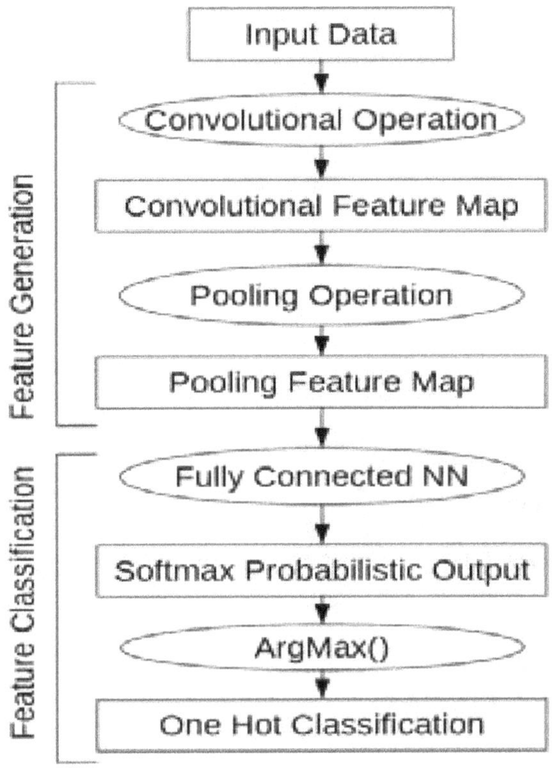

Figure 5. Convolutional Neural Network (CNN) flow diagram.

Our CNN architecture resembles a network that has been recently introduced for hyperspectral data classification in [17]. The network consists of a single convolution layer paired with a pooling layer that feeds a fully connected neural network. Our network was further refined by adding Dropout and Regularization of the fully connected layer [34,35]. Additionally, here we tune the convolutional layers based on a technique that reduces the number of kernels trained within the network by accounting for their impact on the classifier accuracy.

The single convolutional layer accepts the 1-D spectral profile input data and performs the convolutional operation on the input data with each kernel in the architecture (Figure 6). This filtering of input data with each kernel creates the features for classification. In our implementation of CNN, the number of initial kernels to be trained was set at 86, the number of single-season engineered metrics used in our previous research with these hyperspectral data and the RF classifier [6]. In that study, engineered "spectral metrics" targeted vegetation biochemical and structural properties found in reflectance spectra, and perform a similar function as the kernels for this network. Because the learned features within the CNN are not tied to any pre-selected spectral features, rather only spectral properties that best distinguish the classes, the number of kernels in the CNN was drastically reduced by evaluating kernel importance (Section 3.2).

The pooling layer can be thought of as a spectral down-sampling of the convolutional feature map (Figure 6). A Max Pooling operation was utilized here. This layer accepts the convolutional feature map, evaluates pairs of data elements across the spectral dimension of the feature maps and passes the maximum value onto the next layer. This down-samples the data by a factor of two while preserving the maximum excitations from the convolutional feature map. This operation reduces the size of the feature map while preserving the features observed within the convolutional feature map. This provides a level of invariance to the spectral location of feature excitation within the feature generation network and reduces the overall number of connections to the fully connected network; and subsequently, this process will decrease the total number of trainable parameters for the network, thereby reducing overall training time.

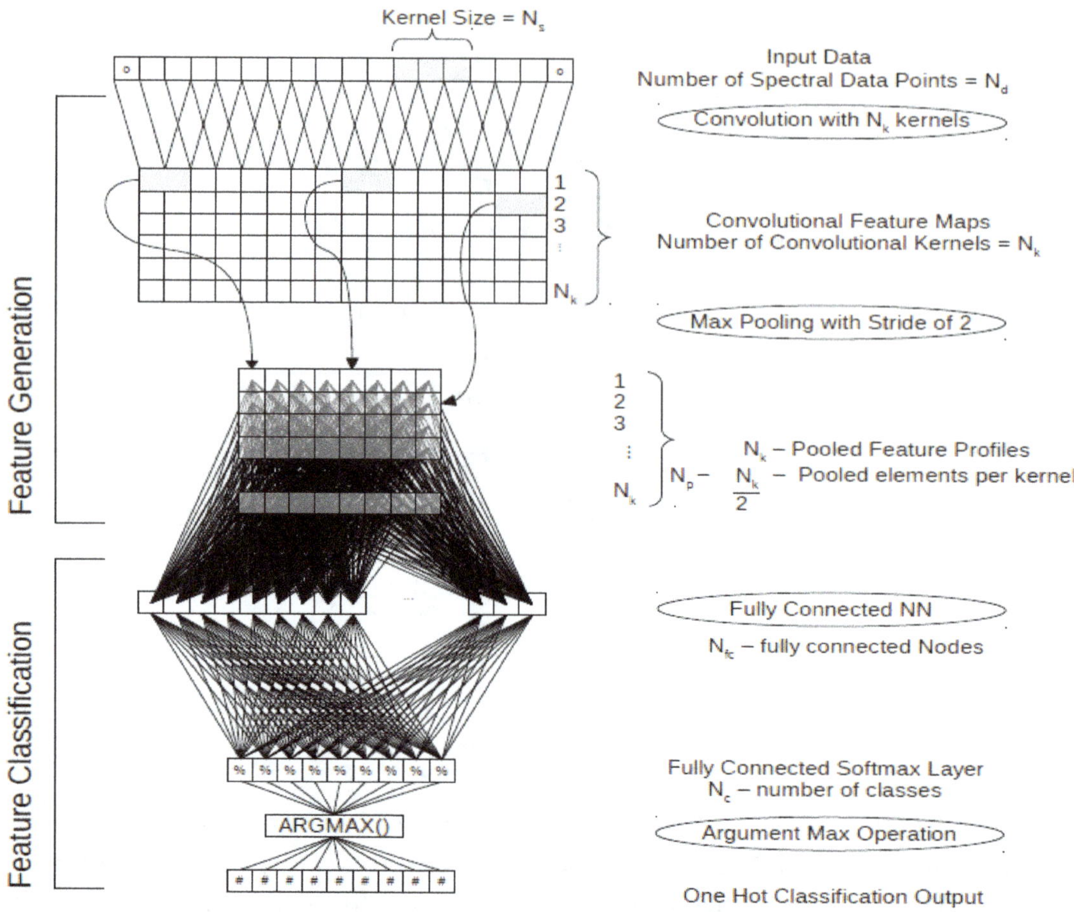

Figure 6. Detailed Convolutional Neural Network (CNN) architecture.

This Pooled Feature map is then provided to the feature classification network (Figure 6). In this architecture the feature classification network consists of a fully connected neural network. This fully connected network layer consists of a hidden layer with 500 nodes. This network's learning is regularized by including standard neural network regularization techniques; specifically, dropout with a 50% dropout level and L2 regularization on the weights connecting the pooled feature map and the fully-connected NN's hidden nodes. These two techniques enable the classification network to learn information throughout the fully connected network and encourage "smaller" weights to be utilized within the classification network. This ensures that individual weights between the two networks (feature generation and feature classification) are not exorbitantly larger than the other weights within the same layer, helping to prevent overfitting of the training data. By applying a penalty for large weights at this layer, the network is encouraged to learn information through all of the kernels and not just a few of them.

The output of the hidden layer is connected to a final Softmax output layer that produces a probabilistic output per class, or a vector of length of the number of classes, with each value representing the probability that the input data belongs to a specific class (Figure 6). This probability or confidence that the Softmax layer calculates is shown in Equation (1).

$$\sigma(z)_j = \frac{e^{z_j}}{\Sigma_{k=1}^{K} e^{z_k}} \; for\, j = 1, \ldots K \tag{1}$$

where Z_j are the inputs from the previous fully-connected layer applied to each Softmax layer node and K is the number of Softmax layer nodes (i.e., the number of classes). Finding the location of the argument with the largest probability value via the Argmax() function provides a one-hot representation of the class (Figure 6).

Stochastic gradient decent was utilized for training of the CNN. During training, this learning method subjects randomized labeled data to the network and calculates a loss function. This loss is then propagated back through the network to modify the network's interconnections, based on the impact the respective weights had on the previous epoch's classification. This is the gradient of the network, and is essentially the degree to which the weights of the network have impacted a given classification and resulted in a respective loss. Modifying the network based on the loss calculated by the classification or misclassification of the labeled data is effectively the back-propagation algorithm. As training progresses, the back-propagation modifies the interconnection of the network to reduce the loss function. This decreasing of the loss function indicates the classified data more closely matches the labeled data. Learning curves are provided in Appendix Figures A2 and A3 for the training and testing data, respectively, and show overfitting is not occurring within the testing data.

The TensorFlow™ platform has built-in implementations for calculating the gradients of the network and the minimization of the loss function to be back-propagated through the network. The Adagrad adaptive learning rate algorithm was utilized from the TensorFlow™ platform. The usage of an adaptive learning rate provided a significant boost to the performance of this architecture on the order of 4% overall classification accuracy. Additionally, the Adagrad adaptive learning rate greatly sped up the training of the network as compared to fixed and the exponential decaying learning rate optimizers. The Adagrad optimizer adjusts the learning rate, the extent the network can be modified, based on the gradient recently seen by the network. As this optimizer is suited for relatively sparse data and the dataset analyzed here is relatively unbalanced (Figure 3), this optimizer seemed to be an appropriate fit for this architecture. Empirical testing proved that this algorithm reached a maximum accuracy with a seed-learning rate of 0.1.

Hyper-Parameter Tuning

Dominant tunable hyper-parameters of this architecture are: K_s the kernel size, N_k the number of kernels, and N_h the number of hidden nodes within the classification network (Figure 6). While other characteristics can be modified within the network, these three hyper-parameters were determined to be the most influential. Other parameters such as dropout level and regularization level on the hidden weights could be adjusted, but our experience was that just having these parameters defined would increase the accuracy of the network. We used values of 50% dropout and 1×10^{-4} for regularization.

The kernel size dictates the size of the feature to capture. It is the size of the local receptive field considered within the spectral dimension for the convolution with the data. A rule of thumb originating from the currently engineered features utilized in [6] is to use local receptive fields that are roughly 10 bands long (10 bands $\times$ 10 nm spacing = 100 nm) for a single season. Informal experiments indicated that K_s could change in size as long as the Number of Kernels, N_k, varied inversely with this value. We believe that this is because the modification of these two parameters effectively varies the capacity of the network to learn the data, as long as the network is sufficiently able to learn the parameters equivalent accuracy should be able to be achieved. Due to familiarity with local receptive fields of 10 in prior work this was used as the kernel size in our CNN implementation.

The number of kernels contained within the network represents the number of features able to be learned. As the capacity of the network to learn features is a combination of the number of kernels and the kernel size. If each of those features are very descriptive, and has a large kernel size, then it would be expected that fewer of them would be required to achieve similar results. With the kernel size set, it was determined that the number of kernels can be reduced until zeroing a kernel from the classifier always has negative effect on overall accuracy. This ensures that the feature maps with respect to each kernel are excited and that each kernel is important to increasing overall accuracy.

Evaluating Kernel Importance

A main goal of training deep network topologies is to appropriately load the weights of the network. If a network has too many nodes or has too many layers, it may not effectively learn the appropriate weights for its interconnections, in effect not learning the decision space. As applied to CNNs, too many kernels can produce excess capacity that does not contribute to an increase in overall accuracy of the classification. With that in mind, we developed a way to verify that each kernel contributes to an increase in accuracy. CNNs can be evaluated with a similar method to the Mean Decrease Accuracy utilized in RF to determine feature importance. By zeroing out each convolutional filter or kernel and assessing the effect on the overall accuracy, information on how important that kernel is to the classification task can be determined. Kernels with significant impact on accuracy can be regarded as containing information of distinguishing characteristic(s) of the data.

In our study, the number of kernels trained was reduced to ensure that all the kernels were important to some degree. Each time the number of kernels was reduced the network was retrained with the training data to appropriately learn the new kernels. It was determined that the number of kernels trained within the network could be reduced until the Kernel Importance Table always showed a negative average percent impact on the classification accuracy for all kernels (Section 3.2). Reducing the number of kernels beyond this started to have a negative impact on the overall classification accuracy. While this was determined experimentally, it proved to be useful in reducing the total size of the network. Additionally, as the learned information is more condensed in fewer trainable kernels, the features that each kernel is extracting should be more expressive. With this kernel reduction performed, we expect that the network does not have too much excess capacity and the likelihood of over training is minimized. In this study, we started with 86 kernels and were able to reduce this drastically. Acceptable parameters for a three-season CNN were: $N_k = 7$ and $K_s = 10$. These parameters were also applied to the single-season CNN.

Hidden Layer Nodes

The number of hidden layer nodes within the classification portion of the network determines the capacity of the classification network to make an accurate "combination" of the features learned. It is important that this network has enough capacity with the upper limit being mainly to keep the size of the network only large enough to not limit the classification accuracy of the network. Here we chose 500 hidden layer nodes, N_h.

Convolutional Filter Visualization

Useful visualizations for CNN are the feature maps generated by the network. Feature maps are created by applying individual pieces of data to the network and extracting the resulting convolutional excitations for each class of data. When a single piece of data is applied to the network, every kernel is convolved with this data point. This creates the $N_d \times N_k$ convolutional feature map. Visualizing this convolutional feature map enables the identification of where within the temporal-spectral signature the data is being excited or filtered by a respective kernel. This is the information that is provided to the feature classification network.

As feature maps are connected to the same spectral locations within the hidden layer, if the feature maps from each of the classes all are excited in the same area then those excitations do not provide any distinctive information to the classification network and would have little effect in determining the class of the data. Thus, for visualization purposes the mean across all class feature maps was removed from individual class feature maps. As any deviation from zero within the modified feature map indicates that the kernel learned something in that region, viewing the magnitude of the mean-removed feature map provides a more useful view. Additionally, averaging feature maps from 75 spectra of the same class provides the average excitations from convolution with those data, showing general trends for each class. The resulting visualization feature maps can then shed some light on the distinguishing characteristics of the class, on a class-by-class basis. To provide context for the feature map, a random piece of data from the feature map's class is provided as a silhouette. This was done to illustrate the typical structure of that class's spectral profile across all three seasons. An example is provided in Figure 7 for Annual Crops.

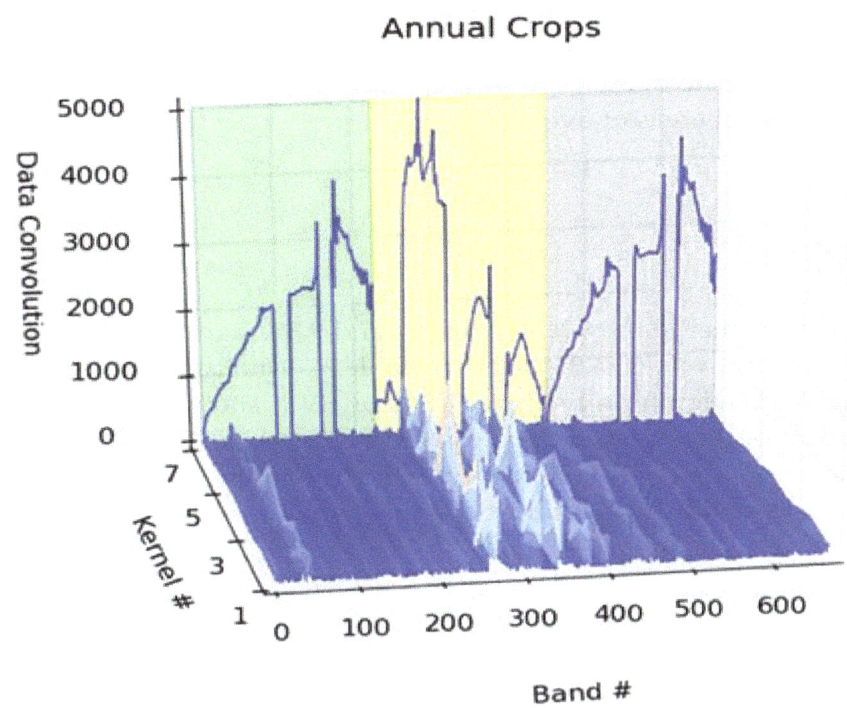

Figure 7. An example convolutional feature map for an Annual Crop three-season spectrum. The example class spectrum is shown in background. Green, yellow and gray background areas represent bands in the spring, summer and fall, respectively. Each season has 186 bands, ordered 370 to 2500 nm. However, in this figure the original 224 bands per season are shown in order to display the bad data gaps (e.g., atmospheric absorption windows).

2.5. Data Pre-Processing

Standardizing data for neural networks is a common practice and enables the network to operate on data that has a similar dynamic range [36]. This frees the network from having to "learn" this dimension or characteristic of the data. The zero mean and standardization of the data with respect to spectra effectively removes common structural content from the data and then scales the data to have an appropriate dynamic range on a per spectra basis. The formula below is the operation performed for this standardization for each spectral band.

$$X_{i\_newspectra} = \frac{x_{i\_rawspectra} - \mu_{i\_rawspectra}}{\sigma_{i\_rawspectra}} \quad for\ i = 1, \ldots N_d \tag{2}$$

where N_d is the number of spectra or data, $\mu_{i\_rawspectra}$ and $\sigma_{i\_rawspectra}$ are the mean and standard deviation of all the training at the spectral index i.

This standardization of the data boosted performance of the CNN roughly 1% in classification accuracy and made the training more stable. These same standardized data were applied to RF and SVM but did not make any noticeable impact on accuracy of these classifiers (Figure 4). As per standard practice, the mean and standard deviations for this standardization were developed from the training dataset and then extended to condition the testing and classification image data. This is to prevent standardization knowledge of the testing data to influence the training of the classifier.

2.6. Classification Post-Processing and Accuracy Assessment

The final classified LCCS map products were created from mosaics of classified runs from respective classifiers (Figure 4). In areas of scene overlap, the map class was determined by selecting the pixels with the maximum number of votes (RF) or highest probability (CNN and SVM). An accuracy assessment was conducted using confusion matrices, overall percent accuracy and kappa statistics based on respective classifiers applied to independent test data. In this case, test data did not include duplicate pixels in areas of scene overlap; that is, the class from the respective mosaicked map was chosen as the reference class in areas of overlap.

3. Results

3.1. Accuracy Assessment

With single-season (summer) data, CNN had 0.5% and 9.3% significantly higher overall accuracy (OA) than SVM and RF, respectively ($Z > 3.2$, $p < 0.01$, Table 2). With three-season (spring, summer, fall) data, CNN had 0.3% and 7.3% significantly higher OA than SVM and RF, respectively ($Z > 2.1$, $p < 0.05$, Table 2). Overall accuracy for three-season data was 1.9 to 3.5% significantly higher than single-season data for all three classifiers ($Z > 2.1$, $p < 0.05$). Given their superior performance, remaining results will focus on three-season CNN and SVM classifications.

Table 2. Overall percent accuracy per classifier using single (summer) or three seasons (spring, summer, fall) of data. Kappa statistics are provided in parentheses.

	CNN	SVM	RF
Single Season	88.0 (0.86)	87.5 (0.86)	78.7 (0.75)
Three Season	89.9 (0.88)	89.5 (0.88)	82.2 (0.80)

The three-season CNN map had average producer accuracies (PA) of 76.7%, and that ranged from 0.0 (Dune Vegetation) to 98.3% (Built-up, Table 3). The CNN map user accuracies (UA) averaged 80.0% and ranged from 0.0 (Dune Veg.) to 97.9% (Tidal Marsh). The SVM map had average PA of 73.5%, and that ranged from 0.0 (Dune Vegetation) to 98.8% (Built-up, Table 4). The SVM map UA averaged 80.0% and ranged from 0.0 (Dune Veg.) to 97.9% (Tidal Marsh). The class accuracy for Dune Vegetation was 0.0% due to low class sample size ($n = 132$ pixels) and spectral confusion with Herbaceous, Annual Crops, Bare, and DBT (Tables 3 and 4). Being located along the coast, Dune Vegetation was not prevalent in the study area and was thus underrepresented in reference data. Annual Crops had 12.3% greater PA with CNN over SVM, with confusion in both classifiers spanning Perennial Crops, Herbaceous and Urban Vegetation. Bare had a 35.3% greater PA in CNN over SVM, with greater confusion with Annual Crops and Built-up in SVM. In both classifiers, deciduous broadleaf forests (DBT) tended to be confused with evergreen broadleaf forests (EBT). Evergreen broadleaf forests were confused with conifers (ENT), deciduous forests (DBT), and shrubs.

The mosaicked classified maps created with RF, CNN and SVM are shown in Figure 8. At this scale, it is visible that all classifiers map land cover relatively well, and are relatively consistent throughout their respective classification. In general, the RF map tended to be more speckled than SVM or CNN maps. The RF map had more obvious problems with classifying shrublands in the southeast corner of the map (Figure 8B), with over-mapping of Perennial Crops; RF also had over-mapping of Built-up in the Tidal Marsh areas along the northeast bay. Some class differences toward the coast, such as near San Francisco, are explained by clouds in the imagery that were not fully masked from the analysis.

Table 3. CNN confusion matrix for the three-season classification.

	Annual Crops	Bare	Built	ENT	DBT	EBT	Shrubs	Dune Veg.	Perennial Crops	Tidal Marsh	Herbaceous	Urban Veg.	Total	User
1. Ann. Crops	3513	62	35	0	0	0	0	45	394	110	41	52	4252	83%
2. Bare	3	479	30	0	0	0	0	22	16	5	1	10	566	85%
3. Built-Up	33	106	13,210	4	2	0	5	0	70	77	6	617	14,130	93%
4. ENT	1	0	2	11,101	0	302	9	0	17	3	0	2	11,437	97%
5. DBT	0	0	0	2	1032	431	16	0	2	4	20	3	1510	68%
6. EBT	2	0	2	394	369	5297	100	0	1	2	3	4	6174	86%
7. Shrubs	11	0	5	262	24	350	7066	0	4	29	156	16	7923	89%
8. Dune Veg.	3	0	0	1	0	0	8	0	0	6	0	0	18	0%
9. Per. Crops	631	0	9	0	10	6	8	0	5359	29	3	37	6092	88%
10. Tidal Marsh	2	0	0	0	0	0	3	1	0	4930	98	2	5036	98%
11. Herbaceous	876	73	52	1	3	21	52	64	148	467	9357	66	11,180	84%
12. Urban Veg.	57	0	92	8	1	17	0	0	0	32	0	1772	1979	90%
Total	5132	720	13,437	11,773	1441	6424	7267	132	6011	5694	9685	2581		
Producer	68%	67%	98%	94%	72%	82%	97%	0%	89%	87%	97%	69%		

Table 4. SVM confusion matrix for the three-season classification.

	Annual Crops	Bare	Built	ENT	DBT	EBT	Shrubs	Dune Veg.	Perennial Crops	Tidal Marsh	Herbaceous	Urban Veg.	Total	User
1. Ann. Crops	2881	256	15	0	0	0	0	21	226	115	38	6	3558	81%
2. Bare	0	225	7	0	0	0	0	25	0	0	0	0	257	88%
3. Built-Up	83	223	13,272	151	1	2	2	7	61	78	65	575	14,520	91%
4. ENT	1	0	1	11,438	1	411	3	0	0	5	0	4	11,864	96%
5. DBT	18	0	4	0	1047	230	11	67	0	2	8	0	1387	75%
6. EBT	3	0	0	131	351	5351	59	0	11	0	0	2	5908	91%
7. Shrubs	0	0	1	34	17	389	7145	0	1	22	129	3	7741	92%
8. Dune Veg.	25	0	0	6	0	1	1	0	0	0	2	0	35	0%
9. Per. Crops	678	5	19	0	20	9	1	0	5560	61	19	65	6437	86%
10. Tidal Marsh	10	0	10	0	0	0	2	1	0	4856	74	5	4958	98%
11. Herbaceous	1124	11	16	1	4	28	43	11	152	524	9307	109	11,330	82%
12. Urban Veg.	309	0	92	12	0	3	0	0	0	31	43	1812	2302	79%
Total	5132	720	13,437	11,773	1441	6424	7267	132	6011	5694	9685	2581		
Producer	56%	31%	99%	97%	73%	83%	98%	0%	92%	85%	96%	70%		

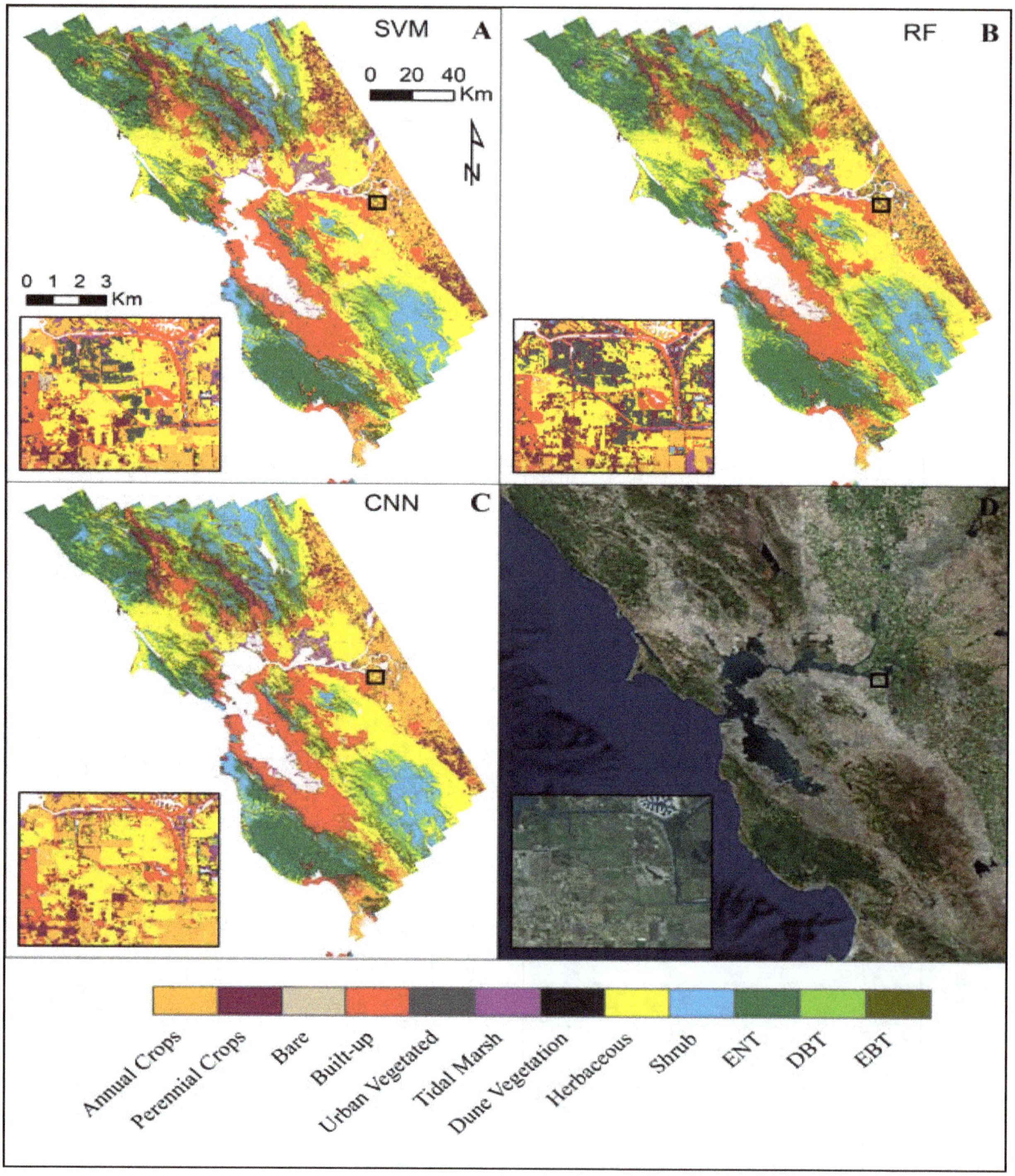

Figure 8. Classified land-cover maps for (**A**) Support Vector Machine; (**B**) Random Forests; and (**C**) Convolutional Neural Networks. White areas indicate pixels that were not classified (e.g., water, clouds, no data); (**D**) Natural color mosaic of imagery from June 2015.

The insets show an example of the classifications in the Bay Area delta agricultural zone. This area is predominantly agricultural land with some urban areas. In this anthropogenic landscape, homogenous geometric shapes are dictated by property lines (Figure 8D). The RF classifier performs the worst with many areas with more speckle (i.e., more heterogeneity) of classes across the landscape. The RF and SVM maps both misclassified several parcels with Urban Vegetated that were mapped as Annual Crops by CNN. This is an area of irrigated sod production, which has temporal and spectral similarities to irrigated golf courses found in the Urban Vegetated class. Despite this spectral similarity, CNN tended to correctly map these areas as agriculture.

3.2. CNN Kernel Importance and Visualization

The three-season CNN kernel importance matrix (Figure 9) shows the impact of zeroing a kernel from the CNN on class producer accuracy. For example, conifer forests (ENT, class index 4) producer accuracy decreases 95.2% if Kernel 7 is removed from the CNN. With Kernel 4 removed, six classes have a 15.9 to 65.31% decrease in accuracy. In contrast, some kernels have a positive impact if removed from the network. For example DBT (class index 5) and Urban Vegetation (class index 12) producer accuracy increases 6% and 10.8% if Kernel 5 is removed from the CNN. On Average, the most important kernel was Kernel 7 while the least important was Kernel 5.

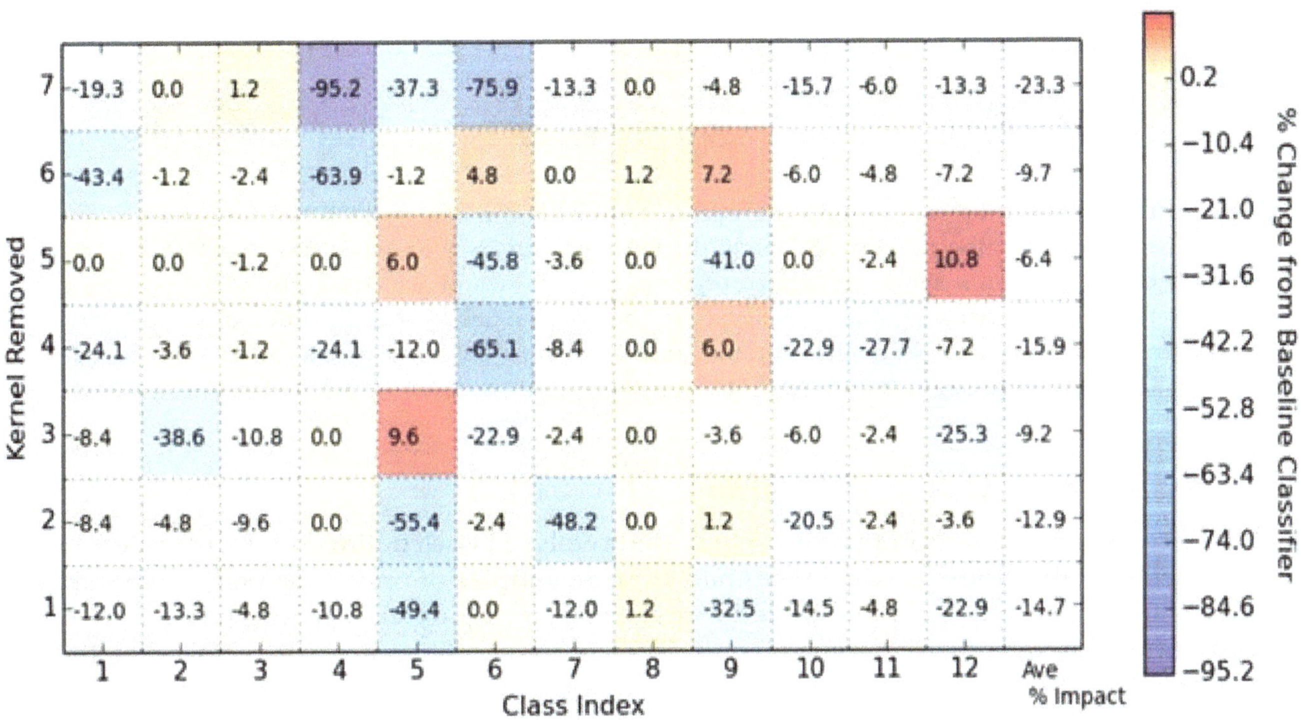

Figure 9. Kernel Importance Matrix. This table shows the percent change in producer accuracy when zeroing a kernel from the CNN. Class index definitions found in Table 3.

3.3. CNN Feature Map Visualization

Convolutional feature maps show, for an example spectrum from each class, kernel excitations subtracted from the average excitation across all classes at a given wavelength (Figure 10). Figure 10 shows the average feature map excitations from 75 randomly selected spectra per class. In general, feature maps show that kernels tend to excite in specific spectral-temporal regions for a given class, forming strips in the feature map. At these regions of activity, the kernels tend to be excited at different magnitudes (e.g., peaks and valleys along a strip).

By observing under which season of the feature map the excitations occur, distinctiveness of phenological spectral variation can be observed within classes. This effectively enables the feature maps to show under which season features are more prominent for respective classes. For example, for the Annual Crop spectrum there is heavy kernel excitation in the summer (Figure 10). The forest spectra (EBT, DBT, ENT) have excitations throughout the seasons. In contrast, Tidal Marsh and Shrubs spectra have relatively low excitation throughout the year.

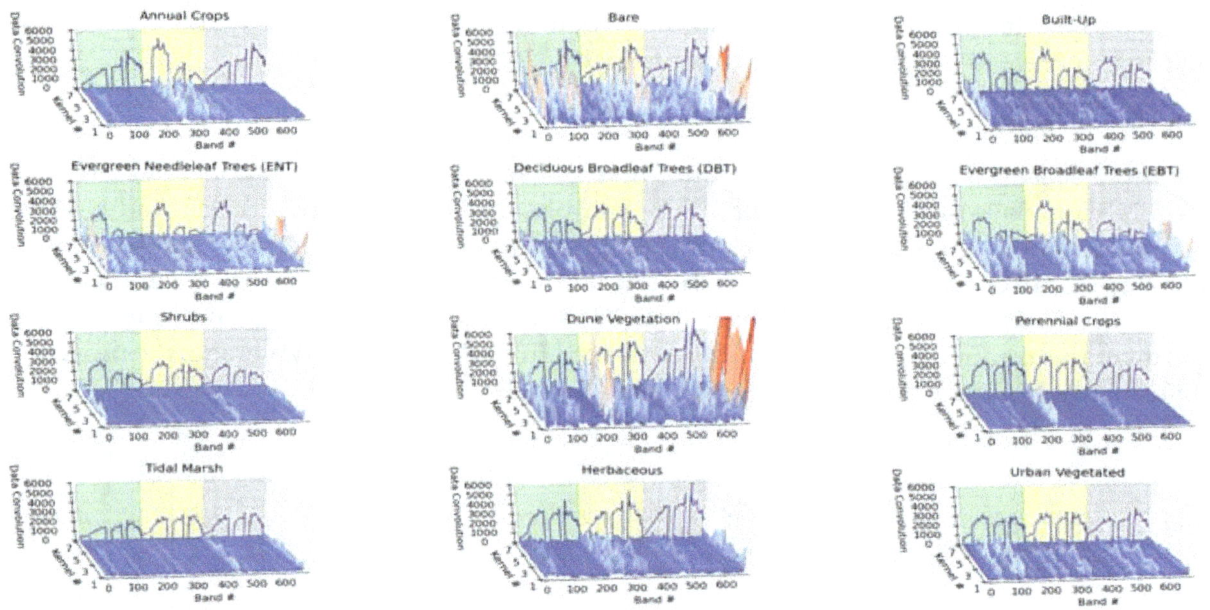

Figure 10. Convolutional feature maps for each class for the three-season CNN. These feature maps are the result of averaging the convolution of the kernels with 75 spectra per class (Section 2.4.3). An example class spectrum is shown in background. Green, yellow and gray background areas represent bands in the spring, summer and fall, respectively. Each season has 186 bands, ordered 370 to 2500 nm. However, in this figure the original 224 bands per season are shown in order to display the bad data gaps (e.g., atmospheric absorption windows).

4. Discussion

4.1. Classified Land-Cover Maps

The classification scheme in this study included twelve classes defined by LCCS rules designed for global-scale applications. Forest classes had closed-canopies of one leaf type (e.g., deciduous or evergreen), and thus excluded more mixed forest or open-canopy classes that have more spectral variability, and subsequent class confusion and decrease in overall accuracy [6]. However, the broad classes include spectral-temporal variability over a large region. This aspect of this study is unique for testing machine-learning classifiers with hyperspectral and multi-seasonal datasets.

We found that the 1-D (per-pixel) SVM and CNN classifiers had over 7% overall accuracy improvement over the RF classifier for these data and a classification scheme. This is similar to a 5% increase in overall accuracy observed for 1-D CNN over RF in a study focused on crop classification in Ukraine with multi-temporal multispectral Landsat and Sentinel-2 images [15]. We found that CNN only had 0.3 (three-season) to 0.5% (summer) greater overall accuracy than the popular SVM classifier. This result is similar to a previous study with 1-D hyperspectral imagery (not multi-temporal) that found CNN to improve overall accuracy by 0.9 to 2.6% relative to SVM [17]. In our study, we found the addition of multi-seasonal data modestly (but significantly) improved classification accuracy by 1.9 to 3.5% over using summer-only data. The Kussal et al. study [15] did not specifically isolate individual seasons for testing multi-temporal data, and to our knowledge there are no studies that compare the relative advantage of temporal data in CNN. However, multi-temporal data have been shown to increase overall accuracy from ~2 to 8% with a variety of classifiers [37]

Our previous research in this study area with 2013 AVIRIS data and the same training and testing reference data was based on the RF classifier [6]. We found reflectance data yielded a 62.9% and 78.8% overall accuracy with single- and three-season data, respectively. The RF classifier in the current study produced 15.8% (single-season) and 3.4% (three-season) higher overall accuracy, respectively. The current study was not designed to replicate our previous research design. Some of the discrepancy in results is explained by a difference in image years, improvements in the atmospheric correction process, and differences in processing code (Python vs. R). Further, our previous work used a RF node size of 10 to avoid overfitting and to train a more generalized classifier, while the current study used a node size of 1. Despite these differences, our results indicate that the CNN and SVM classifiers offer improved performance relative to RF, and both classifiers should be considered in hyperspectral classification applications.

All classifications in our study tended to have class confusion among spectrally similar forest classes, a result in accordance with our previous RF study [6]. Annual Crops were broadly confused with Herbaceous cover, and to lesser extent Urban Vegetated. Not all agriculture in this area is irrigated or plowed, and tends to have similar physiognomy and phenology as grasslands, which are dominated by exotic annual grasses; and thus, spectral-temporal confusion between these two classes is expected.

4.2. CNN Feature Maps and Land Cover

The usage of CNN is promising with respect to classification accuracy as well as visualization of features within the data. From convolutional feature maps, the seasonal and spectral contribution to the CNN for each class is visible as excitations by the kernels (Figure 10). This visualization, paired with the confusion matrices, provide a point of exploration into the distinctive nature of the classes. For example, the example Annual Crop spectrum has heavy kernel excitation within the summer reflectance bands (Figure 7). This indicates that the Annual Crop class contains distinctive features within the summer season as compared to the other classes. This could be due to the lack of chlorophyll absorption features in the spring and fall visible spectrum and presence of these features in summer; the example spectrum appears to be green in the summer. The kernels may thus be sensitive to chemical properties of crops. In contrast, herbaceous cover follows the cycle of precipitation (wet spring, dry summer and fall) and senesces in the early summer and is fully senesced by the fall (Figure 2—note lack of chlorophyll absorption in summer and fall). The Herbaceous example spectrum has convolutional excitations spread more evenly across seasons relative to Annual Crops, with some excitations occurring in fall where annual crops had none (Figure 10). The Bare class by definition has relatively low plant cover and minimal chlorophyll absorption signal, although there is high within-class spectral diversity (Figure 2). With the example spectra, kernels are more uniformly excited across all three seasons of the feature map, indicating that this class is less influenced by seasonal variation than within-season spectral properties. This is visible by the additional excitation around the low-signal area (e.g., low blue, high SWIR) within each season of the class data.

An interesting aspect from feature maps is the kernel activations around areas of low blue, far shortwave infrared and around atmospheric absorption bands. These are spectral regions that tend to have more noise due to lower solar irradiance, atmospheric effects (e.g., scattering, absorption), and spectrometer sensitivity. Engineered metrics used in remote sensing target vegetation chemistry and structure with narrowband ratios (e.g., indices such as NDVI), spectral derivatives, and absorption-feature fitting techniques that span the different spectral ranges with well-known continuous areas [6]. In general, engineered metrics avoid these noise-prone spectral regions. However, because some CNN feature maps are excited in these regions, these results indicate that these regions can be important to classification accuracy and warrant further investigation.

One criticism of neural networks and SVM is that they are relatively "black box"; classification performance may be high, but it is difficult to understand the relative importance of spectral and temporal information in the input vector. The CNN feature maps and kernel importance matrix introduced in this study have potential to provide more insight into the classifier and should be explored further with other datasets. In comparison to RF variable importance, we found these CNN diagnostic tools to be less easy to interpret [6]. With RF variable importance, there is a direct linkage between a predictor variable (e.g., reflectance band) and the impact on class and overall accuracy with its removal. Despite the very different architectures between CNN and RF, we found that RF importance (Figure A1) tended to cluster around some of the same regions as identified by the CNN feature maps, such as blue (spring, summer, fall), high SWIR (spring, summer) and the border of SWIR atmospheric absorption (band index 500, which is 1761 nm from fall imagery).

4.3. Future Work with CNN Hyperspectral Image Classification

The application of CNN to hyperspectral data has only recently been explored and there are thus many avenues for future work. The consideration of the spatial domain is an area of current investigation. The extension of this architecture to include the spatial dimension as in [13] could continue to add more distinctive information that could aid in the classification process. A 3-D convolutional network paired with a respective importance figure (Figure 9) could shed light on to the spatial as well as the spectral features that impact the classification on a per class basis. In a similar manner, the temporal dimension of image data should be investigated. As in this study, this could be done by vertically stacking the temporal season data creating a $N_d \times N_{ns}$ sized input data array, where N_d is the number of bands within a single season of image data and N_{ns} is the number of seasons (rather than horizontally stacked data in temporal sequence, as in our study). This data format would dictate the architecture of networks that could convolve higher dimension kernels across the input data, or perform multiple layered convolutional operations. The additional benefit would be that the kernel's local receptive field could enable the network to capture patterns within the temporal and spectral dimensions in an integrated machine-learning framework. If there is distinctive information within these dimensions, this could increase the accuracy as well as provide more ancillary information about the distinctive aspects of the data through analysis of convolutional feature maps.

It was observed when reviewing the feature maps that the kernels were exciting the convolutional feature maps in similar spectral regions, across all kernels. This appeared as strips of excited kernels. This indicates that the kernels are not capturing unique information. If kernels were "encouraged" to learn information that is "unique" this would enable the feature maps to show different structural features in the data on a per kernel basis. One potential way to do this would be to provide regularization on the kernels during training that would penalize kernels that are not orthogonal. Although this may enhance CNN as a tool for exploring features that discriminate classes, this approach may not improve classification accuracy.

As data from future hyperspectral satellites are more readily available, such as from the Environmental Mapping and Analysis Program (EnMAP) satellite due to launch in 2019, the scalability of classification techniques should be considered. As TensorFlow has been designed to scale from multiple-core, GPU(s), and multiple-computer clustered configurations, the CNN can readily scale for large global and temporal datasets. The Python Scikit-learn library used for RF and SVM in this study does not have this capability, and thus does not offer a direct comparison of GPU processing with CNN and TensorFlow. The computers utilized for this work were an 8 Core Xenon 3.7 Ghz CPU processor with 32 GB ram and a NVIDIA K2200 GPU, as well as a second i7-2640M (2.8 GHz) dual-core CPU with 8 GB of RAM. We tested the CNN architecture with both CPU- and GPU-based configurations. The GPU-based implementation performed all graph computations within the GPU, and was 5 to 8 times faster than CPU-based processing, depending on the learning batch size (larger batch size increased performance within the GPU space). With our data, the GPU-based CNN was slower to train than SVM; however, image classification was 9 times faster with CNN (Table A2). For applications that use large datasets, particularly with the inclusion of spatial and temporal information in the classification design, the CNN with TensorFlow or similar machine learning platform may have a considerable advantage in processing time while offering relatively high classification accuracy.

5. Conclusions

A broad goal of this work was to implement and explain how a Convolutional Neural Network (CNN) with a one-dimensional architecture could be applied to multi-seasonal hyperspectral images. The results show that CNN can be applied to simulated spaceborne hyperspectral reflectance data (HyspIRI) to achieve high classification accuracy rates comparable to that from Support Vector

Machine (SVM), and surpassing the Random Forest (RF) classifier. Highest overall accuracies were with three-season data (spring, summer, fall), with 89.9% for CNN, 89.5% for SVM and 82.2% for RF. Single-season (summer) classifications had overall accuracies that were 1.9 to 3.5% lower than with three-season data. Spectral and temporal information is readily visible through CNN feature map visualizations and their respective importance is traceable back to kernel importance. In summary, the CNN is a promising classifier for future hyperspectral classification tasks and this study identifies future work to increase CNN performance, scalability and incorporation of spatial and temporal information.

Acknowledgments: Funding for this project was provided by NASA HyspIRI Preparatory Airborne Activities and Associated Science Research, grant NNX12AP09G. Simulated HyspIRI-like reflectance products were provided by NASA Jet Propulsion Lab (JPL), with key involvement by Bo Cai Gao, David R Thompson, Sarah Lundeen and Robert Green, as well as Philip Dennison of Univ. of Utah for AVIRIS-to-HyspIRI resampling (NASA grant NNX12AP08G). Kelsey Dunn helped prepare maps for this paper. Bala Ravikumar and Farid Farahmand of Sonoma State University provided initial review and input into this study.

Author Contributions: D.G. and M.C. conceived and designed the experiments; D.G. performed the experiments; D.G. and M.C. analyzed the data; M.C. contributed materials and analysis tools; D.G. and M.C. wrote the paper.

Appendix A

Table A1. The Land-Cover Classification System (LCCS) rules applied to percent cover estimates of land cover in reference samples in order to create discrete class labels. Open-canopy (open) trees and shrubs have >10–65% trees or shrubs, respectively. Closed-canopy (closed) trees and shrubs have >65% trees or shrubs, respectively. This study focused on closed-canopy tree and shrub classes only.

Rules Based on Land-Cover Abundance	LCCS Class
1. If >10% Vegetated	
2. If >50% Natural/Semi-natural Vegetation	
3. If >50% tidal salt marsh	**Tidal Marsh**
3. If >10% of cover is woody vegetation (trees + shrubs) and >10% of woody vegetation is trees	
4. If ≥75% of relative tree cover is needleleaf trees	**Evergreen Needleleaved Trees (ENT)**
4. If >75% of relative tree cover is broadleaf trees	
5. ≥75% of tree cover is evergreen	**Evergreen Broadleaved Trees (EBT)**
5. ≥75% of broadleaf tree cover is deciduous	**Deciduous Broadleaved Trees (DBT)**
3. If >10% of cover is woody vegetation (trees + shrubs) and >10% of woody vegetation is shrubs	**Shrubs**
3. Else herbaceous cover	
6. ≥75% of herbaceous cover is upland grasses and forbs	**Herbaceous**
6. ≥75% of herbaceous cover is dune vegetation	**Dune Vegetation**
2. Else Cultivated/Managed Vegetation	
7. >50% perennial crops	**Perennial Crops**
7. >50% annual crops	**Annual Crops**
7. >50% urban landscape	**Urban Vegetated**
1. Else Not Vegetated	
8. >50% impervious surface	**Built-up**
8. >50% non-vegetated	**Bare**

Table A2. Classifier training and classification times in seconds for three season data. For CNN, (1) initial steady-state validation accuracy; and (2) a 4-mil epoch standard training session. The CNN classifier utilized GPU-based processing, while SVM and RF were restricted to CPU-based processing.

	CNN-1 (s)	CNN-2 (s)	SVM (s)	RF (s)
Training Time	1500	21,146	358	3452
Classification Time per GB	65	65	590	768
Classification Time Per Pixel	0.0008	0.0008	0.00773	0.00959

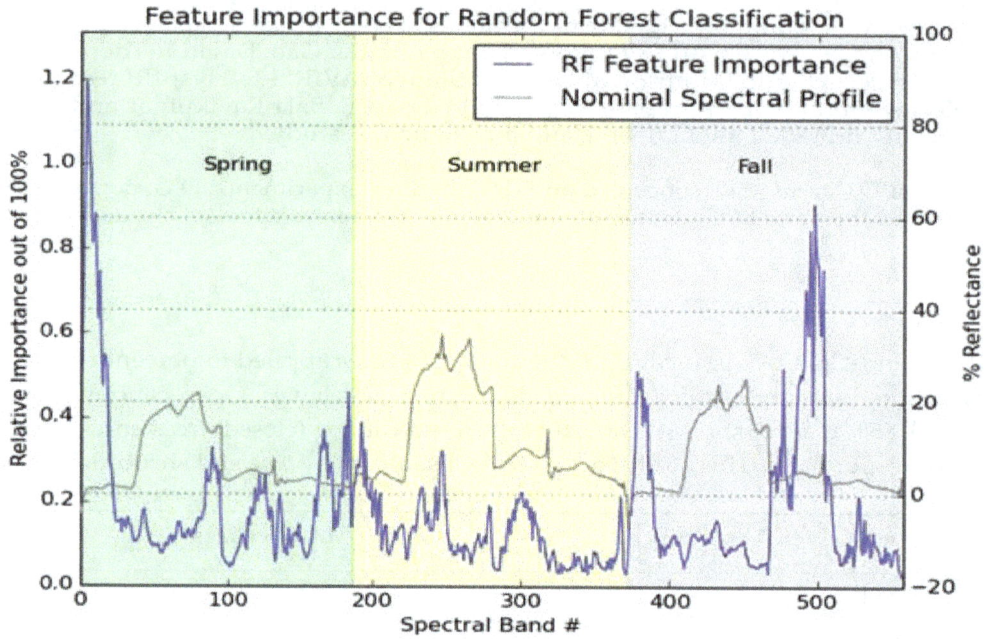

Figure A1. Feature importance for the three-season Random Forests classifier. Nominal spectral profile for ENT shown for context.

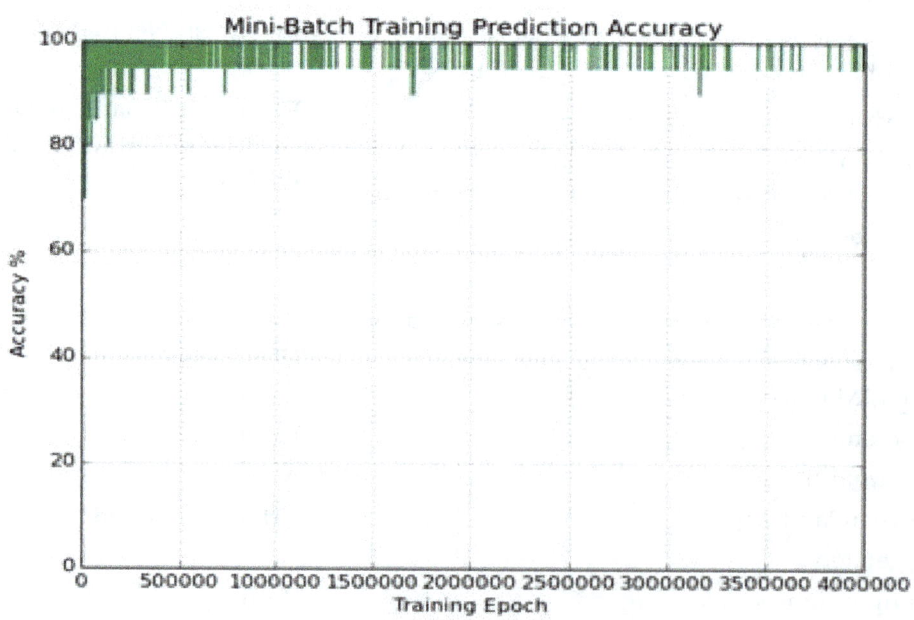

Figure A2. Mini-batch training accuracy curve.

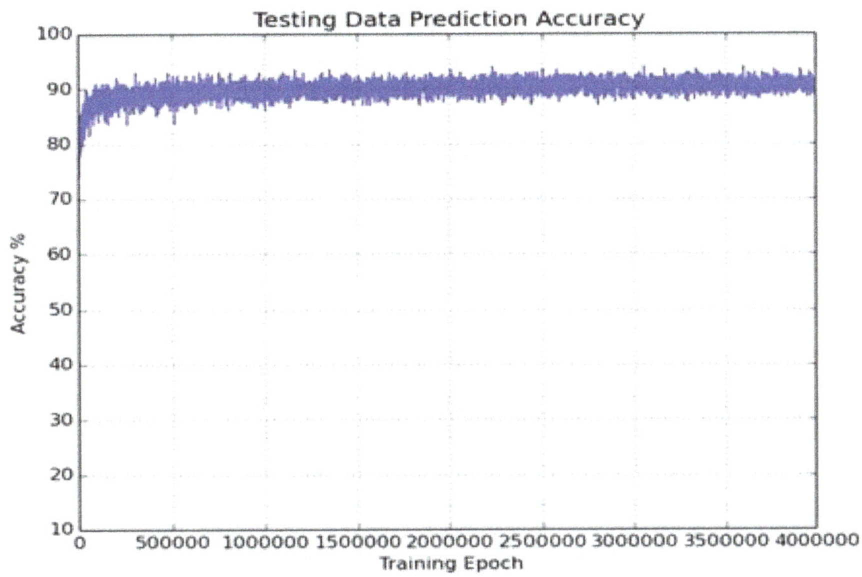

Figure A3. Testing accuracy curve.

References

1. Lee, C.M.; Cable, M.L.; Hook, S.J.; Green, R.O.; Ustin, S.L.; Mandl, D.J.; Middleton, E.M. An Introduction to the NASA Hyperspectral InfraRed Imager (HyspIRI) Mission and Preparatory activities. *Remote Sens. Environ.* **2015**, *167*, 6–19. [CrossRef]
2. Somers, B.; Asner, G.P.; Tits, L.; Coppin, P. Endmember Variability in Spectral Mixture Analysis: A Review. *Remote Sens. Environ.* **2011**, *117*, 1603–1616. [CrossRef]
3. Roberts, D.A.; Gardner, M.; Church, R.; Ustin, S.; Scheer, G.; Green, R.O. Mapping Chaparral in the Santa Monica Mountains Using Multiple Endmember Spectral Mixture Models. *Remote Sens. Environ.* **1998**, *65*, 267–279. [CrossRef]
4. Franke, J.; Roberts, D.A.; Halligan, K.; Menz, G. Hierarchical Multiple Endmember Spectral Mixture Analysis (MESMA) of Hyperspectral Imagery for Urban Environment. *Remote Sens. Environ.* **2008**, *113*, 1712–1723. [CrossRef]
5. Camps-Valls, G. Machine Learning in Remote Sensing Data Processing. Proceedings of IEEE International Workshop on Machine Learning for Signal Processing (MLSP), Grenoble, France, 2–4 September 2009; pp. 1–6.
6. Clark, M.L.; Kilham, N.E. Mapping of Land Cover in Northern California with Simulated HyspIRI imagery. *ISPRS J. Photogramm. Remote Sens.* **2016**, *119*, 228–245. [CrossRef]
7. Mountrakis, G.; Im, J.; Ogole, C. Support Vector Machines in Remote Sensing: A Review. *ISPRS J. Photogramm. Remote Sens.* **2011**, *66*, 247–259. [CrossRef]
8. Fernandex-Delgado, M.; Cernadas, E.; Barro, S.; Amorim, D. Do We Need Hundreds of Classifiers to Solve Real World Classification Problems? *J. Mach. Learn. Res.* **2014**, *15*, 3133–3181.
9. Krizhevsky, A.; Sutskever, I.; Hinton, G. Imagenet Classification with Deep Convolutional Neural Networks. Proceedings of 25th International Conference on Neural Information Processing Systems, Lake Tahoe, NV, USA, 3–6 December 2012; pp. 1097–1105.
10. Szegedy, C.; Liu, W.; Jia, Y.; Sermanet, P.; Reed, S.; Anguelov, D.; Erhan, D.; Vanhoucke, V.; Rabinovich, A. Going Deeper with Convolutions. In Proceedings of the IEEE Conference on Computer Vision and Pattern Recognition, Boston, MA, USA, 7–12 June 2015.
11. Abdel-Hamid, O.; Mohamed, A.; Jiang, H. Convolutional Neural Networks for Speech Recognition. *IEEE/ACM Trans. Audio Speech Lang. Process.* **2014**, *22*, 1533–1545. [CrossRef]
12. Castelluccio, M.; Poggi, G.; Sansone, C.; Verdoliva, L. Land Use Classification in Remote Sensing Images by Convolutional Networks. Available online: http://arxiv.org/abs/1508.00092 (accessed on 8 May 2017).

13. Li, Y.; Zhang, H.; Shen, Q. Spectral-Spatial Classification of Hyperspectral Imagery with 3D Convolutional Neural Network. *Remote Sens.* **2017**, *9*, 67. [CrossRef]
14. Langkvist, M.; Kiselev, A.; Alirezaie, M.; Loutfi, A. Classification and Segmentation of Satellite Orthoimagery Using Convolutional Neural Networks. *Remote Sens.* **2016**, *8*, 329. [CrossRef]
15. Kussul, N.; Lavreniuk, M.; Skakun, S.; Shelestov, A. Deep Learning Classification of Land Cover and Crop Types Using Remote Sensing Data. *IEEE Geosci. Remote Sens. Lett.* **2017**, *14*, 5. [CrossRef]
16. Chen, C.; Li, W.; Su, H.; Liu, K. Spectral-Spatial Classification of Hyperspectral Image Based on Kernel Extreme Learning Machine. *Remote Sens.* **2014**, *6*, 5795–5814. [CrossRef]
17. Hu, W.; Huang, Y.; Wei, L.; Zhang, F.; Li, H. Deep Convolutional Neural Networks for Hyperspectral Image Classification. *J. Sens.* **2015**, *2015*, 258619. [CrossRef]
18. Makantasis, K.; Karantzalos, K.; Doulamis, A.; Doulamis, N. Deep Supervised Learning for Hyperspectral Data Classification through Convolutional Neural Networks. In Proceedings of the 2015 IEEE International Symposium Geoscience and Remote Sensing (IGARSS 2015), Milan, Italy, 26–31 July 2015; pp. 4959–4962.
19. Li, W.; Fu, H.; Yu, L.; Cracknell, A. Deep Learning Based Oil Palm Tree Detection and Counting for High-Resolution Remote Sensing Images. *Remote Sens.* **2017**, *9*, 22. [CrossRef]
20. Wang, J.; Luo, C.; Huang, H.; Zhao, H.; Wang, S. Transferring Pre-Trained Deep CNNs for Remote Scene Classification with General Features Learned from Linear PCA Network. *Remote Sens.* **2017**, *9*, 225. [CrossRef]
21. Hu, F.; Xia, G.-S.; Hu, J.; Zhang, L. Transferring Deep Convolutional Neural Networks for Scene Classification of High-Resolution Remote Sensing Imagery. *Remote Sens.* **2015**, *7*, 14680–14707. [CrossRef]
22. Mesay, B.B.; Zeggada, A.; Nouffidj, A.; Melgani, F.; A Convolutional Neural Network Approach for Assisting Avalanche Search and Rescue Operations with UAV Imagery. *Remote Sens.* **2017**, *9*, 100.
23. Liang, H.; Li, Q. Hyperspectral Imagery Classification using Sparse Representations of Convolutional Neural Network Features. *Remote Sens.* **2016**, *8*, 99. [CrossRef]
24. Zhao, W.; Du, S. Spectral-Spatial Feature Extraction for Hyperspectral Image Classification: A Dimension Reduction and Deep Learning Approach. *IEEE Trans. Geosci. Remote Sens.* **2016**, *54*, 8. [CrossRef]
25. Wang, Q.; Lin, J.; Yuan, Y. Salient Band Selection for Hyperspectral Image Classification via Manifold Ranking. *IEEE Trans. Neural Net. Learn. Syst.* **2016**, *27*, 6. [CrossRef] [PubMed]
26. Yuan, Y.; Lin, J.; Wang, Q. Hyperspectral Image Classification via Multitask Joint Sparse Representation and Stepwise MRF Optimization. *IEEE Trans. Neural Netw. Learn. Syst.* **2016**, *46*, 12. [CrossRef] [PubMed]
27. Di Gregorio, A. *Land Cover Classification System: Classification Concepts and User Manual*; LCCS (No.8); Food and Agriculture Organization: Rome, Italy, 2005.
28. Green, R.O.; Eastwood, M.L.; Sarture, C.M.; Chrien, T.G.; Aronsson, M.; Chippendale, B.J.; Faust, J.A.; Pavri, B.E.; Chovit, C.J.; Solis, M.S.; et al. Imaging Spectroscopy and the Airborne Visible/Infrared Imaging Spectrometer (AVIRIS). *Remote Sens. Environ.* **1998**, *65*, 227–248. [CrossRef]
29. Thorpe, A.K.; Frankenberg, C.; Aubrey, A.D.; McFadden, J.P. Mapping Methane Concentrations from a Controlled Release Experiment using the Next Generation Airborne Visible/Infrared Imaging Spectrometer (AVIRIS-NG). *Remote Sens. Environ.* **2016**, *179*, 104–115. [CrossRef]
30. Thompson, D.R.; Gao, B.C.; Green, R.O.; Roberts, D.A.; Dennison, P.E.; Lundeen, S.R. Atmospheric Correction for Global Mapping Spectroscopy: ATREM advances for the HyspIRI preparatory campaign. *Remote Sens. Environ.* **2015**, *167*, 64–77. [CrossRef]
31. Breiman, L. Random forests. *Mach. Learn.* **2001**, *45*, 5–32. [CrossRef]
32. Ensemble Methods. Available online: http://scikit-learn.org/stable/modules/ensemble.html#forest (accessed on 8 May 2017).
33. RBF SVM Parameters. Available online: www.scikit-learn.org/stable/auto_examples/svm/plot_rbf_parameters.html (accessed on 8 May 2017).
34. Srivastava, N.; Hinton, G.; Krizhevsky, A.; Sutskever, I.; Salakhutdinov, R. Dropout: A Simple Way to Prevent Neural Networks from Overfitting. *J. Mach. Learn. Res.* **2014**, *15*, 1929–1958.
35. Nielsen, M. Chapter 3, Improving the Way Neural Networks Learn, Neural Networks and Deep Learning. Available online: http://neuralnetworksanddeeplearning.com/chap3.html (accessed on 8 May 2017).
36. CS231n Convolutional Neural Networks for Visual Recognition. Available online: http://cs231n.github.io/convolutional-networks/ (accessed on 8 May 2017).

10

Refinement of Hyperspectral Image Classification with Segment-Tree Filtering

Lu Li [1,2], Chengyi Wang [1,*], Jingbo Chen [1] and Jianglin Ma [1]

1 Institute of Remote Sensing and Digital Earth, Chinese Academy of Sciences, Datun Road North 20A, Beijing 100101, China; lilu@radi.ac.cn (L.L.); chenjb@radi.ac.cn (J.C.); majianglin2003@gmail.com (J.M.)
2 The University of Chinese Academy of Sciences, Beijing 100049, China
* Correspondence: wangcycastle@163.com.

Academic Editors: Qi Wang, Nicolas H. Younan, Carlos López-Martínez, Lenio Soares Galvao and Prasad S. Thenkabail

Abstract: This paper proposes a novel method of segment-tree filtering to improve the classification accuracy of hyperspectral image (HSI). Segment-tree filtering is a versatile method that incorporates spatial information and has been widely applied in image preprocessing. However, to use this powerful framework in hyperspectral image classification, we must reduce the original feature dimensionality to avoid the Hughes problem; otherwise, the computational costs are high and the classification accuracy by original bands in the HSI is unsatisfactory. Therefore, feature extraction is adopted to produce new salient features. In this paper, the Semi-supervised Local Fisher (SELF) method of discriminant analysis is used to reduce HSI dimensionality. Then, a tree-structure filter that adaptively incorporates contextual information is constructed. Additionally, an initial classification map is generated using multi-class support vector machines (SVMs), and segment-tree filtering is conducted using this map. Finally, a simple Winner-Take-All (WTA) rule is applied to determine the class of each pixel in an HSI based on the maximum probability. The experimental results demonstrate that the proposed method can improve HSI classification accuracy significantly. Furthermore, a comparison between the proposed method and the current state-of-the-art methods, such as Extended Morphological Profiles (EMPs), Guided Filtering (GF), and Markov Random Fields (MRFs), suggests that our method is both competitive and robust.

Keywords: hyperspectral image classification; SELF; SVMs; Segment-Tree Filtering

1. Introduction

Hyperspectral image (HSI) classification is important for urban land use monitoring, crop growth monitoring, environmental assessment, etc. Various machine learning algorithms that process high-dimension data can be employed in pixel-wise classification, such as Support Vector Machines (SVMs) [1], Logistic Regression [2,3], Artificial Neural Networks (ANNs) [4], etc. However, these conventional approaches do not consider spatial HSI information between neighboring pixels, which can lead to noisy classification output. Including the spatial relationships between pixels can enhance the classification accuracy. For example, there is a high probability that a pixel shares the same class as its neighboring pixels if the similarity measure between them is high. Otherwise, if the similarity measure is low, this probability decreases. Therefore, HSI classification could be improved further by combining spatial and spectral features.

Many spatial-spectral methods have been proposed to incorporate spatial or contextual information. For example, spatial information was represented using Markov Random Fields (MRFs) in [5,6], and classification has been performed using α–Expansion [7] and Belief Propagation [8], which are commonly used max-flow/min-cut algorithms in MRF optimization. Another method is presented in [9], namely Extended Morphological Profiles (EMPs). After the first two principal components of the HSI are computed using the Principal Component Analysis (PCA) method, spatial features are extracted by morphological operations. Together with spectral information, they are concatenated for HSI classification. As morphological operations such as opening and closing involve neighboring pixel calculations, contextual information is naturally utilized in this manner. Another example of employing contextual information is via texture analysis. In [10], a Gray Level Co-occurrence Matrix (GLCM) is used to extract this type of contextual information, which is then employed to concatenate spectral features used for classification. In [11], segmentation is employed to represent spatial information based on a minimum spanning tree method, and majority voting is used to assign a class label to each region. Similar to [11], methods based on segmentation [12–15] have attracted increased attention because they produce satisfactory results. However, there are some drawbacks to algorithms based on hard segmentation. For example, they assume that all pixels in the same region are homogenous. After segmentation is completed, the relationship between pixels in different regions is fully disconnected; thus, if the segmentation is incorrect, the accuracy decreases dramatically. Although an over-segmentation approach is applied in [11,15] to improve the similarity in a region, the computational complexity increases considerably. Therefore, these algorithms are not efficient because of the complex voting processes in thousands of regions. In [16], super-pixel segmentation is applied to feature extraction and then classification is conducted in a novel framework via multiple kernels, which avoid voting but the super-pixel method still needs over-segmentation. To make classification more efficient, Edge-Aware Filtering and Edge-Preserving Filtering (EAF and EPF) methods [17,18] can be applied. These methods have been adopted successfully in many computer vision applications, such as stereo matching [19], optical flow [20], image fusion [21], etc. Unlike image segmentation, the most prominent merit of the EAF method is that in homogeneous image areas, EAFs can generate smooth output, while in inhomogeneous image areas, they can adaptively preserve boundaries, even in challenging situations. In this paper, we implement a tree-structure EAF for HSIs that is based on the segment-tree algorithm [22,23] and combines the advantages of segmentation and EAF. Unlike other EAF methods, the window size does not need to be set in this scheme. It is difficult to establish a proper window size for bilateral and guided filters [17], largely because objects of interest display the most prominent features at different scales. Another merit of this scheme is that the segment-tree filter is more efficient than other EAFs because of its tree structure [24]. By traversing the tree in two sequential passes, from the leaves to the root and from the root to the leaves, every pixel in HSI can be filtered and labeled.

However, the Segment-Tree Filter cannot be used for original HSI directly because the computational cost of this method is extremely high and the Hughes phenomenon always makes the classification accuracy unsatisfactory. The hyperspectral bands that are contaminated by noise may destroy the true connection between neighboring pixels. To avoid this problem, there are several literature that introduce how to choose the bands of original HSI [25,26] or produce new salient features. In this paper, the Semi-supervised Local Fisher (SELF) discriminant analysis method [27] is employed to reduce dimensionality. The SELF method of feature reduction is used because it retains prior knowledge from training sets and statistical distribution of clusters, unlike methods such as PCA, Linear Discriminant Analysis (LDA)/Fisher Discriminant Analysis (FDA) [28], and Local FDA (LFDA) [29]. In practice, the segmentation will be less sensitive to the number of training samples based on the SELF method. Additionally, we can construct the Segment-Tree Filter using a limited number of bands, which can reduce the computational cost of segmentation. Because the extracted SELF bands can reduce the effect of noise, a limited number of bands can well represent the inherent spatial structure of the image, which can lead to better output.

The remainder of this paper is organized as follows. In Section 1, we discuss some related methods and processes, including initial classification, SELF, Graph-based Segment-Tree Filter construction, and filtering. In Section 2, the proposed HSI classification scheme is described in detail. The experimental results are presented in Section 3. Finally, we draw our conclusions and present our outlooks for future research in Section 4.

2. HSI Classification Refinement Using Segment-Tree Filtering

A schematic diagram of the proposed method is shown in Figure 1.

1. Step 1: Construct the Segment-Tree Filter, which involves feature extraction using the SELF method followed by building a tree-structure filter for an HSI based on dimensionality reduction.
2. Step 2: Use a Multi-class SVM method to obtain the initial classification map.
3. Step 3: Perform Segment-Tree Filtering based on the Multi-class SVM, pixel-based initial classification map. By combining this initial classification map and the Segment-Tree Filter, we can incorporate spatial information and spectral features, adaptively. Finally, the HSI classification map can be derived from the result of Segment-Tree Filtering.

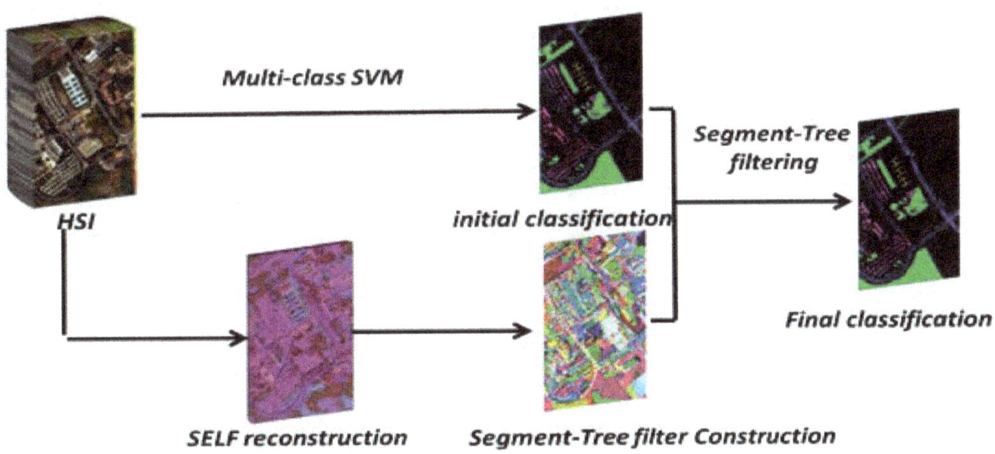

Figure 1. Workflow of Segment-Tree Filtering for HSI classification.

2.1. Initial Classification

In this paper, a Multi-class SVM classifier is adopted in the initial classification step. SVMs are widely used classifiers in remote sensing image classification. These supervised learning models are used for classification and regression in binary classification problems. In this paper, we utilize a Multi-class SVM from LIBSVM library with a radial basis function (RBF) kernel. In this method, a "one against one" strategy [30] is employed to extend the binary SVM to multi-class cases. The punishment parameter C and the spread of the kernel gamma are optimally determined by cross-validation.

In most cases, the output of the initial classification is a probability map, which can be represented as a tensor [18] as follows:

$$\{\mathbf{M}_p^k \mid p(i,j); i = 1,2,\ldots,H; j = 1,2,\ldots,W; k = 1,2,\ldots,S\} \quad (1)$$

where (i,j) is the position of sample p in the image; H and W are the height and width of the image, respectively; k is the label of the sample; S is the total number of classes in the classification; and $\mathbf{M}_p^k$ is the probability that the sample p belongs to the kth class. Based on the Multi-class SVM classifier, $\mathbf{M}_p^k$ is either 0 or 1 depending on whether the sample belongs to the kth class.

$$\mathbf{M}_p^k = \begin{cases} 1\ if & class(p) = k \\ 0 & otherwise \end{cases} \quad (2)$$

2.2. Semi-Supervised Local Fisher Discriminant Analysis

In this section, we review SELF briefly. The SELF method seeks an embedding transformation such that the local, between-class scatter is maximized and the local, within-class scatter is minimized in both the training and test sets. We assume that HSI $\mathbf{X}$ has m hyperspectral bands and n samples, and the training set $\mathbf{X}'$ has n' samples. Then, the test set has $n - n'$ samples. There are three pre-defined input parameters: the trade-off parameter β, the dimensionality of the reconstruction space r, and the KNN parameter K. The five steps in this process are as follows:

1. Local scaling coefficient σ_i is pre-computed for each sample in the training set, which is equal to the Euclidean distance between the sample x_i and its Kth nearest neighbor x_i^K among all samples in both the training and test sets.
$$\sigma_i = \|x_i - x_i^K\| \tag{3}$$
2. Local between-class weight matrix $\mathbf{W}^{lb}$ and local within-class weight matrix $\mathbf{W}^{lw}$ are computed as Equations (4) and (5), respectively. In this step, if two samples have the same label in the training set, σ_i is used to scale the local geometric structure with heat kernel weighting.
$$\mathbf{W}_{i,j}^{lb} = \begin{cases} (\frac{1}{n'} - \frac{1}{n'_{y_i}})\exp(\frac{-\|x_i - x_j\|^2}{\sigma_i\sigma_j}) & if \quad y_i = y_j \\ \frac{1}{n'} \quad otherwise \end{cases} \tag{4}$$
$$\mathbf{W}_{i,j}^{lw} = \begin{cases} \frac{1}{n'_{y_i}}\exp(\frac{-\|x_i - x_j\|^2}{\sigma_i\sigma_j}) & if \quad y_i = y_j \\ 0 \quad otherwise \end{cases} \tag{5}$$
3. The local between-class scatter matrix $\mathbf{S}^{lb}$ and local, within-class scatter matrix $\mathbf{S}^{lw}$ are calculated by Equations (6) and (7):
$$\mathbf{S}^{lb} = \mathbf{X}'\{diag(\mathbf{W}^{lb}\mathbf{1}_{n'})\}\mathbf{X}'^T; \tag{6}$$
$$\mathbf{S}^{lw} = \mathbf{X}'\{diag(\mathbf{W}^{lw}\mathbf{1}_{n'})\}\mathbf{X}'^T; \tag{7}$$
Note that $\mathbf{1}_{n'}$ is a unit column vector of size $n' \times 1$. Steps 1–3 are the same as those used in the LFDA procedure. In our procedure, only samples in the training set have been used at this point, and the statistical distribution of clusters has not been assessed or applied.
4. The covariance matrix $\mathbf{S}^t$ is computed based on all samples in both the training and test sets as below:
$$\mathbf{S}^t = \frac{\sum_{i=1}^{n}(x_i - \overline{x})(x_i - \overline{x})^T}{n}; \tag{8}$$
where $\overline{x}$ is the mean of all samples. Then, the regularized, local, between-class scatter matrix $\mathbf{S}^{rlb}$ and the regularized, local, within-class scatter matrix $\mathbf{S}^{rlw}$ are derived by Equations (9) and (10), respectively.
$$\mathbf{S}^{rlb} = (1-\beta)\mathbf{S}^{lb} + \beta\,\mathbf{S}^t; \tag{9}$$
$$\mathbf{S}^{rlw} = (1-\beta)\mathbf{S}^{lw} + \beta\,\mathbf{I}_m; \tag{10}$$
β is the trade-off parameter based on prior knowledge from the training set and the statistical distribution of clusters. Therefore, SELF maintains the advantages of LFDA and PCA. Note that $\mathbf{I}_m$ is an $m \times m$ identity matrix.
5. Transformation matrix $\mathbf{T}$ can be computed based on generalized eigenvalue decomposition. $\mathbf{T}$ consists of weighted eigenvectors corresponding to the r largest eigenvalues. After $\mathbf{T}$ is determined, all the pixels in the HSI can be reprojected to a new low-dimensional space.

The parameter β plays an important role in the algorithm. When it is relatively small (e.g., $\beta = 0.01$), SELF is nearly identical to LFDA, and SELF becomes increasingly similar to PCA when β approaches 1. β balances prior knowledge regarding the labels in the training set and the statistical distribution of clusters in the test set. In our experiment, we set β to 0.6, as we found that this value fully utilizes the advantages of the algorithm.

As a Semi-Supervised Learning (SSL) method, SELF can adapt based on the number of training samples. When the number of samples in a training set is small and prior knowledge about labels is limited and/or noisy, SELF can use the statistical distribution of clusters in the test set to offset these issues. We demonstrate this advantage in Figures 2 and 3. Figure 2 illustrates the reconstructed image of the Indian Pines dataset using PCA, LDA, LFDA, and SELF when the training samples accounted for only 1% of all samples. Figure 3 shows the same reconstructed image when the training percentage increases to 20%. In both figures, the color images (R, G, and B) are composed of the first three bands extracted using the corresponding feature-transformation methods. In Figure 2a, because the reconstruction based on LDA only relies on a small number of training set samples, the image exhibits considerable noise and error. When the number of training samples increases, LDA reconstruction is improved (e.g., there are fewer fractions and more homogenous areas in Figure 3a than in Figure 2a). Additionally, LFDA is more robust than LDA, even if the number of training samples is limited. However, as the number of samples increases, numerous false edges and fractions can be observed in the reconstructed image based on LFDA, as shown in Figures 2b and 3b. The boundary of the purple trapezoid in the bottom-left portion of Figure 2b was correctly extracted; however, it was incorrectly extracted in Figure 3b. As shown in Figures 2d and 3d (see the red rectangular region), SELF is robust regardless of the number of samples in the training set. Conversely, PCA does not depend on the training set; therefore, Figures 2c and 3c display the same reconstruction result. However, the comparison between PCA and SELF in the rectangular region in Figure 3d shows that PCA creates more segments and fractions because it aligns the boundaries of objects rather than classes, which causes more errors in subsequent processing steps compared to using SELF. In our experiment, the best reconstructed images based on SELF had 10 bands. Thus, the spectral dimension of the original HSI was dramatically reduced, but the most discriminatory information within the spectral bands was retained.

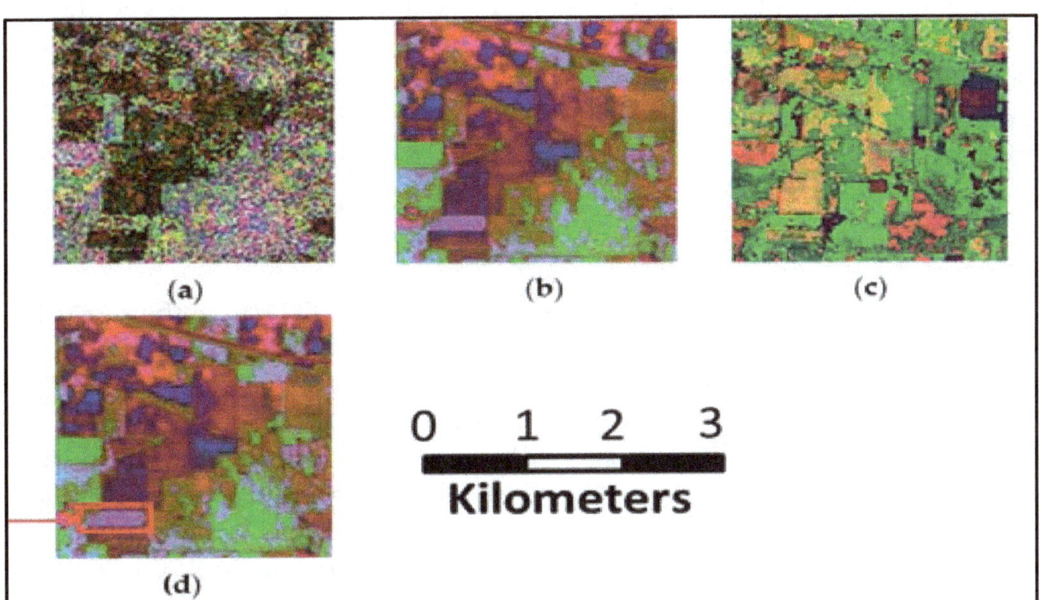

Figure 2. Indian Pines reconstructed using different methods of dimensional reduction: (**a**) LDA; (**b**) LFDA; (**c**) PCA; (**d**) SELF. The number of samples in the training set accounts for only 1% of all samples.

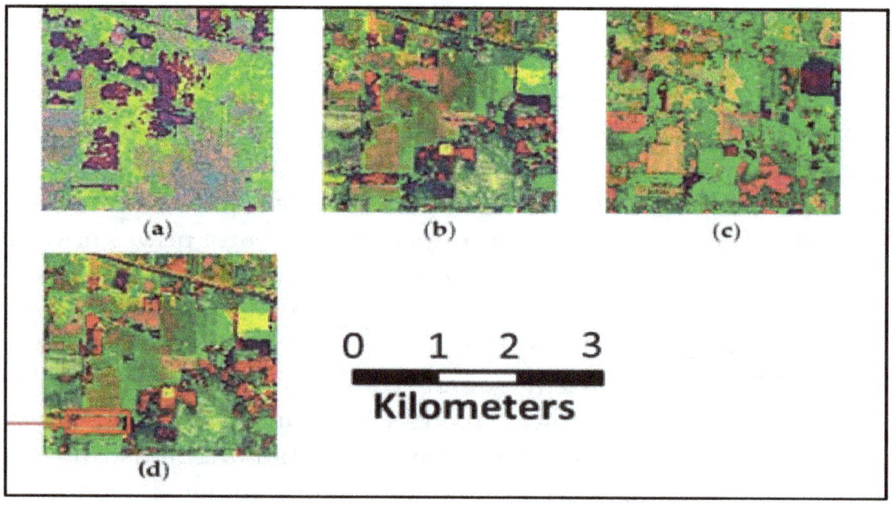

Figure 3. Indian Pines reconstructed using different methods of dimensional reduction: (**a**) LDA; (**b**) LFDA; (**c**) PCA; (**d**) SELF. The number of samples in the training set accounts for 20% of all samples.

2.3. Segment-Tree Filter Construction

The image transformed using SELF is used as input for the graph-based Segment-Tree Filter. The implementation of Segment-Tree Filtering is based on the methodology presented in [22,23], which use the Kruskal algorithm to construct a Minimum Spanning Tree (MST). The general workflow is summarized as follows. First, a graph $G = \{V, E\}$ is constructed for an image ($m \times n$ pixels), where V represents the vertices and each pixel is a vertex. E represents an edge that links four neighbors, and there are $m(n-1) + n(m-1)$ edges in total. A weight w_e is assigned for each edge E to represent the dissimilarity between the linked vertices. Several dissimilarity measures, such as the L1-norm, L2-norm, L∞-norm, and Spectral Angle Mapper (SAM), have been proposed in the literature [11]. In our experiment, SAM is used as the dissimilarity measure.

1. All the edges are sorted in ascending order according to their weights. This step can be performed efficiently using a quicksort algorithm [31], even if the number of edges is very large.
2. For each vertex, we initialize a tree $T_i(V_i, E_i)$.
3. A subtree is then built for each segment. Then, subtrees are merged based on the order of sorted edges. Segment-Tree Filtering is a variant of the conventional MST approach that considers an extra criterion to merge trees [22,32], as shown in Equation (11):

$$w_e \leq \min(\max(w_{Tp}) + \frac{k}{|Tp|}, \max(w_{Tq}) + \frac{k}{|Tq|}) \quad (11)$$

   where w_e is the weight of the edge between subtrees T_p and T_q, $|T_p|$ is the number of vertices in the subtree T_p, and k is a constant. In our experiments, k is set to five times the standard deviation of all weights in the graph. If criterion (3) is satisfied, subtrees T_p and T_q are merged. Criterion (3) establishes a trade-off between the edge weights and the numbers of pixels in the subtrees. Initially, merging subtrees is easy because the number of pixels in each subtree is small. As the number of pixels increases, the criterion becomes increasingly rigorous; therefore, it is adaptive.
4. All the remaining edges that are not part of any subtree are sorted again. If the number of vertices in a subtree is smaller than a threshold T_0, then the subtrees should be merged. In our experiment, $T_0 = 6$. This processing step is based on the improvement presented in [23] to omit small fractions caused by noise. The obtained subtrees are illustrated in Figure 4, in which each color represents an obtained subtree. As shown, constructed subtrees can be used to segment HSIs adaptively.

5. Finally, subtrees are merged until all vertices are included in the trees. For each tree, all the connected vertices exhibit the highest similarity and are within the shortest possible distance. As shown in Figure 5b, the edges of the final tree minimally cross the boundaries between two regions.

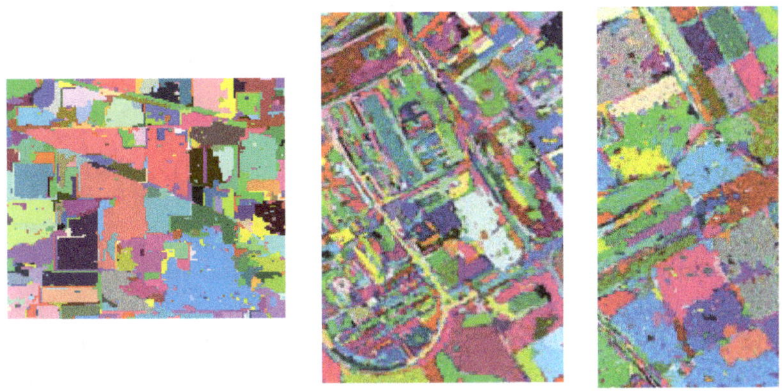

(a) (b) (c)

Figure 4. Subtrees constructed for different standard benchmarks: (**a**) Indian Pines; (**b**) University of Pavia; (**c**) Salinas. Each color segment represents a subtree.

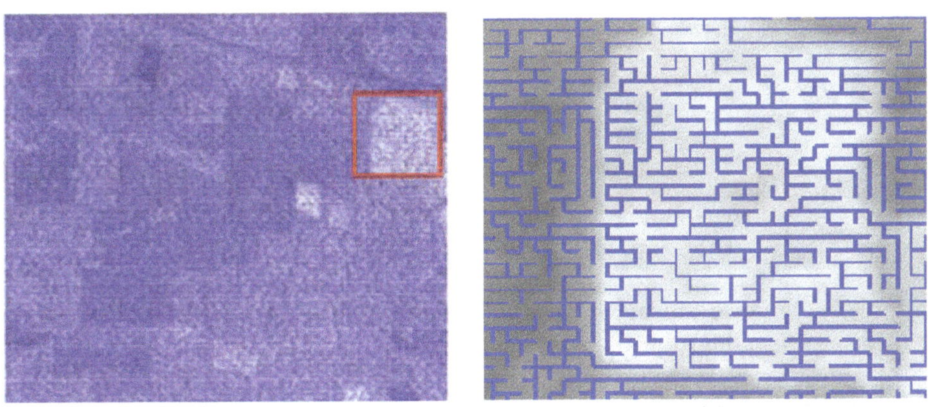

(a) (b)

Figure 5. Tree structure of the Segment-Tree Filter for the Indian Pines dataset: (**a**) Image of the segment tree; (**b**) Close-up of the red rectangular region in (**a**).

2.4. Segment-Tree Filtering

The final step is to filter the initial probability maps using the tree-structure filter. The objective of the filtering process is to compute the aggregated probabilities. In the proposed approach, all vertices contribute to the aggregated probabilities, unlike using local neighbor methods. The non-local, aggregated probabilities $\overline{\mathbf{M}}_p^d$ can be defined as follows:

$$\overline{\mathbf{M}}_p^d = \sum_{q \in I} S(p,q)\mathbf{M}_q^d \tag{12}$$

where $\mathbf{M}_p^d$ is defined in Equation (1). $S(p,q)$ is a weighting function that denotes the weight contribution of pixel q to p:

$$S(p,q) = \exp(-\sum_{i \in path(p,q)} w_{ei}/\gamma) \tag{13}$$

where w_{ei} is the weight of an edge in the tree structure connecting p and q and γ is a constant parameter.

Due to the tree structure, all the aggregated probabilities of class d in the image can be computed efficiently through traversing the tree in two sequential passes. In the first pass, forward filtering occurs from the leaf pixels to the root:

$$\mathbf{M}_p^{d\uparrow} = \mathbf{M}_p^d + \sum_{q \in c(p)} S(p,q)\mathbf{M}_q^{d\uparrow} \tag{14}$$

where $c(p)$ represents all the children of vertex $c(p)$. In the second pass, backward filtering occurs from the root to the leaf pixels:

$$\overline{\mathbf{M}}_p^d = S(pa(p),p)\overline{\mathbf{M}}_{pa(p)}^d + (1 - S^2(pa(p),p))\mathbf{M}_q^{d\uparrow} \tag{15}$$

where $pa(p)$ represents the parent of vertex p.

As Figure 6 shows, vertex V_4 aggregates the probabilities of V_5, V_6, V_7, and itself during the forward filtering step using Equation (16). During the backward filtering step, the probabilities of V_1, V_2, and V_3 contribute to V_4 based on Equation (17). After only two filtering steps, the aggregated probabilities of all vertices are computed, which reflects an extremely low computational complexity. Finally, the classification map is obtained using a simple Winner-Take-All (WTA) rule.

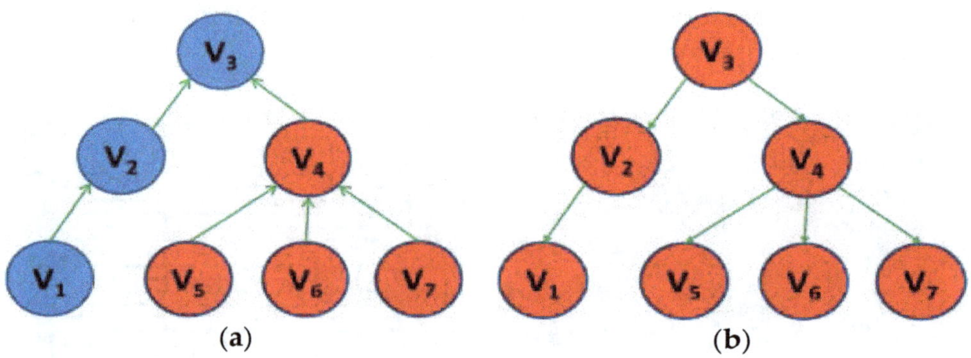

Figure 6. Segment-Tree Filtering in two sequential passes: (**a**) Forward filtering from the leaves to root; (**b**) Backward filtering from the root to leaves.

3. Experiments and Results

The proposed method has been implemented in C++ with the OpenCV library and Lapack library. The implemented code is available by contracting author. Evaluations were performed using three hyperspectral benchmark datasets as below:

1. The first HSI is a 2 × 2 mile portion of agricultural area over the Indian Pines region in Northwest Indiana, which was acquired by NASA's Airborne Visible/Infrared Imaging Spectrometer (AVIRIS) sensor. This scene with a size of 145 × 145 pixels, comprises 202 spectral bands in the wavelength range from 0.4 to 2.5μm, with spatial resolution of 20 m. The ground truth of scene (see Figure 7a) contains 16 classes of interest and total 10,366 samples. Due to the imbalanced number of available labeled pixels and a large number of mixed pixels per class, this dataset creates a challenge in HSI classification.
2. The second HSI is a 103-band image acquired by Reflective Optics Spectrographic Image System (ROSIS-03) sensor over the urban area of the University of Pavia, Italy. The spatial resolution is 1.3 m and the scene contains 610 × 340 pixels and nine classes. The number of samples is 42,776 in total. The ground truth of the scene is shown in Figure 8a.
3. The third HSI is also derived by AVIRIS sensor over Salinas Valley, California. This scene with a size of 512 × 217 pixels, and 204 spectral bands is used for classification. There are 16 classes in the ground truth image, which is shown in Figure 9a.

The overall accuracy (OA), average accuracy (AA), Kappa coefficient, and producer accuracy (PA) are used to assess the classification accuracy.

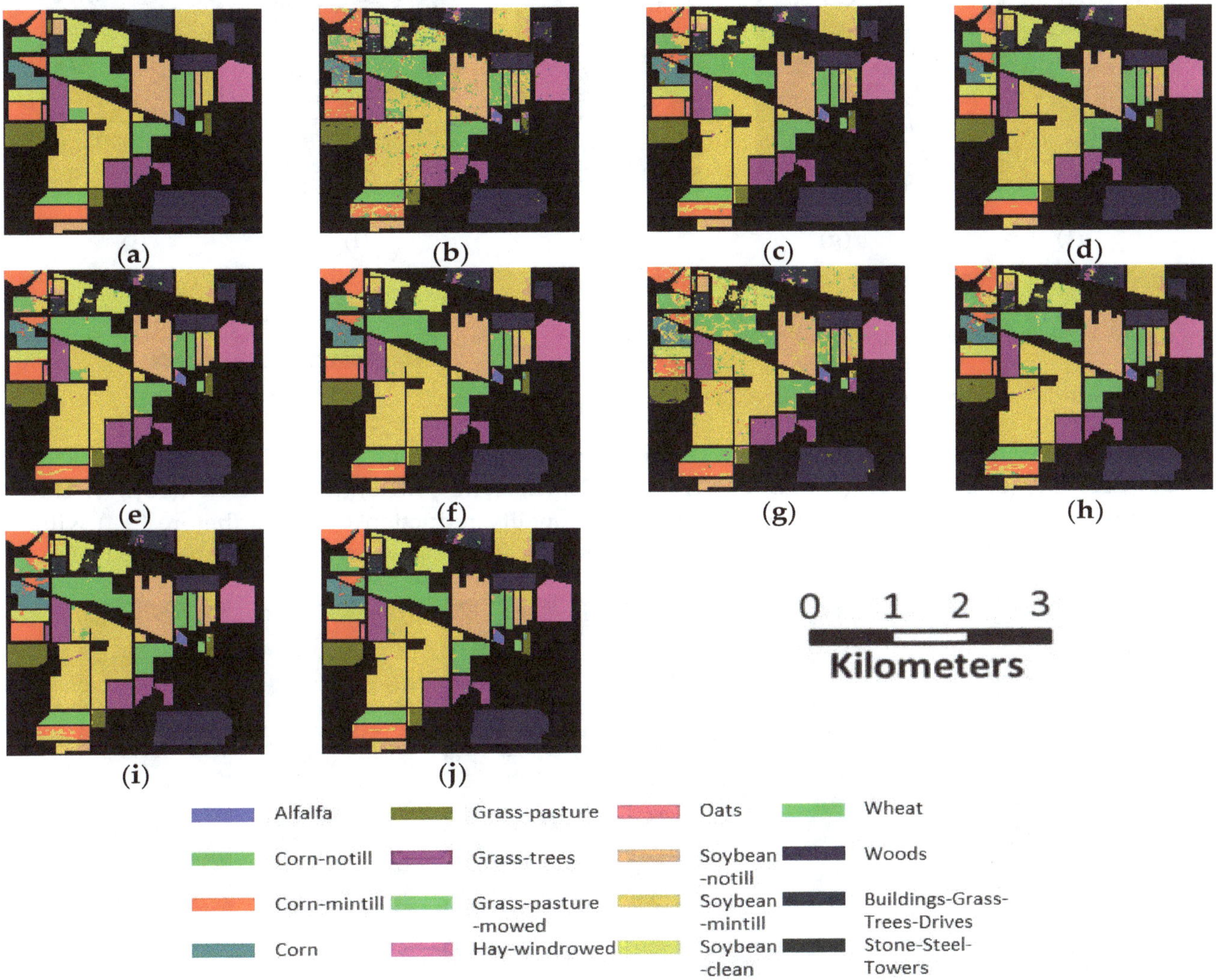

Figure 7. Classification results for Indian Pines: (**a**) Actual values; (**b**) Multi-class SVM; (**c**) ST; (**d**) PCA + ST; (**e**) LDA + ST; (**f**) LFDA + ST; (**g**) EMPs; (**h**) SVM + MRF; (**i**) PCA + GF ([18]); (**j**) The proposed method; (**d–f**) combine different dimensionality reduction methods (before "+") with Segment-Tree Filtering (after "+"); (**g–i**) are other methods of spatial-spectral classification for HSIs.

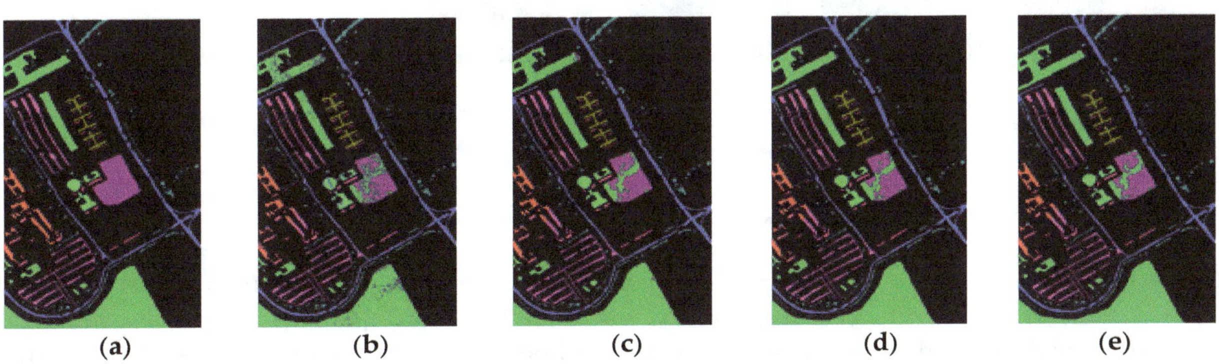

Figure 8. *Cont.*

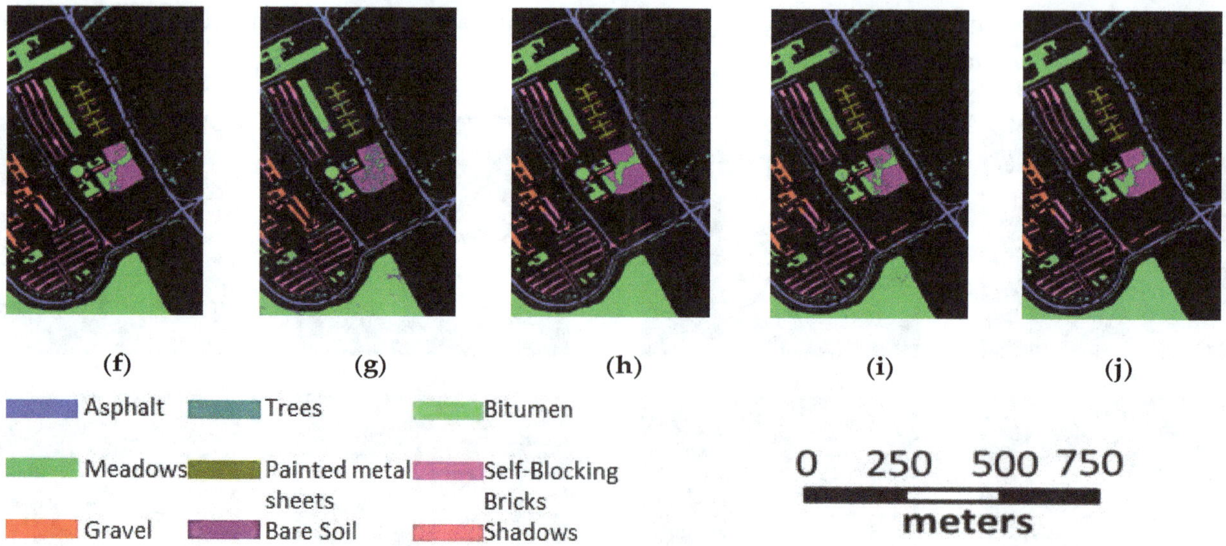

Figure 8. Classification results for Pavia University: (**a**) Actual values; (**b**) Multi-class SVM; (**c**) ST; (**d**) PCA + ST; (**e**) LDA + ST; (**f**) LFDA +ST; (**g**) EMPs; (**h**) SVM + MRF; (**i**) PCA + GF ([18]); (**j**) The proposed method. (**d–f**) combine different dimensionality reduction methods (before "+") with Segment-Tree Filtering (after "+"); (**g–i**) are other methods of spatial-spectral classification for HSIs.

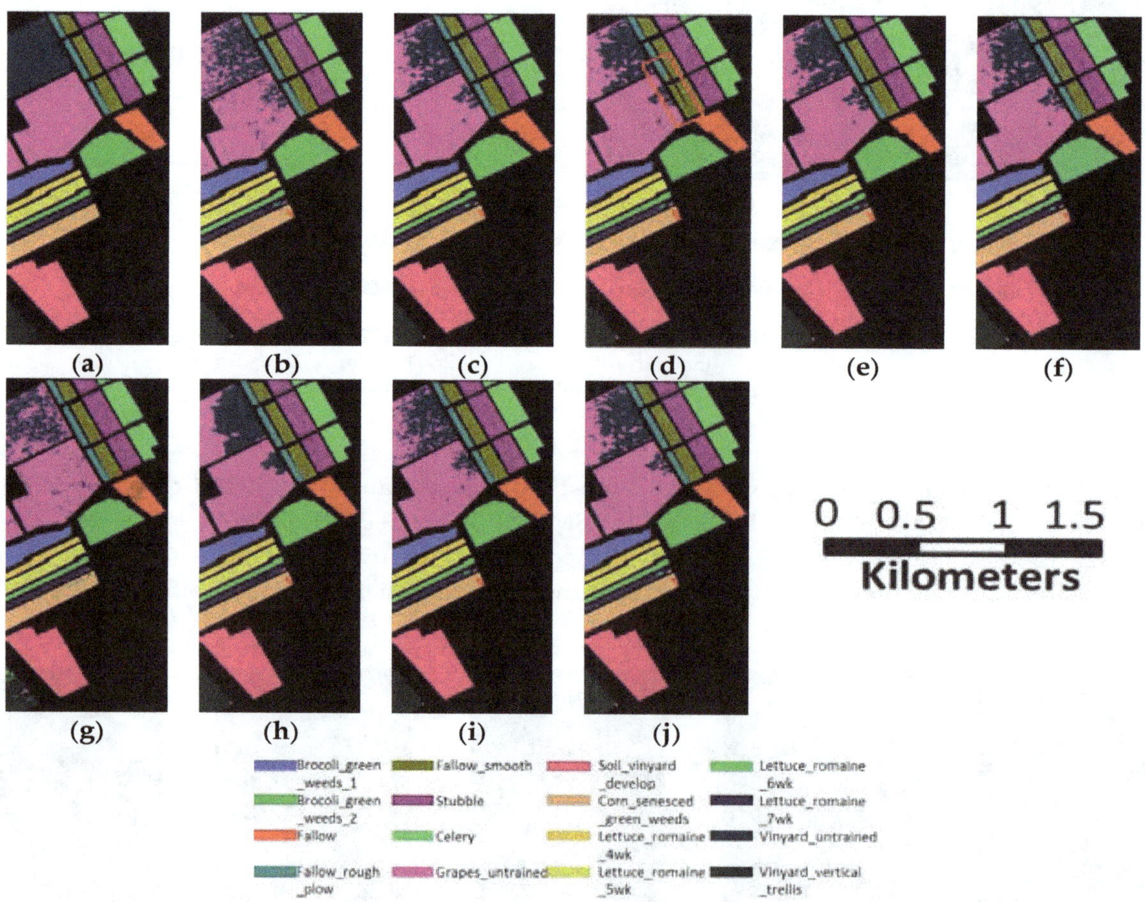

Figure 9. Classification results for Salinas; (**a**) Actual values; (**b**) Multi-class SVM; (**c**) ST; (**d**) PCA + ST; (**e**) LDA + ST; (**f**) LFDA + ST; (**g**) EMPs; (**h**) SVM + MRF; (**i**) PCA + GF ([18]); (**j**) The proposed method; (**d–f**) combine different dimensionality reduction methods (before "+") with Segment-Tree Filtering (after "+"); (**g–i**) are other methods of spatial-spectral classification for HSIs.

3.1. *Influence of Different Parameters*

Some parameters in our proposed method may affect the classification accuracy, such as β, K, and r. Therefore, the Indian Pines dataset is used to test the importance of these parameters to the classification. In this case, the training samples account for 15% of all samples, regardless of their class. When one of the parameters is measured, the other parameters are fixed. Five-fold cross validation is used to tune all parameters. The influences of β, K, and r on the classification accuracy are shown in Figures 10–12, respectively. The influences of β and r are less than 1%, while the influence of K is greater than 1%. Figures 10–12 illustrate that the classification accuracy is the most sensitive to the influence of parameter K. Different dissimilarity measures are adopted during graph-based Segment-Tree Filter construction in our experiments, including the Minkowski distance (from 1 to 6 and infinity) and SAM, as shown in Figure 13. All the parameters affect the classification accuracy by approximately 1% to 2%, and the SAM dissimilarity measure was the largest in our Segment-Tree Filtering approach.

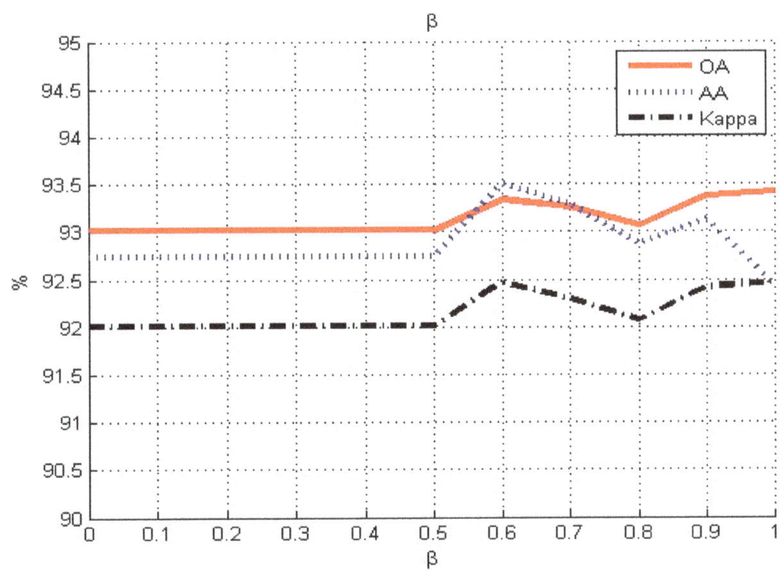

Figure 10. Influence of parameter β on the classification accuracy.

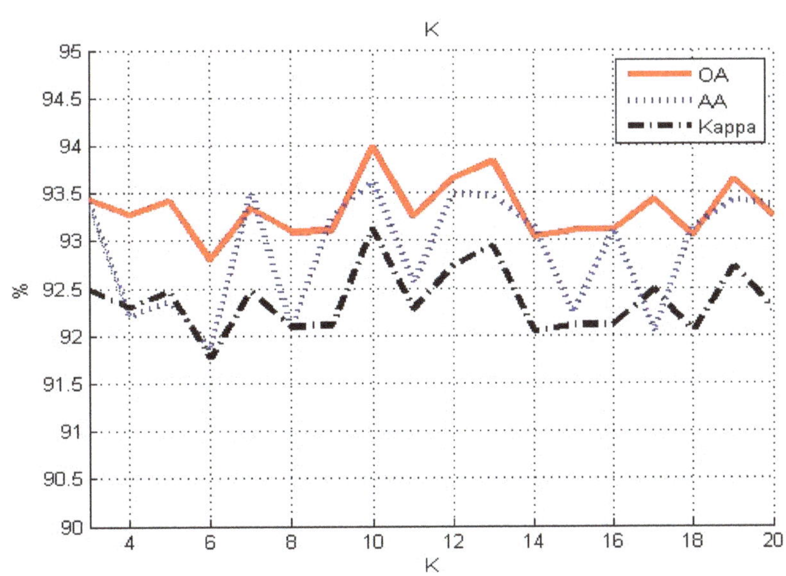

Figure 11. Influence of parameter K on the classification accuracy.

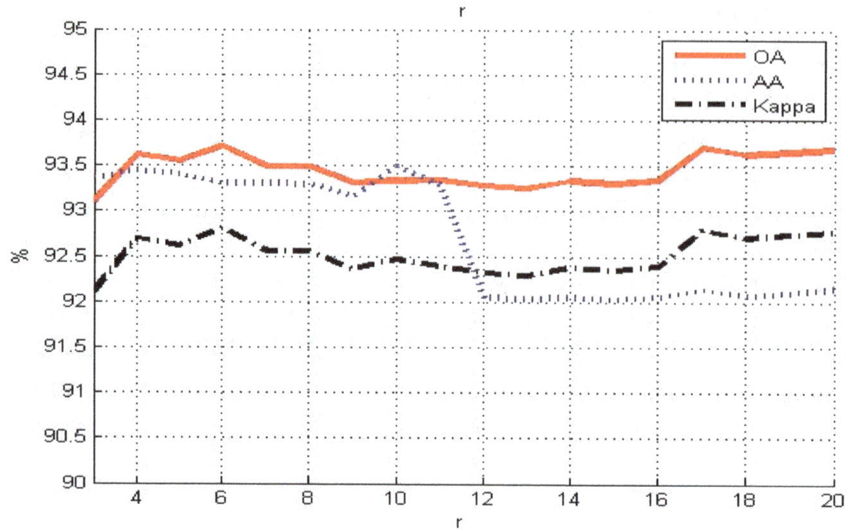

Figure 12. Influence of parameter r on the classification accuracy.

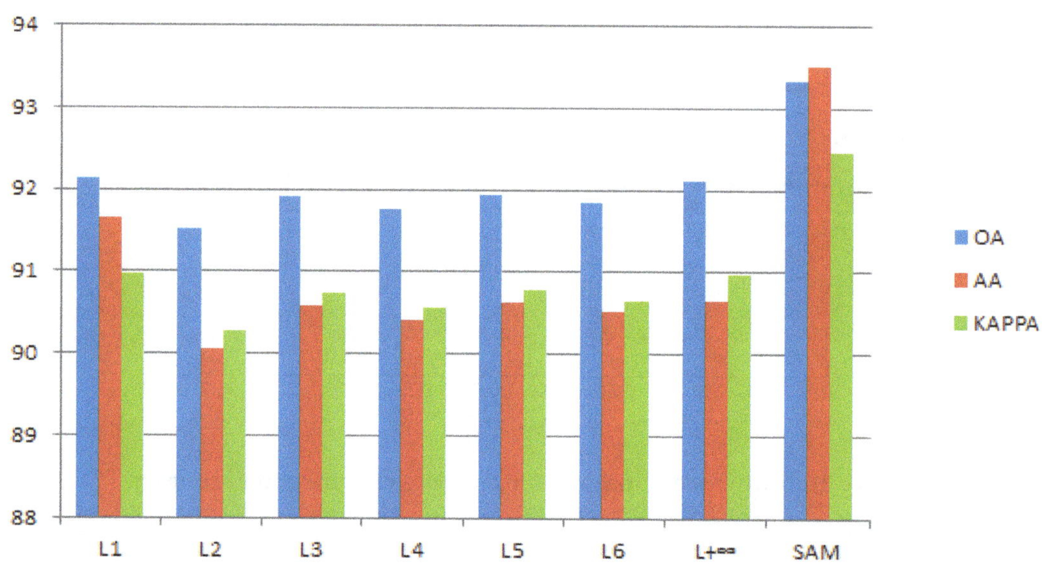

Figure 13. Influences of different dissimilarity measures on the classification accuracy.

3.2. Classification Accuracy Analysis

In this experiment, the training samples account for 15% of all the available samples, regardless of their class. The parameter settings are summarized as follows. In the initial classification step, the parameters were based on observed values, as discussed in Section 2.1. In the SELF transformation step, $\beta = 0.6$, $K = 7$, and $r = 10$. In the graph-based Segment-Tree construction step, SAM is used as the dissimilarity measure, and in the Segment-Tree Filtering step, γ is set as three times the standard deviation of w_e.

All experiments would be repeated five times according to different sampling training-set. The average of the five classification accuracies is recorded. The results of our proposed method based on analyses of the Indian Pines, University of Pavia, and Salinas datasets are shown in Tables 1–3, respectively. The visual results of the initial classification of each HSI are shown in Figure 7b, Figure 8b, and Figure 9b, respectively. Although SVMs are powerful classifiers, the results of the initial classification based solely on spectral features contain substantial noise. However, our proposed method greatly improves the classification accuracy after Segment-Tree Filtering, as shown in Figure 7j, Figure 8j, and Figure 9j. The OA increases by 8.56% for the Indian Pines dataset, 4.64% for the Pavia University dataset, and 1.22% for the Salinas dataset. The largest PA increase is 32.60% for the Indian Pines dataset, followed by 22.71% for the Pavia University dataset and 3.41% for the Salinas dataset.

Table 1. Number of training and test samples from the Indian Pines dataset and the classification accuracies (in percentages) of different methods (the bolded item in each line means the best accuracy).

Class	Training/Test	SVM	ST	PCA + ST	LDA + ST	LFDA + ST	EMP	SVM + MRF	PCA + GF	Proposal
Alfalfa	7/46	58.70	71.74	**91.30**	76.09	82.61	82.61	83.30	80.43	**91.30**
Corn-N	214/1428	81.65	**92.86**	89.22	90.68	89.64	71.57	88.51	89.29	92.37
Corn-M	125/830	75.90	83.25	96.51	91.81	92.65	76.99	80.84	81.20	**92.16**
Corn	36/237	64.56	85.23	76.79	91.56	**93.67**	62.87	87.34	89.87	88.61
Grass-Pa	73/483	89.23	90.89	92.75	91.10	92.13	78.05	91.72	93.17	**93.37**
GrassT	109/730	96.99	**99.73**	99.45	99.59	99.45	98.36	99.18	100	**99.73**
GrassP	5/28	89.29	96.43	92.85	**100**	96.42	89.29	96.43	96.43	96.43
Hay-W	72/478	98.74	99.79	**100**	**100**	**100**	98.95	**100**	**100**	**100**
Oats	5/20	75.00	**100**	**100**	90.00	**100**	90.00	0.00	0.00	**100**
Soy-N	146/972	82.10	**89.20**	86.31	88.17	88.16	84.77	87.14	84.26	88.17
Soy-M	368/2455	85.62	**94.70**	95.62	95.11	94.50	93.40	94.87	95.52	94.58
Soy-C	89/593	74.20	94.77	**99.66**	96.46	97.64	71.16	93.76	95.45	97.98
Wheat	31/205	96.59	99.51	99.51	99.51	99.51	97.56	99.51	**100**	99.51
Woods	190/1265	95.42	97.23	98.50	97.94	98.26	98.26	96.60	**98.66**	98.10
Building	58/386	61.14	59.07	60.62	58.55	60.33	**85.23**	62.79	76.42	64.88
Stone-ST	14/93	87.10	87.10	**98.92**	87.10	**98.92**	60.22	88.17	94.62	**98.92**
OA		84.78	92.11	93.32	92.83	93.01	70.48	91.22	92.20	**93.34**
AA		82.01	90.09	92.44	90.85	92.74	67.26	89.07	85.96	**93.50**
Kappa		82.60	90.97	**92.47**	91.80	92.00	67.28	90.11	91.08	**92.47**

Table 2. Number of training and test samples from the Pavia University dataset and the classification accuracies (in percentages) of different methods (the bolded item in each line means the best accuracy).

Class	Training/Test	SVM	ST	PCA + ST	LDA + ST	LFDA + ST	EMP	SVM + MRF	PCA + GF	Proposal
Asphalt	995/6631	93.03	96.95	95.97	96.03	95.99	93.53	**97.10**	96.44	96.67
Meadows	2797/18,649	95.06	99.33	99.88	99.81	99.88	95.69	**100**	99.18	99.79
Gravel	315/2099	66.32	72.46	71.46	72.74	71.46	76.13	69.03	**73.80**	71.08
Trees	460/3064	93.05	94.65	94.39	94.09	94.45	98.63	90.31	**97.03**	93.79
P-M-S	202/1345	99.70	99.55	99.78	99.63	99.78	84.31	**100**	**100**	99.70
Bare Soil	754/5029	66.65	70.33	71.01	70.01	71.12	76.89	70.87	68.50	**71.30**
Bitumen	200/1330	77.21	96.99	97.07	95.78	97.74	84.81	83.38	97.22	**99.92**
Self-B B	552/3682	91.55	98.37	97.99	98.09	97.83	89.19	98.29	99.76	**98.80**
Shadows	142/947	100	98.94	96.09	91.12	96.09	77.82	98.10	**100**	98.32
OA		89.25	93.74	93.76	93.51	93.78	90.75	93.21	93.78	**93.89**
AA		86.95	91.95	91.52	90.81	91.59	86.34	89.68	**92.43**	92.15
Kappa		85.60	91.58	91.59	91.26	91.62	87.72	90.82	91.62	**91.77**

Table 3. Number of training and test samples from the Salinas dataset and the classification accuracies (in percentages) of different methods (the bolded item in each line means the best accuracy).

Class	Training/Test	SVM	ST	PCA + ST	LDA + ST	LFDA + ST	EMP	SVM + MRF	PCA + GF	Proposal
g_w_1	301/2009	98.80	**100**	**100**	**100**	**100**	98.95	**100**	99.80	**100**
g_w_2	491/3276	99.97	**100**	99.97	**100**	99.97	99.68	**100**	99.97	99.97
Fallow	296/1976	98.68	**100**	**100**	99.75	**100**	90.13	**100**	99.89	**100**
Fallow_r_p	209/1394	99.35	98.21	60.04	99.50	98.92	99.14	99.43	**100**	98.92
Fallow_s	401/2678	98.28	98.81	**99.22**	98.62	98.62	96.83	99.10	98.92	98.62
Stubble	593/3959	99.97	99.92	99.90	99.97	99.87	99.75	**100**	99.97	99.87
Celery	536/3579	99.35	99.61	**99.66**	99.52	99.64	98.46	99.89	99.78	99.64
Grapes_u	1690/11,271	92.11	95.03	95.20	95.16	95.51	95.40	94.58	95.11	**95.52**
Soil_v_d	930/6203	99.43	99.50	**99.87**	99.48	99.69	96.45	99.79	99.47	99.68
C_s_g_w	491/3278	93.62	96.06	94.63	95.18	95.79	94.97	**96.86**	95.73	95.64
L_r_4	160/1068	96.44	99.25	96.72	97.94	**99.63**	94.66	98.69	99.25	**99.63**
L_r_5	289/1927	99.53	**100**	95.69	**100**	**100**	99.79	**100**	**100**	**100**
Le_r_6	137/916	98.03	97.82	98.25	97.49	98.14	95.09	98.25	**98.80**	98.14
L_r_7	160/1070	92.05	93.74	93.74	93.93	93.48	96.07	**96.26**	93.64	93.49
V_u	1090/7268	56.89	57.33	56.70	56.84	57.75	50.30	57.04	57.66	**57.85**
V_v	271/1807	98.83	99.06	**99.11**	99.00	99.06	89.60	99.39	99.45	99.06
OA		91.56	92.59	91.35	92.49	92.77	90.32	92.76	92.72	**92.78**
AA		95.08	95.89	93.05	95.77	96.00	93.45	**96.17**	96.09	96.00
Kappa		90.57	91.73	90.33	91.60	91.91	89.17	**91.92**	91.85	**91.92**

3.3. Influences of Different Techniques for Dimensionality Reduction

In the above section, we illustrated that incorporating spatial information can improve the classification accuracy. In the following section, we will evaluate how different techniques for dimensionality reduction can affect the classification accuracy. First, we assess the classification accuracy with/without dimensionality reduction. If no dimensionality reduction method is used, the Segment-Tree Filter is constructed using the original HSI. Figure 7c, Figure 8c, and Figure 9c show the results of classification using this scheme. Because redundant bands negatively affect segmentation, the classification accuracy using the original bands in the three HSI datasets is less than that produced using the proposed method. The fourth column in Tables 1–3 illustrates that reducing the dimensionality is necessary to increase the classification accuracy, and the average OA increased by approximately 0.53%. In addition, dimensionality can considerably improve the computational speed.

Next, we examine how different methods of dimensionality reduction can affect the classification accuracy. Figures 7–9 show the classification results for various methods, including the PCA, LDA, and LFDA methods; however, the classification accuracy produced by SELF is better than the accuracies of those methods.

Figure 7d–f, Figure 8d–f, and Figure 9d–f show the classification results for PCA, LDA, and LFDA, respectively. As expected, the OA and Kappa coefficient of SELF is the highest among these methods and the AA is the highest for the Indian Pines dataset and the second highest for the Pavia University and Salinas datasets. Although PCA sometimes performed better than SELF in some PAs, PCA with Segment-Tree Filtering is not a robust algorithm and can easily over-smooth spatial information, as illustrated in Figure 9d. The results illustrated in the red rectangle exhibit considerable classification error-based PCA, and the PA of Fallow_r_p decreases from 99.35% to 60.04% in the fourth row of Table 3.

3.4. Comparison to Other Methods of Spectral-Spatial Classification

As discussed in Section 1, spectral-spatial classification is a powerful method of combining contextual information. Therefore, we compared the proposed method to other common methods of spectral-spatial classification. We implemented the following spectral-spatial classification algorithms in our analysis.

1. The first algorithm is based on EMPs [9]. In [9], a neural network classifier was applied; however, an SVM is used instead of a back-propagation neural network to create a fair comparison. The EMPs are shown in Figure 7g, Figure 8g, and Figure 9g.
2. The second approach uses MRFs [6] with Multi-class SVM, which is, to the best of our knowledge, the state of the art method for spatial-spectral image classification based on remote sensing. Multi-class SVM is used as the initial classifier, and the spatial optimization is performed using the max-flow/min-cut algorithms. In our experiment, α-expansion is adopted, and the regularization coefficient is fixed to 0.5. The results of the SVM with MRFs are shown in Figure 7h, Figure 8h, and Figure 9h.
3. The third approach is based on a Guided Filter with PCA [18]. The size and blur degree in Guided Filtering are tuned adaptively by cross-validation. The results of this classification are shown in Figure 7i, Figure 8i, and Figure 9i.

The proposed method produced results that were more accurate than those of the EMP-based method for the Indian Pines dataset; however, the results were similar for the other datasets. This result suggests that the proposed method is more suitable for different datasets compared to the EMP-based approach.

The classification accuracies of the proposed method and the SVM method with MRFs were nearly equal. This result indicates that the proposed method achieved an accuracy comparable to that of a state of the art method for spatial-spectral HSI classification. Furthermore, when the number of training samples within a class is very small, e.g., the PAs of the "Oats" class in Table 1, the classification accuracy of the proposed approach is perfect, while the SVM with MRFs method fails. This occurs because MRFs over smooth spatial features; thus, the regularization parameter requires complex tuning steps. Therefore, our proposed approach is more robust than the SVM with MRFs method.

The classification accuracy of the proposed method is slightly higher than that of the Guided Filter based on the OAs of the three datasets.

However, we computed the computational times associated with the three HSIs based on GF with PCA [18] and our method. We assume that N pixels, M bands, D classes, and local window size R are used for the reconstruction of the HSIs. The complexity of Segment-Tree Filtering is O (ND), and that of Guided Filtering is O (NDM). For the Indian Pines dataset, GF needed 2.6138 s to process 10 bands in the compressed dataset, while the proposed method required only 0.0536 s (all programs were executed using an Intel(R) Xeon(R) CPU E5-2620 with 24 GB of RAM.). Guided Filtering is slower because it computes the inverse covariance matrix for each sample. In extreme cases, Guided Filtering using an original HSI as the guide image can be time consuming and ineffective.

3.5. Effect of the Training Set on Classification

In this section, we assess how the number of training samples affects the classification accuracy of the proposed method. Thus, we varied the training sample size from 1% of all samples to 20% of all samples. We found that when the training sample size increases, the classification accuracy also increases. Therefore, we only illustrate how the number of training samples affects the classification accuracy of the Indian Pines dataset, as shown in Figure 14. The classification accuracy improves considerably until the number of pixels in the training set reaches 5% of the total pixel number. Then, the accuracy continues to improve but at a lower rate.

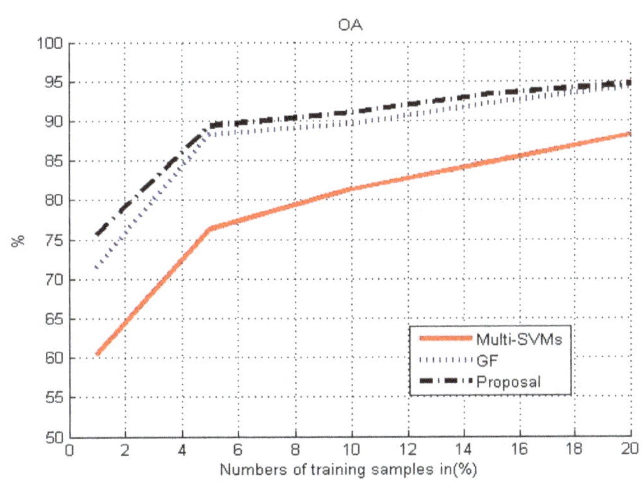

Figure 14. Classification accuracy based on the number of training samples for the Indian Pines dataset.

We also evaluate the effects of the training sample size on the classification accuracies of the Guided Filter [18] and Multi-class SVM methods. As shown in Figure 14, the proposed method and Guided Filter approach improve the classification accuracy regardless of the size of the training set, and the proposed method yields better results. Furthermore, the advantage is larger when the size of the training set is small.

4. Conclusions

A novel and efficient approach based on a Segment-Tree Filter has been proposed for hyperspectral image classification. Our proposed approach is based on spatial-spectral filtering, which is a special EAF. This filter construction utilizes both spectral features using a SELF transformation and spatial information using a Segment-Tree algorithm. After an initial classification map is generated by Multi-class SVM, we can filter the map using the Segment-Tree Filter. One advantage of our proposed approach is that the classification accuracy has been improved dramatically. Experimental results show that the proposed method produced a high classification accuracy for hyperspectral image benchmark sets, including 93.34% for the Indian Pines dataset, 93.89% for the Pavia University dataset, and 92.78% for the Salinas dataset. Compared to other spatial-spectral methods, another advantage of the proposed method is that it provides a more robust classification approach for different datasets and training sets of different sizes.

In the future, two major aspects of our approach could be improved. First, dimensionality reduction using could be performed in SELF to reconstruct HSIs with nonlinear projections. Second, other classifiers, including fuzzy classifiers, could be applied to improve the classification accuracy.

Acknowledgments: The authors would like to thank. D. Landgrebe and P. Gamba for providing the Indian Pines and Pavia University respectively hyperspectral data set available to the community, and Website http://www.ehu.eus/ccwintco/index.php?title=Hyperspectral_Remote_Sensing_Scenes for downloading all the three hyperspectral data sets. This research was supported in part by the Natural Science Foundation of China with Project IDs 41301499 and 41401474, and the National Science and Technology Major Project with Project ID 30-Y20A03-9003-15/16.

Author Contributions: Lu Li and Chengyi Wang conceived and designed the experiments; Lu Li performed the experiments; Lu Li and Jingbo Chen analyzed the data; Lu Li and Jianglin Ma wrote the paper.

References

1. Melgani, F.; Bruzzone, L. Classification of hyperspectral remote sensing images with support vector machines. *IEEE Trans. Geosci. Remote Sens.* **2004**, *42*, 1778–1790. [CrossRef]
2. Böhning, D. Multinomial logistic regression algorithm. *Ann. Inst. Stat. Math.* **1992**, *44*, 197–200. [CrossRef]
3. Li, J.; Bioucas-Dias, J.; Plaza, A. Semi-supervised hyperspectral image segmentation using multinomial logistic regression with active learning. *IEEE Trans. Geosci. Remote Sens.* **2010**, *48*, 4085–4098.
4. Ratle, F.; Camps-Valls, G.; Weston, J. Semi-supervised neural networks for efficient hyperspectral image classification. *IEEE Trans. Geosci. Remote Sens.* **2010**, *48*, 2271–2282. [CrossRef]
5. Li, W.; Prasad, S.; Fowler, E. Hyperspectral image classification using Gaussian mixture models and Markov random fields. *IEEE Lett. Geosci. Remote Sens.* **2014**, *11*, 153–157. [CrossRef]
6. Li, J.; Bioucas, M.; Antonio, P. Spectral-spatial hyperspectral image segmentation using subspace multinomial logistic regression and Markov random fields. *IEEE Trans. Geosci. Remote Sens.* **2012**, *50*, 809–823. [CrossRef]
7. Boykov, O.; Veksler, O.; Zabih, R. Fast approximate energy minimization via graph cuts. *IEEE Trans. Pattern Anal. Mach. Intell.* **2001**, *23*, 1222–1239. [CrossRef]
8. Li, J.; Bioucas, M.; Antonio, P. Spectral-spatial classification of hyperspectral data using loopy belief propagation and active learning. *IEEE Trans. Geosci. Remote Sens.* **2013**, *51*, 844–856. [CrossRef]
9. Benediktsson, A.; Palmason, A.; Sveinsson, R. Classification of hyperspectral data from urban areas based on extended morphological profiles. *IEEE Trans. Geosci. Remote Sens.* **2005**, *43*, 480–491. [CrossRef]
10. Zortea, M.; de Martino, M.; Serpico, S. A SVM Ensemble approach for spectral-contextual classification of optical high spatial resolution imagery. In Proceeding of the IEEE International Conference on Geoscience and Remote Sensing Symposium, Boston, MA, USA, 23–28 July 2007.
11. Yuliya, T.; Jocelyn, C.; Jón Atli, B. Segmentation and classification of hyperspectral images using minimum spanning forest grown from automatically selected markers. *IEEE Trans. Syst. Man Cybern. B Cybern.* **2010**, *40*, 1267–1279. [CrossRef]
12. Fu, W.; Li, S.; Fang, L. Spectral-spatial Hyperspectral image classification via superpixel merging and sparse representation. In Proceeding of the IEEE Proceedings on Geoscience and Remote Sensing Symposium, Milan, Italy, 26–31 July 2015.
13. Duan, W.; Li, S.; Fang, L. Spectral-spatial hyperspectral image classification using superpixel and extreme learning machines. In *Pattern Recognition, Proceedings the 6th Chinese Conference, Changsha, China, 17–19 November 2014*; Springer: Berlin/Heidelberg, Germany, 2014; Volume 483, pp. 159–167. [CrossRef]

14. Weihua, S.; David, M. Trilateral filter on multispectral imagery for classification and segmentation. *Proc. SPIE* **2011**. [CrossRef]
15. Tarabalka, Y.; Benediktsson, A.; Chanussot, J. Spectral-spatial classification of hyperspectral imagery based on partitional clustering techniques. *IEEE Trans. Geosci. Remote Sens.* **2009**, *47*, 2973–2987. [CrossRef]
16. Fang, L.; Li, S.; Duan, W. Classification of hyperspectral images by exploiting spectral-spatial information of superpixel via multiple kernels. *IEEE Trans. Geosci. Remote Sens.* **2015**, *53*, 6663–6673. [CrossRef]
17. He, K.; Sun, J.; Tang, X. Guided image filtering. *IEEE Trans. Pattern Anal. Mach. Intell.* **2013**, *35*, 1397–1409. [CrossRef] [PubMed]
18. Kang, X.; Li, S.; Jón Alti, B. Spectral-spatial hyperspectral image classification with edge-preserving filtering. *IEEE Trans. Geosci. Remote Sens.* **2014**, *52*, 2666–2677. [CrossRef]
19. Hosni, A.; Rhemann, C.; Bleyer, M.; Rother, C.; Gelautz, M. Fast cost-volume filtering for visual correspondence and beyond. *IEEE Trans. Pattern Anal. Mach. Intell.* **2013**, *35*, 504–511. [CrossRef] [PubMed]
20. Hosni, A.; Rhemann, C.; Bleyer, M.; Gelautz, M. Temporally consistent disparity and optical flow via efficient spatio-temporal filtering. In Proceedings of the 5th Pacific Rim Symposium, Gwangju, Korea, 20–23 November 2011; Springer: Berlin, Germany, 2011; pp. 165–177.
21. Shutao, L.; Xudong, K.; Jianwen, H. Image fusion with guided filtering. *IEEE Trans. Image Process.* **2013**, *22*, 2864–2875. [CrossRef] [PubMed]
22. Felzenszwalb, F.; Huttenlocher, P. Efficient graph-based image segmentation. *Int. J. Comput. Vis.* **2004**, *59*, 167–181. [CrossRef]
23. Mei, X.; Sun, X.; Dong, W.; Wang, H.; Zhang, X. Segment-tree based cost aggregation for stereo matching. In Proceeding of the 2013 IEEE Conference on Computer Vision and Pattern Recognition, Portland, OR, USA, 23–28 June 2013.
24. Yang, X. A Non-local cost aggregation method for stereo matching. In Proceeding of the 2012 IEEE Conference on Computer Vision and Pattern Recognition, Providence, RI, USA, 16–21 June 2012.
25. Wang, Q.; Lin, J.; Yuan, Y. Salient band selection for hyperspectral image classification via manifold ranking. *IEEE Trans. Neural Netw. Learn. Syst.* **2016**, *27*, 1279–1289. [CrossRef] [PubMed]
26. Yuan, Y.; Lin, J.; Wang, Q. Dual clustering based hyperspectral band selection by contextual analysis. *IEEE Trans. Geosci. Remote Sens.* **2016**, *54*, 1431–1445. [CrossRef]
27. Sugiyama, M.; Idé, T.; Nakajima, S.; Sese, J. Semi-supervised local Fisher discriminant analysis for dimensionality reduction. *Mach. Learn.* **2010**, *78*, 35–61. [CrossRef]
28. Fisher, R.A. The use of multiple measurements in taxonomic problems. *Ann. Eugen.* **1936**, *7*, 179–188. [CrossRef]
29. Sugiyama, M. Dimensionality reduction of multimodal labeled data by local fisher discriminant analysis. *J. Mach. Learn. Res.* **2007**, *8*, 1027–1061.
30. Huang, T.K.; Weng, R.C.; Lin, C.J. Generalized bradley-terry models and multi-class probability estimates. *J. Mach. Learn. Res.* **2006**, *7*, 85–115.
31. Cormen, T.; Leiserson, C.; Rivest, R.; Stein, C. *Introduction to Algorithms*, 3rd ed.; MIT Press: London, UK, 2005; pp. 170–174.
32. Shi, J.; Malik, J. Normalized cuts and image segmentation. *IEEE Trans. Pattern Anal. Mach. Intell.* **2000**, *22*, 888–905. [CrossRef]

11

Cost-Effective Class-Imbalance Aware CNN for Vehicle Localization and Categorization in High Resolution Aerial Images

Feimo Li [1,2,*], Shuxiao Li [1,2,*], Chengfei Zhu [1,2], Xiaosong Lan [1,2] and Hongxing Chang [1,2]

1 Institute of Automation Chinese Academy of Sciences, Beijing 100190, China; chengfei.zhu@ia.ac.cn (C.Z.); lanxiaosong2012@ia.ac.cn (X.L.); hongxing.chang@ia.ac.cn (H.C.)
2 University of Chinese Academy of Science, Beijing 100049, China
* Correspondence: lifeimo2012@ia.ac.cn (F.L.); shuxiao.li@ia.ac.cn (S.L.);

Academic Editors: Qi Wang, Nicolas H. Younan, Carlos López-Martínez, Gonzalo Pajares Martinsanz, Xiaofeng Li and Prasad S. Thenkabail

Abstract: Joint vehicle localization and categorization in high resolution aerial images can provide useful information for applications such as traffic flow structure analysis. To maintain sufficient features to recognize small-scaled vehicles, a regions with convolutional neural network features (R-CNN) -like detection structure is employed. In this setting, cascaded localization error can be averted by equally treating the negatives and differently typed positives as a multi-class classification task, but the problem of class-imbalance remains. To address this issue, a cost-effective network extension scheme is proposed. In it, the correlated convolution and connection costs during extension are reduced by feature map selection and bi-partite main-side network construction, which are realized with the assistance of a novel feature map class-importance measurement and a new class-imbalance sensitive main-side loss function. By using an image classification dataset established from a set of traditional real-colored aerial images with 0.13 m ground sampling distance which are taken from the height of 1000 m by an imaging system composed of non-metric cameras, the effectiveness of the proposed network extension is verified by comparing with its similarly shaped strong counter-parts. Experiments show an equivalent or better performance, while requiring the least parameter and memory overheads are required.

Keywords: vehicle localization; vehicle classification; high resolution; aerial image; convolutional neural network (CNN); class imbalance

1. Introduction

For most of the sliding window-based vehicle detection methods involving localization and categorization, predictions are often performed in a separated manner, where the categories are estimated after the positional information is obtained. In the localization process—also called vehicle detection in its narrow sense—the positional existence of vehicles is estimated by analyzing the features extracted from the sliding window that moves across the region of interest with a pre-defined route and stepping pattern. The features used for vehicle detection can either be hand-crafted shallow descriptors or the deep features generated by convolutional neural network (CNN). Shallow features such as Haar [1], histogram of oriented gradients (HOG) [2,3], and local binary pattern (LBP) [3], etc.—although they are less robust and accurate as the deep ones—can make a good compromise between speed and efficiency when the computational resources or the quantity of training samples are very limited. However, once these limitations no longer exist, the detection methods based on deep features are often superior with strong resistance to disturbances in scale, lighting condition, and shadow, and their supreme performances have been repeatedly verified in many studies [4–8]. For these CNN-based methods, their underlying structures generally follow the regions with convolutional neural network features (R-CNN) [9] or its accelerated variants [10–13] with region of interest (ROI)-pooling [14]. More specifically, the R-CNN detector—whose features are calculated from the full-scale input image without sub-sampling—despite being primitive, turns out to be informative for recognizing small objects. Because of this, in large aerial images with small-scaled vehicles, R-CNN-like structure [5–8] is often preferred over those with ROI-pooling [4,15], which is also used in this article. Moreover, it can be accelerated by lossless preprocessing means such as saliency detection [16,17] and objectness filtering [8].

Once the vehicle locations are obtained, they are fed to the subsequent categorization process as positional indications to extract features. Similar to the localization process, features for classification can be produced by either shallow or deep models. At present, limited by the number of publicly available high-resolution aerial image datasets, only a small number of vehicle detection methods involve a classification procedure [18–20]. Among these limited publications, authors in both [19] and [20] tried to categorize vehicles by the "SVM + feature" strategy, while in [20] the strong influence of the class-imbalance issue on classification accuracies has been observed.

The separated estimation scheme discussed above is quite natural, and has been adopted for the positional classification of many general objects [10,12,13,21]. However, it could be troublesome for classifying targets as small as vehicles. Considering a private car only six pixels in width, any location error greater than four pixels will miss the main body of the vehicle and make the following categorization meaningless. Detecting objects in dense scenes can be untangled via density estimation [22] or object counting [23], which has already been validated for congested traffic scene classification [24]. While in this article, without loss of generality, taking the R-CNN detector as a common CNN-based classifier as in [7,8], the previously mentioned cascaded localization error can be avoided by treating the samples with deviation as a negative class and classifying them alongside the accurately centered but differently typed positives.

This arrangement primarily solves the problems caused by the small target scale, and strictly constrains type classification to those accurately located situations. Except for that, however, the introduction of a large quantity of negatives further skews the unbalanced categorical distributions between vehicle types. To address this problem, a bi-partite network extension driven by a class-imbalance-aware cost function is proposed. This cost function is designed based on the idea of providing the two network components with different training losses, intentionally correlating the extended component to the minority classes which are badly classified. Moreover, to reduce the extension costs, the extended components are built with feature maps from lower convolutional layers selected by a novel importance measurement. Notably, compared with other similarly-shaped structures, this proposed modification scheme is capable of achieving equal or better performance with much less extension overhead.

The rest of the paper is arranged as follows: Related and similar works are discussed in Section 2. The CNN basics and the semantic interpretation of convolutional kernels are given in Section 3. The proposed extension and its details are introduced in Section 4. Dataset preparation, experiment setup, and analysis of experimental results are presented in Section 5. Conclusive discussions on the experiments are given in Section 6. Section 7 concludes the paper.

2. Related Work

Class imbalance is a ubiquitous issue existing in nearly every real-life classification problem. As it has been intensively studied for more than two decades, many comprehensive and insightful reviews have been published to generalize the methods on this topic [25–28]. According to [28], these proposed treatments generally fall within three categories: data-level, algorithm-level, and hybrid treatments. The data-level methods focus on balancing the training samples, modifying their distributions via over-sampling or under-sampling. Typical techniques would include synthetic minority over-sampling technique (SMOTE) [29] and many of its variants, such as adaptive synthetic sampling (ADASYN) [30] and cluster-based oversampling (CBO) [31]. The algorithm-level methods—which are mostly based on the cost-sensitive principle [32,33]—alleviate the bias with majority classes by assigning greater penalties for the minority ones in training. The hybrid methods (e.g., the ensemble style classifiers [34]) take the advantages of the previous two for further performance enhancement, which is common, as mentioned in [25,28].

All of the previously mentioned means are for "shallow" models, but their class-imbalance-addressing principles still apply to the deep learning-based classifiers [35]. For instance, the re-sampling tricks work fine [36,37], although some more advanced dealing methods (e.g., the generative adversarial network (GAN) [38]) should be used to avoid noise and over-fitting in the re-sampling. Similarly, algorithm-level cost function reformation is also widely applicable [39–41], where the softmax loss [42,43], cross-entropy loss [39], and logistic regression [44] are mostly taken as the basis format. More recently, a new branch of cost-sensitive methods based on improving the underlying micro feature space structure have appeared, and they have achieved a significant improvement by constraining the relative sample distances [35,45–47]. Representative methods in this category include the triplet loss [43,45], quintuplet loss [35], and the center loss [47], which are now hotly debated in the academy.

Although the proposed method in this article generally follows the algorithm-level principle, it is more concerned about achieving a robust performance improvement with less or no influence on the original structure. This goal is achieved by re-balancing the classification bias with the assistance of an extra network component, where the structural expansion cost is kept at a minimum by the incorporative usage of feature map selection.

Plain extension of the convolutional kernel was theoretically analyzed in [48] without involving the class-imbalance issue. Structural extension is a common method for network performance enhancement whose underlying intentions focus either on feature space enhancement [41,48–50] or strong prior generation [51,52], and it has been applied to numerous topics, including classification [49,51], tracking [52], edge detection [41], etc. Specifically, only one paper [53] has been found to directly address the class-imbalance issue by combining the feature vectors outputted from a dual arrangement of auto-encoders, where the issue of cost-efficiency has not been emphasized.

Feature map selection can be viewed as a special case of feature selection based on the CNN structure. Consistent with the feature selection methods, it also has two categories with three types [54]: the first category includes the filters [55–57], where ranks of the features are obtained without the help of classifiers; the second category employs the predictor, and for the included types, wrappers [58] explicitly score the feature, while the embedded methods [59–61] do it implicitly in the training process. Mostly, feature map selection is used for the enhancement of network performance. However, for the purpose of structural simplification, the wrappers principle would be more appropriate in our case.

Due to such specialized requirement, per the brief review above, few studies have tried to make a combinatorial usage of these two methods to seek effective network performance improvement with optimized expansion costs. So, the method proposed in this article acts as a novel approach to the class-imbalance problem with convenient usage, where no tricky hard negative mining or parameter selection is involved.

3. Background

3.1. Basic Knowledge of Convolutional Neural Networks

Convolutional neural networks (CNNs) currently dominate computer vision studies, with constant state-of-the-art performance in almost every topic to which they are applied. CNNs are a special kind of deep belief network (DBN) with components called convolutional layers, composed of units called kernels or filters. Due to space limitations, a very brief introduction based on [62] is given for the principles of DBN and CNN and to help with the clarity of symbols. Firstly, a normal

DBN can be viewed as a stack of fully-connected layers, where each layer has a set of learnt parameters θ composed of connection weights $\mathbf{W}$ and bias $\mathbf{b}$. During the forward propagation, every input vector $\mathbf{x}$ will be processed by an affine transformation to get the output $\mathbf{z}$, as in Equation (1).

$$\mathbf{z} = \mathbf{W}^T\mathbf{x} + \mathbf{b} \tag{1}$$

In practice, the output $\mathbf{z}$ will be further corrected by a nonlinear function such as $\mathbf{h} = g(\mathbf{z})$ to overcome the XOR problem, where the rectified linear unit (or ReLU) [63,64] will always be chosen as the $g(\cdot)$. At the final stage of forward propagation, an output vector from the topmost fully-connected layer would be transformed by a probability distribution function (e.g., the softmax function) before being outputted. The softmax function defined in Equation (2) is one of the most commonly used Bernoulli distribution outputs calculated through normalized exponential transformation.

$$\begin{aligned} P(y = i|x) &= \text{softmax}(z_i) \\ &= \frac{\exp(z_i)}{\sum_j \exp(z_j)} \end{aligned} \tag{2}$$

To obtain the highest probability on the correct class label y on the input $\mathbf{x}$, this output for softmax function is minimized by its negative log-likelihood format, which is defined in Equation (3).

$$\begin{aligned} J(\theta; \mathbf{x}, y) &= L(\hat{y}, y) + \lambda \cdot \Omega(\theta) \\ L(\hat{y}, y) &= -\log(\text{softmax}(\mathbf{z})_i) \\ &= \log \sum_j \exp(z_j) - z_i \end{aligned} \tag{3}$$

Here, $J(\cdot)$ is the loss to be minimized during training, and $L(\hat{y}, y)$ is the softmax-based loss term in which y and $\hat{y}$ are the true and estimated labels for input $\mathbf{x}$. $\Omega(\cdot)$ is some regularization term with restrictions defined on the network parameters θ (e.g., the weights $\mathbf{W}$ or biases $\mathbf{b}$). More often than not, gradient descent-based optimization is employed to reduce the value of $J(\cdot)$, where the updating gradient from the softmax loss is $\mathbf{g} = \nabla_{\hat{y}} J$, based on the estimated label. Similarly, the updating gradients for $\mathbf{W}$ and $\mathbf{b}$ are defined in Equation (4), calculated by the chain rule.

$$\nabla_{\mathbf{W}^{(k)}} J = \mathbf{h}^{(k-1)T}\mathbf{g},\ \nabla_{\mathbf{b}^{(k)}} J = \mathbf{g} \tag{4}$$

$$\mathbf{W}^{(k)} \leftarrow \mathbf{W}^{(k)} + \alpha \cdot \nabla_{\mathbf{W}^{(k)}} J,\ \mathbf{b}^{(k)} \leftarrow \mathbf{b}^{(k)} + \alpha \cdot \nabla_{\mathbf{b}^{(k)}} J \tag{5}$$

$\mathbf{W}^{(k)}$ and $\mathbf{b}^{(k)}$ are the weights and bias for the fully connected layer at level k, whose rectified output is denoted as $\mathbf{h}^{(k-1)}$. During the back propagation, at each layer, the weights and bias are updated by adding the deviations $\nabla_{\mathbf{W}^{(k)}} J$ and $\nabla_{\mathbf{b}^{(k)}} J$, with the latter ones multiplied by a learning rate α to control the convergence rate, as in Equation (5).

Those are the cases for the DBN, while all things are almost identical in the case of CNN, except for the part involving the convolutional layers. Convolutional layers can be treated as a special kind of fully-connected layer with shared connection weights held by kernels. Take the network in Figure 1 for illustration; considering a 4-D kernel tensor $\mathbf{K}^{(k)}$ from the kth convolutional layer, during the back propagation, the input signal data $\mathbf{V}^{(k-1)}$ is convoluted with $\mathbf{K}^{(k)}$ with step s to get the output $\mathbf{Z}^{(k)}$. The produced activation map $\mathbf{Z}^{(k)}$ is also called the *feature map*, which will always be under-sampled in practice by an operation called *pooling* to get the input data for the next layer, denoted as $\mathbf{Z}^{(k)} \rightarrow \mathbf{V}^{(k+1)}$. After the input image $\mathbf{V}^{(0)}$ has gone through all five convolutional layers in Figure 1, the final feature map $\mathbf{V}^{(5)}$ will be flattened into a 1-D vector $\mathbf{h}^{(5)}$ to be fed to the fully-connected trailing layer FC6 and the following FC7, FC8 to get the final predicted probabilities. Likewise, in the back-propagation, the 1-D difference $\mathbf{g}^{(5)}$ from the FC6 layer is reshaped into 3-D as $\mathbf{G}^{(5)}$ to update the feature maps. Assuming the objective function value is $J(\mathbf{V}, \mathbf{K})$ on the feature maps $\mathbf{V}$ and kernels $\mathbf{K}$, its back-propagated differences from the upper layer should be calculated as $\mathbf{G}^{(k)} = \nabla_{\mathbf{V}^{(k)}} J\left(\mathbf{V}^{(k)}, \mathbf{K}^{(k)}\right)$ and $\nabla_{\mathbf{K}^{(k)}} J\left(\mathbf{V}^{(k)}, \mathbf{K}^{(k)}\right)$. Then, the feature maps and kernels are updated in a manner identical to Equation (5), where the convolutional kernels and feature maps are updated by adding with the derivations multiplied by a learning rate coefficient α, as in Equation (6).

$$\begin{aligned} \mathbf{V}^{(k)} &\leftarrow \mathbf{V}^{(k)} + \alpha \cdot \nabla_{\mathbf{V}^{(k)}} J\left(\mathbf{V}^{(k)}, \mathbf{K}^{(k)}\right) \\ \mathbf{K}^{(k)} &\leftarrow \mathbf{K}^{(k)} + \alpha \cdot \nabla_{\mathbf{K}^{(k)}} J\left(\mathbf{V}^{(k)}, \mathbf{K}^{(k)}\right) \end{aligned} \tag{6}$$

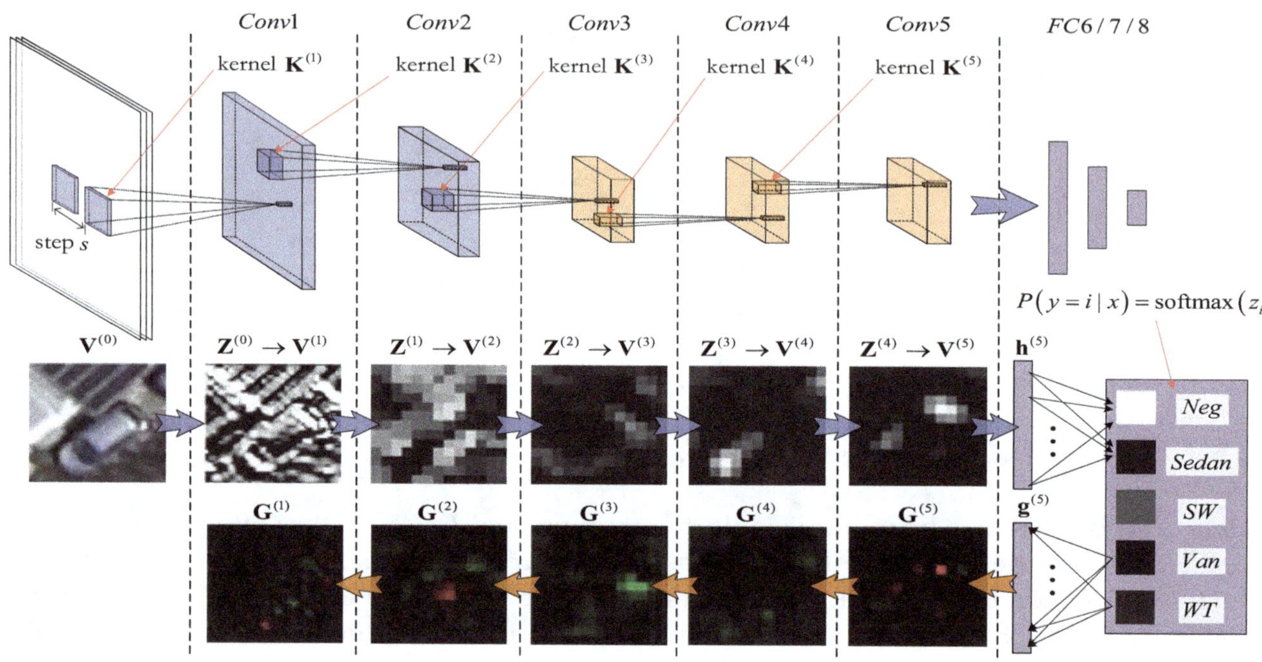

Figure 1. A typical convolutional neural network (CNN) structure, with feature and difference maps produced by the forward and backward propagations. SW: station wagon; WT: working truck.

3.2. The Semantic Texture Encoding Pattern for Convolutional Kernels

Despite of all the symbols and equations listed above, studies like [65] sought to produce more interpretable results, helping to better understand and improve the network. One of the important functional components of the DeepVis toolbox proposed in [65] is to find and show the image crops causing the top-most activations by each kernel. This kind of data-centric visualization measure [66–68] differs from other means such as deconvolution [69] or image synthesis [70], showing the correlations between kernel and image samples more directly.

The manifestation effectiveness of the previously mentioned data-centric max-activation illustration method is shown in Figure 2. Therein, six kernels from the CONV5 layer are arranged into two separated groups, denoted as $\{S_i \,|i = 1, 2, 3\}$ and $\{W_i \,|i = 1, 2, 3\}$ by their correlation strengths with the input image $\mathbf{x}$ shown in the column *Raw Image*. Under the column *Top Activation Image Crops*, the max-activation image crops are listed for each kernel, from which a stable image content can be observed, and that represents the textural pattern being encoded. Finally, for each kernel, the correlation between its texture and the input image can be measured by the corresponding feature maps being listed under the *Feature Map* column. Clearly, the feature maps from the kernels $\{S_i\}$ have greater activation values, while those belonging to $\{W_i\}$ are almost black. Considering that these pixel-wise activations will be fed to the trailing fully-connected layers to produce the class-wise likelihoods, strongly activated feature maps from $\{S_i\}$ indicate that they have stronger correlations with the input image.

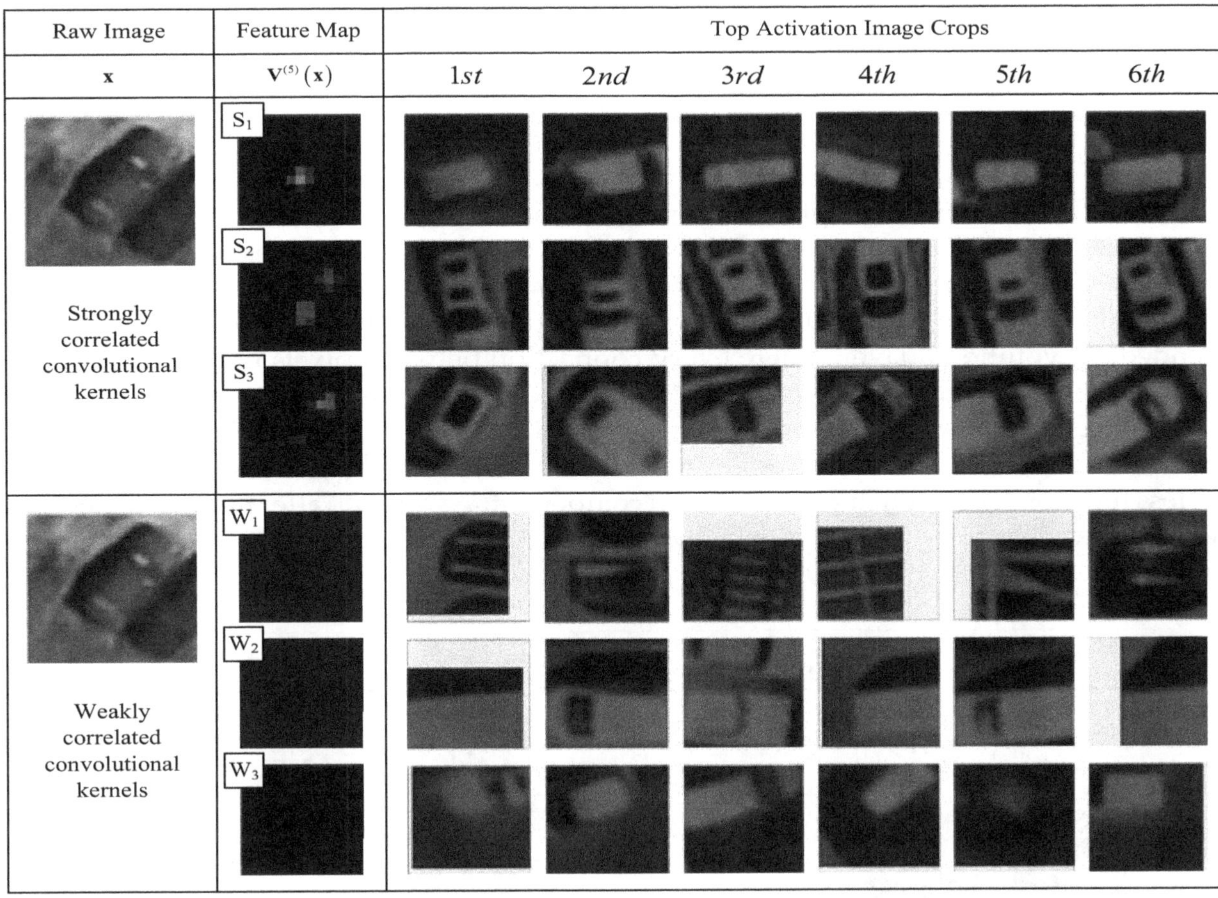

Figure 2. Illustration of the semantic meaning of the convolutional kernels. The raw input image is displayed in the *Raw Image* column; the six feature maps produced by six different kernels at the CONV5 layer are shown in the *Feature Map* column; and six arrays of local image crops on which the top six feature map activations are produced are shown in the *Top Activation Image Crops* column.

In fact, the way in which the high activations in feature maps from the last convolutional layer help with efficient classification can be exemplified by using Equation (5). Considering two activations $h_{l_1}^{(k-1)}$ and $h_{l_2}^{(k-1)}$ at the same position i, m, n from two feature maps $Z_{q1}^{(k-1)}$ and $Z_{q2}^{(k-1)}$ at layer level $k-1$, with that $h_{l_1}^{(k-1)} = Z_{q1,i,m,n}^{(k-1)}$ and $h_{l_2}^{(k-1)} = Z_{q2,i,m,n}^{(k-1)}$. The connection weights bounded with these two activations are $W_{l_1}^{(k)}$ and $W_{l_2}^{(k)}$ in a single trailing fully-connected layer with its final categorical probabilities generated by transformation $z_j^{(k)} = \sum_l W_l^{(k)} h_l + b_j^{(k)}$. Then, by Equation (4), the updating differences for $W_{l_1,j}^{(k)}$ and $W_{l_2,j}^{(k)}$ can be calculated by Equation (7), where $g_j^{(k)} = \frac{\partial}{\partial z_j^{(k)}} J$.

$$\nabla_{\mathbf{W}_{l_1,j}^{(k)}} = g_j^{(k)} \cdot h_{l_1}^{(k-1)}, \; \nabla_{\mathbf{W}_{l_2,j}^{(k)}} = g_j^{(k)} \cdot h_{l_2}^{(k-1)} \tag{7}$$

So, when there is $h_{l_1}^{(k-1)} > h_{l_2}^{(k-1)}$, greater updating differences will be generated for the $W_{l_1}^{(k)}$ as $\nabla_{\mathbf{W}_{l_1,j}^{(k)}} > \nabla_{\mathbf{W}_{l_2,j}^{(k)}}$. Assuming activations $h_{l_1}^{(k-1)}$ and $h_{l_2}^{(k-1)}$ are all beneficial for the final probabilistic estimation on class j, the weighted connection $W_{l_1}^{(k)}$ will grow faster and larger with respect to $W_{l_2}^{(k)}$. This means that the feature map $Z_{q1}^{(k-1)}$ produced by convolutional kernel $\mathbf{K}_{q1}^{(k-1)}$ is more effective for recognizing samples from class j.

4. Methods

4.1. Overview of the Proposed CNN Extension Scheme

So, being aware of the fact that the modeling power of a CNN is strongly correlated with the diversity of feature maps at the last convolutional layer, this article sets out to tackle the problem of class-imbalance by adopting a cost-effective imbalance-aware feature map extension. Commonly, two kinds of overheads will be introduced when new feature maps are added: the *convolution overhead* and the *connection overhead*. Specifically, the *convolution overhead* refers to the extra convolution operation and extra feature map storage. The *connection overhead* happens in the fully-connected layer right above the extended convolutional layer, where every connection between pixels in the new feature map and the hidden-neurons in the fully-connected layers should be added. In order to reduce these two overheads, two general measures are adopted, which are illustrated in Figure 3: (1) the selective feature map extension by a newly derived class-importance measurement; (2) a class-imbalance-sensitive softmax loss function for optimizing the extended component. As a result, after these two modifications, the original network is turned into a bi-partite structure with enhanced sensitivities to the samples in the minority classes.

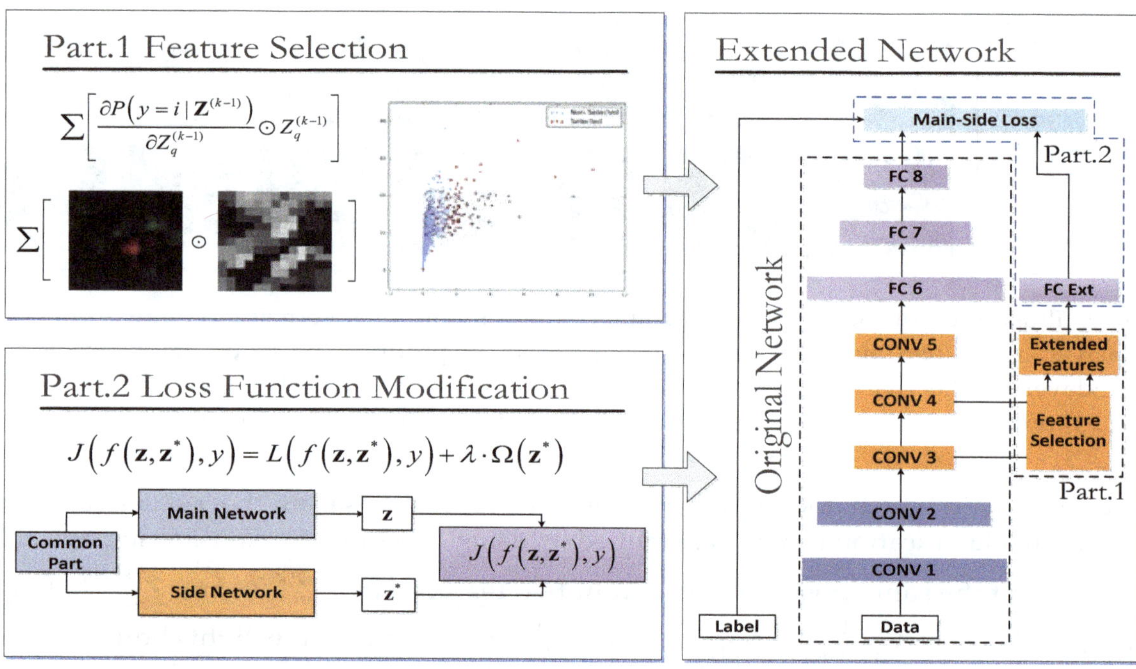

Figure 3. The general structure of the proposed network enhancement method.

(1) Part 1: The selective feature map extension by class-importance measurement. This measure aims to reduce the *convolution overhead* by reusing feature maps selected from the preceding layers. The criteria adopted in the feature selection process—named feature map class-importance—are similar to that in [58], but are further extended for a multi-class problem with slight modification. Additionally, according to [58], these selected feature maps are further filtered by an extra convolutional layer to reduce noise before being used as the *Extended Features* component in Figure 3.

(2) Part 2: the class-imbalance-sensitive softmax loss function. This measure aims at reducing the *connection overhead* and increasing the class-imbalance awareness of the improved structure. Firstly, the extended network components holding the *Extended Features* are isolated from the main part of the Original Network by a single-layered fully-connected (FC) layer *FC Ext*. This FC layer has hidden neurons only as few as the number of output classes; thus, the additional connection

quantity for the new maps is largely reduced. Secondly, as shown in the right-most text-box of Figure 3, a new loss function named main-side loss is adopted in place of the original softmax loss to raise the sensitivities of the *Extended Features* to the minority classes.

For the rest of Section 4, the proposed extension is described in detail based on a network prototype miniature visual geometry group (VGG-M) shown in Figure 3, which is very similar to AlexNet [71], but has slight improvements on the local convolutional parameters. This illustrative network has five convolutional layers (denoted as CONV1 to CONV5) and three fully-connected layers (denoted as FC6 to FC8), and feature maps for extension are selected from layers CONV3 and CONV4. All of these terms will be used in the following explanations.

4.2. The Network Extension by Selected Feature Maps

The idea of using quadratic expansion of the loss function to reduce less-effective network connections is not new—similar studies can be seen in [72], dating back to 1989. However, loss function-based feature map significance cannot be used to make class related pruning. Instead, the class-wise importance measurement for the feature maps is not hard to obtain—it can be produced by using a similar expansion technique on the output class likelihoods from the output neurons. Considering a general case where $\mathbf{Z}^{(k-1)}$ is the collection feature maps at the $k-1$th layer generated from input image $\mathbf{x}$, and the predicted probability for class i is $P\left(y=i|Z_q^{(k-1)}\right)$. Then, the contribution of feature map $Z_q^{(k-1)}$ to the estimated likelihood on class i can be approximated by Equation (8).

$$\begin{aligned} P\left(y=i|Z_q^{(k-1)}\right) &\approx P\left(y=i|\mathbf{Z}^{(k-1)}\right) - P\left(y=i|\mathbf{Z}^{(k-1)}{}_{/q}\right) \\ &= \sum\left[\frac{\partial P\left(y=i|\mathbf{Z}^{(k-1)}\right)}{\partial Z_q^{(k-1)}} \odot Z_q^{(k-1)}\right] + R_2\left(Z_q^{(k-1)}\right) \end{aligned} \tag{8}$$

In Equation (8), $\mathbf{Z}^{(k-1)}{}_{/q}$ is the collection of feature maps $\mathbf{Z}^{(k-1)}$ without $Z_q^{(k-1)}$, and $R_2\left(Z_q^{(k-1)}\right)$ denotes the other higher-order expansions based on $Z_q^{(k-1)}$. In the first expansion term, $\frac{\partial P\left(y=i|\mathbf{Z}^{(k-1)}\right)}{\partial Z_q^{(k-1)}}$ is the feature map differences back-propagated from the probability value at the ith output neuron, and $\odot$ is the element-wise multiplication between matrices. In practice, this difference can be efficiently obtained by back-propagation. By summing the pixel-wise production of the feature map and its differences, the class-importance for the feature map on class i can be obtained. This is vividly shown in Figure 4.

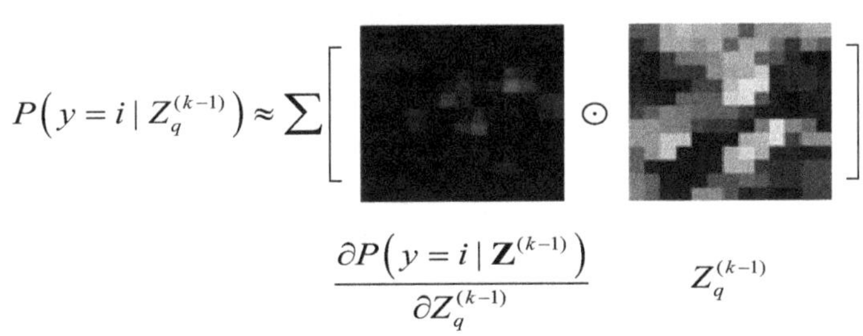

Figure 4. The first-order term of the Taylor expansion in Equation (8). $\frac{\partial P\left(y=i|\mathbf{Z}^{(k-1)}\right)}{\partial Z_q^{(k-1)}}$ denotes the feature map difference, positive, negative, and zero values marked as green, red, and black.

The class-important measure is validated in Figure 5, where the correlations for the maximal feature map activation and maximal feature significance to the probability values on negative samples are presented. Specifically, the x-axis max activations in Figure 5a means the topmost activation value measured from all the feature maps from CONV5. This is also the case for the x-axis class importance in Figure 5b. As can be seen, the data points in Figure 5b are much tighter and dense, roughly distributed on a curve with shape $y = K \cdot \frac{x}{a+x}$, where $a > 0$. Such strong correlation also indicates that the final categorical estimation is mostly based on a single feature map, which again emphasizes the importance of effective feature map selection.

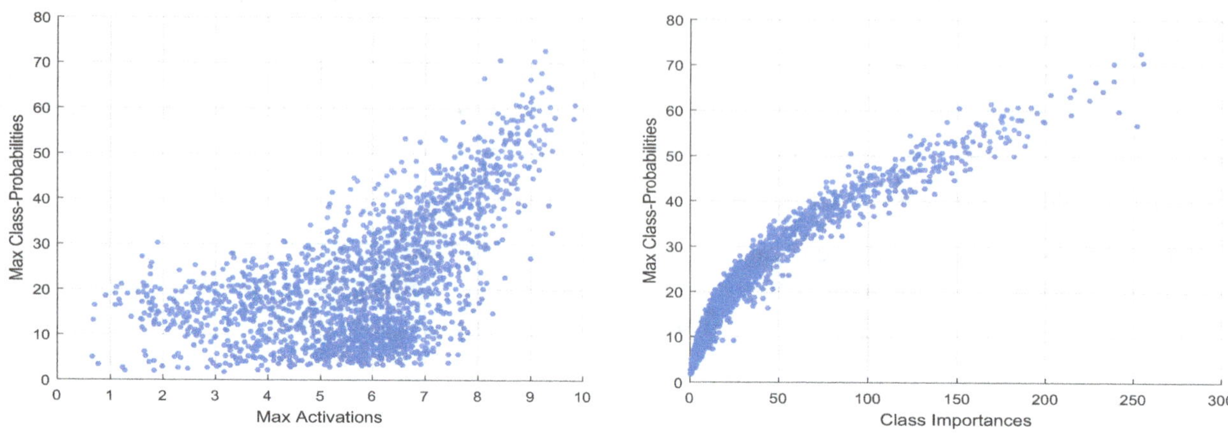

(a) (b)

Figure 5. Correlations of the max-activations and class-importance with the class probability of the negative class. (**a**) Max-activation vs. class probability. (**b**) Max class-importance vs. class probability.

Figure 6a shows the distribution pattern of feature maps from the CONV3 and CONV4 layers in the max class-importance vs. max-activation space. From Figure 6a, it can be determined that feature maps from the CONV4 are slightly more significant than those from CONV3, with elements in the high class-importance section distributed closer to the x-axis. The categorical inclination of a specific feature map Z_q can be calculated by getting the index i of its largest class importance as $i = \arg\max_j P\left(y = j|Z_q\right)$, and their categorical distributions are shown in Figure 6b for five vehicle classes. As can be observed, feature maps belonging to all five classes have similar distributions in the importance section either high and low. Accordingly, in picking the most relevant feature maps for extension, it would be reasonable to select the ones with highest importance scores from each class and control that class-wise quantity according to their classification deficiencies, as in Algorithm 1.

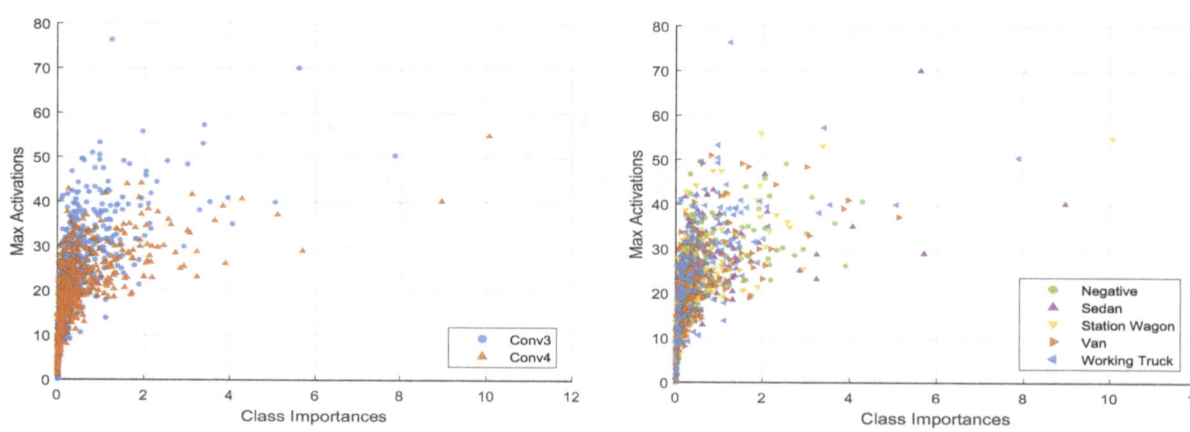

(a) (b)

Figure 6. Scatter plots showing the distribution of the feature maps $\{Z_q\}$ from CONV3 and CONV4 in the class-importance vs. max-activation space. (**a**) The distributions of CONV3 and CONV4 feature maps. (**b**) Feature maps correlated to the five classes by the class-importance measurement.

Algorithm 1 Class Imbalance-Aware Extension Feature Map Selection

Input: Classification accuracies $\{ACC\ (j)\}$, class-importance $\{P\ (y = i|Z_q)\}$ for feature maps $\{Z_q\}$ from the CONV3 and CONV4 layers, and the total number of maps to be selected N_{sel}.

Output: Selected feature map indexes $\{i\}_{CONV3,CONV4}$ on CONV3 and CONV4

1: Calculate the number of extension maps needed for each class. For instance, for class j, denote the required extension quantity as $N_{sel}^{(j)}$, then there is $N_{sel}^{(j)} = \left\lceil \frac{1-ACC(j)}{\sum_i (1-ACC(i))} \right\rceil \cdot N_{sel}$.

2: For each class j, sort the CONV3 and CONV4 feature maps $\{Z_q\}$ by their class importance values $\{P\ (y = i|Z_q)\}$ in descending order, with the indexes denoted as $\{m_i\}_{CONV3,CONV4}^{DESC(j)} = \left\{m_i\ |P\ (y = j|Z_{m_1}) \geq ... \geq P\left(y = j|Z_{m_{N_{all}}}^{(k-1)}\right)\right\}$, where $N_{all} = |\{Z_q\}|$.

3: For each class j, get the top $N_{sel}^{(j)}$ map indexes from the descending order set as $\{m_i\}_{CONV3,CONV4}^{TOP\left(N_{sel}^{(j)}\right)} = \left\{n_i \middle| i = 1, ..., N_{sel}^{(j)}, n_i \in \{m_i\}_{CONV3,CONV4}^{DESC(j)}\right\}$.

4: Merge the class-wise top indexes $\{m_i\}_{CONV3,CONV4}^{TOP(j)}$ from the previous step, and get the output feature map index set as $\{i\}_{CONV3,CONV4} = \bigcup_j \{m_i\}_{CONV3,CONV4}^{TOP\left(N_{sel}^{(j)}\right)}$.

More specifically, as in Algorithm 1, the selection ratio for each class is measured by their pro rata accuracy deficiencies $\frac{1-ACC(j)}{\sum_i (1-ACC(i))}$, so the class-wise selection quantity is $N_{sel}^{(j)} = \left\lceil \frac{1-ACC(j)}{\sum_i (1-ACC(i))} \right\rceil \cdot N_{sel}$. Two exemplified CONV3 and CONV4 feature map selections are illustrated in Figure 7, where the total selection quantities are $N_{sel} = 64$ and $N_{sel} = 160$. Therein, the extended feature map candidates mainly reside in the high class-importance region.

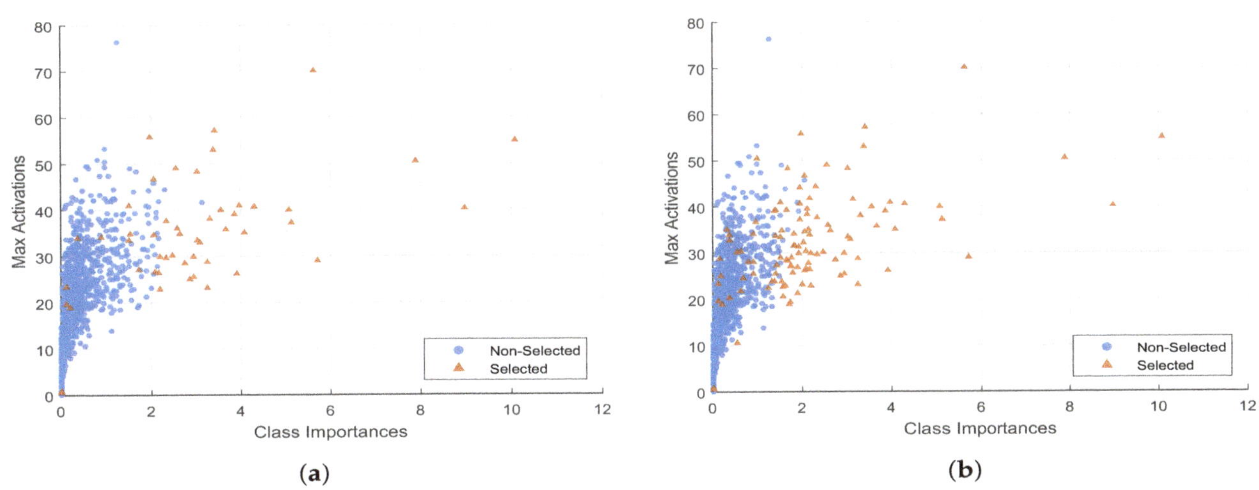

Figure 7. (**a**) The 50 selected maps for $N_{sel} = 64$. (**b**) The 109 selected maps for $N_{sel} = 160$.

4.3. *Class Imbalance-Sensitive Softmax Loss Function*

Up to the present, explanations for the cost-effective network extension have been focused on the feature map selection process which reduces the *convolution overhead*. However, the *connection overhead* is also significant if the newly extended feature maps are encoded directly by the trailing fully-connected layer FC6. Typically, for a network with structure similar to that in Figure 3, there can be as many as 4096 hidden neurons in FC6. Supposing the feature maps from CONV5 are of shape

13×13, then as many as $4096 \times 13 \times 13$ real-valued connection weights will be introduced for every newly added feature map. This kind of overhead can be greatly reduced if these extended feature maps are encoded by a single-layered fully-connected layer independent of the original network, which has hidden neurons with quantity equal to the number of output classes. As shown in Figure 8, the resulting bi-partite network is generalized as composed by three structural components: the common part, the main network, and the side network. For the eight-layered network in Figure 3, the common part refers to the shared layers CONV1 and CONV2, the Main Network refers to layers CONV3 through FC8 in the Original Network, and the Side Network refers to the *Extended Features* along with the isolated *FC Ext* layer.

In this structure, the output values from *FC Ext* can be viewed as an extra categorical estimation based purely on the newly added feature maps, whereas the final categorical prediction from the extended network can be calculated as the summation of these two. Taking the predicted likelihoods from the Main and Side Network components as $\mathbf{z}$ and $\mathbf{z}^*$, this kind of summarization-based likelihood mixture can be viewed as applying a hard connection on these two likelihoods as $f(\mathbf{z}, \mathbf{z}^*) = 1 \cdot \mathbf{z} + 1 \cdot \mathbf{z}^*$, in which both predictions are equally weighted. However, according to the analysis in Section 3.2, this straightforward means does not promise that the extended part will be more correlated with minority class samples. Take the $h_l^{(k-1)}$ as some activation from a feature map in the Side Network component at layer $k-1$, and its connection weights to a majority class i and a minority class j are denoted as $W_{l,i}^{(k)}$ and $W_{l,j}^{(k)}$. Then, according to Equation (9), in the case of using softmax loss, the updating differences $g_i^{(k)}$ and $g_j^{(k)}$ from upper layer will be almost equal.

$$\nabla_{\mathbf{W}_{l,i}^{(k)}} = g_i^{(k)} \cdot h_l^{(k-1)}, \; \nabla_{\mathbf{W}_{l,j}^{(k)}} = g_j^{(k)} \cdot h_l^{(k-1)} \tag{9}$$

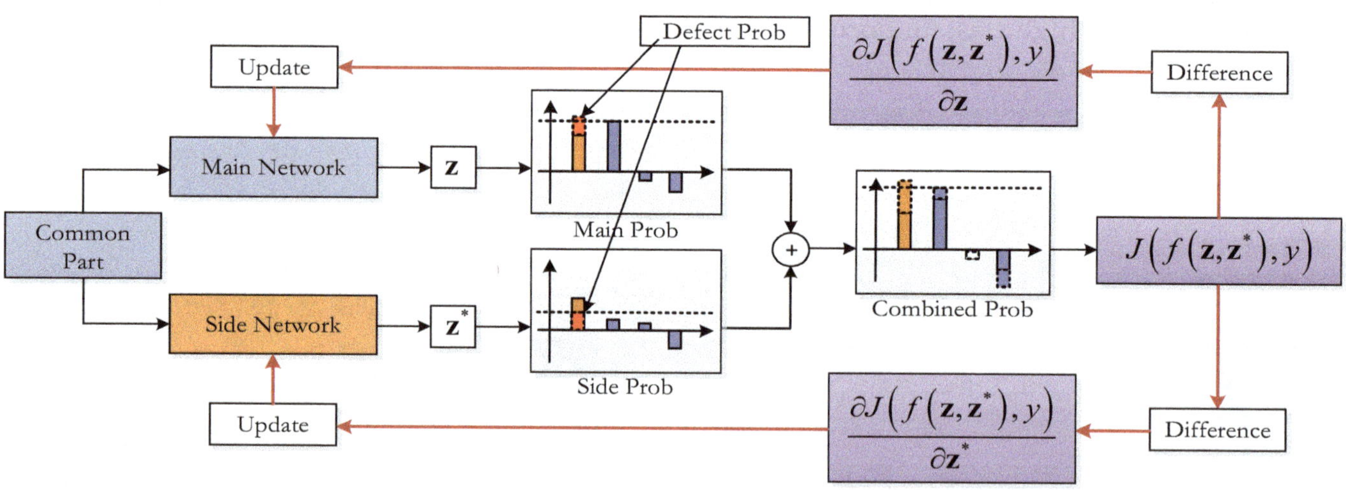

Figure 8. Principle structure of the class-imbalance aware Main-Side Network.

So, in order to achieve class-imbalance sensitivity, the loss function should differ the back-propagated values for the Side Network between majority and minority classes. This setting is manageable. By Figure 8, considering the likelihood summation format as employing the $\mathbf{z}^*$ from Side Network to rectify the estimates $\mathbf{z}$ from the Main Network, then the $\mathbf{z}^*$ acts as a filling for the probability deficiencies of $\mathbf{z}$ marked by the red bar in Figure 8. Then, if this kind of likelihood amendment is intentionally diminished for the majority classes and encouraged for the minority classes, the predictions from the Side Network are more likely to be correlated with samples from the minority classes.

More specifically, as in Equation (10), the newly introduced Main-Side loss is denoted as $J(f(\mathbf{z},\mathbf{z}^*),y)$ at the right-side of the picture, which takes the softmax loss $L(f(\mathbf{z},\mathbf{z}^*),y)$ as its main component. Then, in order to make these two updating values vary from each other, an extra regularization only relevant with $\mathbf{z}^*$ is added to the loss function, which is denoted as $\Omega(\mathbf{z}^*)$ with a global penalization coefficient λ. Since $\Omega(\mathbf{z}^*)$ is only dependent on the Side Network output $\mathbf{z}^*$, the back-propagated differences for the Main and Side Network components $\frac{\partial J(f(\mathbf{z},\mathbf{z}^*),y)}{\partial \mathbf{z}}$ and $\frac{\partial J(f(\mathbf{z},\mathbf{z}^*),y)}{\partial \mathbf{z}^*}$ will be different, as in Equation (11).

$$\begin{aligned} J(f(\mathbf{z},\mathbf{z}^*),y) &= L(f(\mathbf{z},\mathbf{z}^*),y) + \lambda\cdot\Omega(\mathbf{z}^*) \\ L(f(\mathbf{z},\mathbf{z}^*),y) &= -\log\left[\mathrm{softmax}(\mathbf{z}+\mathbf{z}^*)_{y=i}\right] \end{aligned} \tag{10}$$

$$\begin{aligned} \frac{\partial J(f(\mathbf{z},\mathbf{z}^*),y)}{\partial \mathbf{z}} &= \mathrm{softmax}(\mathbf{z}+\mathbf{z}^*)_{y=i} - \mathbf{1}(y=i) \\ \frac{\partial J(f(\mathbf{z},\mathbf{z}^*),y)}{\partial \mathbf{z}^*} &= \mathrm{softmax}(\mathbf{z}+\mathbf{z}^*)_{y=i} - \mathbf{1}(y=i) + \lambda\cdot\frac{\partial\Omega(\mathbf{z}^*)}{\partial \mathbf{z}^*} \end{aligned} \tag{11}$$

Recalling that the softmax loss term $L(f(\mathbf{z},\mathbf{z}^*),y)$ should be diminished during training, this Side Network correlated regularization $\Omega(\mathbf{z}^*)$ should produce small penalty values for the minority classes, but large values for the majority classes. The simplest way to achieve this is to assign varied penalty coefficients for class-wise likelihood values in $\mathbf{z}^*$, and the classification accuracies for these classes measured on the cross-validation dataset serves such needs. So, as in Equation (12), the additional loss function regularization term $\Omega(\mathbf{z}^*)$ is defined as the Norm-2 of the element-wise multiplication of $\mathbf{z}^*$ and the class-wise accuracies measured on the Main-Network.

$$\begin{aligned} \Omega(\mathbf{z}^*) &= \|\mathrm{B}\odot\mathbf{z}^*\|_2 = \sqrt{\textstyle\sum_j\left(\beta_j\cdot z_j^*\right)^2} \\ \beta_j \propto ACC(\mathbf{X})_j, ACC(\mathbf{X})_j &= \frac{TP(\mathbf{X})_j}{TP(\mathbf{X})_j+FP(\mathbf{X})_j} \end{aligned} \tag{12}$$

The $ACC(\mathbf{X})_j$ in Equation (12) is the averaged accuracy for the given image set $\mathbf{X}$ on class j measured by $\mathbf{z}$ from the Main Network, B is the categorical penalization coefficient applied on $\mathbf{z}^*$, and $\odot$ means element-wise multiplication between two vectors. Following this definition, for a majority class i already having very high accuracy $ACC(\mathbf{X})_i$, its penalization will be higher than a minority class j with lower $ACC(\mathbf{X})_i$, and vice versa. Besides, due to the flexibilities in choosing the set of input images $\mathbf{X}$, three penalization modes can thus be derived, here denoted as Global, Local, and Batch-wise, as shown in Figure 9.

Conceptually, these three penalization modes have specific pros and cons of their own. According to Figure 9, the global penalization β based on the overall sample set O stays unchanged for all training subsets, and thus is insensitive to abnormalities in local space. The local penalization $\beta_{i,k}$ partially improves the flexibility by using accuracy local cluster S_i for each training sample that k belongs to, but such accuracy must still be measured beforehand and could be obsolete during the training. Instead, for the batch-wise mode, a real-time tracking of accuracy can be acquired from the training mini-batch B_i, while the additional price is the increased non-linearity in convergence.

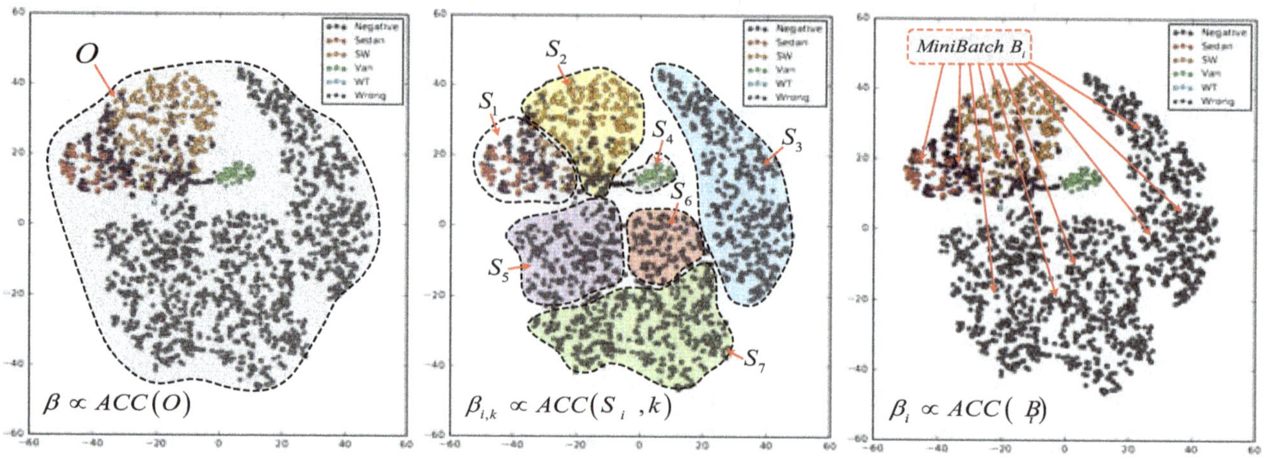

(a) (b) (c)

Figure 9. The t-Distribution stochastic neighbor embedding (t-SNE) -based visualization [73] of the negatives and vehicle types in the FC8 output space, and the three penalization modes used for B: (**a**) global, (**b**) local, and (**c**) batch-wise.

5. Experiments and Analysis

5.1. Data Set Description and Experiment Setup

5.1.1. DLR 3K Aerial Image Dataset

The DLR 3K aerial image dataset is an aerial image dataset made publicly available online by the Germany Aerospace Center, which has been studied in [19]. It contains 20 aerial images with resolution of 5616 × 3744 captured over the city of Munich by a low-cost airborne imaging system called DLR 3K+ Cam, which is composed of three non-metric Canon Eos 1Ds Mark III cameras with Zeiss lenses. This system is intended to be fixed on an airplane or a glider with a ZEISS shock mount, where images are taken at the height of 1000 m with real-time ortho-rectification made either on-board or at ground station. All pictures are in RGB real-colored spectral bands (each being digitalized by 8 bits) with ground sampling distance (GSD) at 13 cm.

Although there is no modern skyscraper, these pictures contain quantities of medium-height residential buildings, workshops, trees, lawns, railway track, and streets filled with kinds of vehicles either wide or narrow. All of these put together form a rich set of scenarios which would include most of the typical conditions that cause false detections. Figure 10 shows some of the image samples. The four sub-figures marked as $b_1 \sim b_4$ are cropped from spots marked by yellow squares in the main image a on the left, which represent classical detection disturbances: tight parking (b_1), shadows from trees and houses (b_1, b_2, and b_3), and partial occlusion (b_4). In addition, buildings and man-made facilities in this area have complex textures similar to vehicles, which further increases the localization and categorization difficulties.

Instead of using the original vehicle classes in the dataset, we defined a new set of classes, with the main focus being put on small and medium-sized vehicles; that is, Sedan, Station Wagon (i.e., private SUV), Van, and Working Truck. Quantitative distributions and averaged scales of these vehicle types are listed in Table 1, by which a highly skewed inter-class distribution of samples can be clearly observed. In it, the Station Wagon class has the top-most quantity with the Sedan and Van lagging far behind. The quantity of Working Truck is trivial, with occupation ratios at merely 0.8% in the training set and 0.6% in the testing set.

Figure 10. (**a**) A typical frame from the training sample. (**$b_1 \sim b_4$**) Typical difficult detection cases. (**c**) The close-to-vehicle region (shaded blue) and categorical sampling positions.

Table 1. The vehicle types defined in this paper and the basic statistics.

Type	Samples	Training Set			Testing Set		
		L (px)	W (px)	N	L (px)	W (px)	N
Sedan	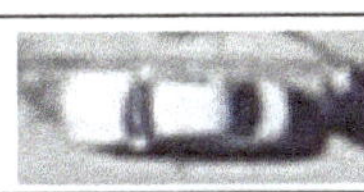	21.45	10.47	776	20.83	10.16	1075
Station Wagon	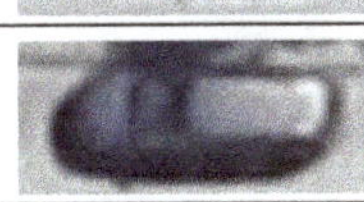	19.99	9.76	2302	19.14	9.32	4178
Van	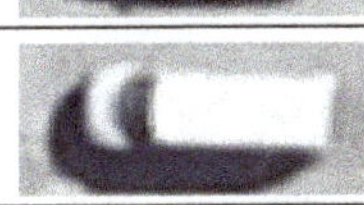	24.65	12.06	312	24.14	11.83	512
Working Truck	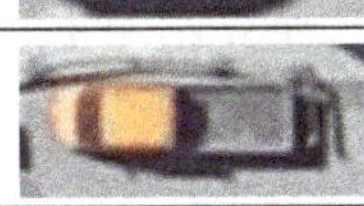	27.17	13.31	29	26.58	13.02	34

Note: **L (px)** and **W (px)** denote the length and width of vehicles in pixels, and **N** denotes the quantity.

5.1.2. Training and Testing Preparation as a Classification Problem

Since the R-CNN detection structure is employed, it can be regarded as a common CNN-classifier making categorization on full-scaled input images. To facilitate the analysis and verification, 48×48 sized patches are uniformly extracted from the original image for a simplified experimental environment. Furthermore, to reduce the quantity of unnecessary negative samples with redundant textural patterns, these image patches are produced from three different regions by their distance to vehicle centers $Dist_V$, which are shown in Figure 11:

- The *Centered* category: position marked by yellow square in sub-figure *c* in Figure 10 with $Dist_V$ no more than 3 pixels;
- The *Close Range* category: positions marked by red squares within the blue shaded region in sub-figure *c* in Figure 10, whose $Dist_V$ are in range from 4 to 20 pixels;
- The *Far Range* category: in sub-figure *c* from Figure 10, positions marked by green squares outside the blue shaded region with $Dist_V$ more than 20 pixels.

Centered			Close Range			Far Range		
$Dist_V$	$N_{aug.}$	$\Delta\theta$	$Dist_V$	$N_{aug.}$	$\Delta\theta$	$Dist_V$	$N_{aug.}$	$\Delta\theta$
≤ 3	$16\times$	$22.5°$	$4 \sim 20$	$8\times$	$45°$	> 20	$1\times$	-

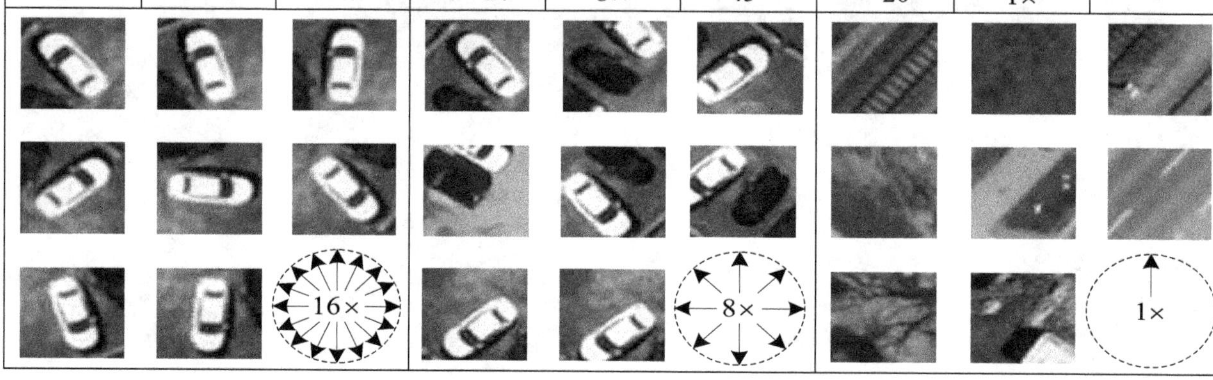

Figure 11. The sample categories used on the three regions: Centered, Close Range, and Far Range.

Only samples in the *Centered* category are treated as positives, which will be further categorized into different vehicle types. Samples from the *Close Range* and *Far Range* categories are taken as negatives with different classification difficulties. By Figure 11, data enhancements are performed on these samples by rotation with different times $N_{aug.}$, where rotated angle spacing $\Delta\theta$ is calculated by $\Delta\theta = 360°/N_{aug.}$. Since samples in the *Centered* and *Close Range* categories are less populated, their rotate angle spacings are 22.5° and 45° with augmentation times at $16\times$ and $8\times$. Finally, sample quantities in these categories are kept equal (approximately 33.3% for each), which results in 210,944 training samples and 534,624 testing samples.

5.1.3. The Baseline Network Structure and Extension Styles for Analysis

To manifest the effectiveness of feature selection and Main-Side loss-based fine-tuning, the VGG-M [74] network is employed as the baseline for the optimized extension. The VGG-M and its full version the 16-layered VGG [75] are powerful holistically structured networks, and have achieved a top-5 error at only 13.7% and 7.4% on the ILSVRC-2012-val dataset, which are the best scores until 2014. Different from its ancestor AlexNet [71], VGG-M uses small kernels of size 3×3 with 1 pixel-sized padding, making them ideal for encoding the local structural differences. After that, the development on CNN have either sought greater depth by shortcut connection [76–78] or more miscellaneous structural complexities [21,50,79].

Three typical kinds of extension structures based on VGG-M are illustrated in Figure 12, which will be studied in the following subsections. Figure 12a shows the case when the network is extended with blank randomly initialized kernels, and Figure 12b shows the case when selected feature maps from the preceding layers are used for extension. Figure 12c shows the case when both the feature selection and Main-Side loss techniques are employed for extension.

During the experimental analysis in the rest of the section, six kinds of network extension in total are involved for general or specific analysis, and their principle structures are shown in tables in Figure 13. Therein, the miniature VGG-M and full-sized VGG-16 are abbreviated as *Orig.M* and *Orig.16* in Figures 13a,b. Plain network extension by blank kernels and selected feature maps with the original softmax loss are abbreviated as *New Ext.* and *Select Ext.*, shown by Figures 13c,d. Blank kernel-based and selected feature map-based extension with the class-imbalance-sensitive Main-Side loss are denoted as *New S-Ext.* and *Select S-Ext.*, shown by Figures 13e,f.

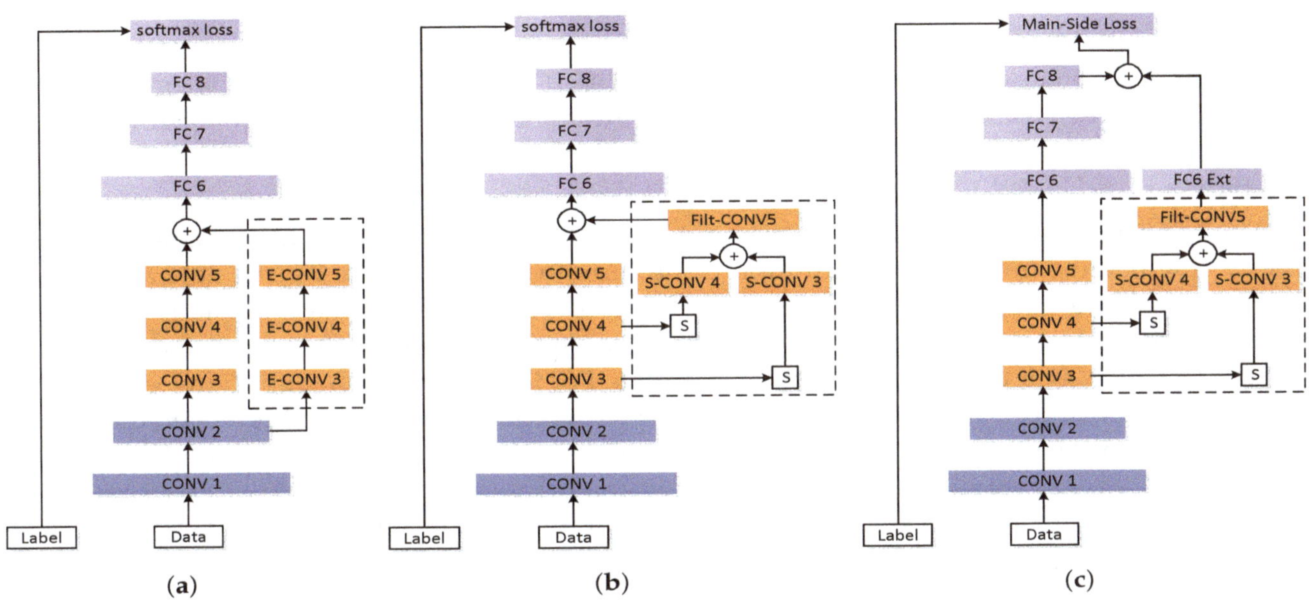

Figure 12. Three typical extension schemes. (**a**) Plain extension with blank kernel generated feature maps; (**b**) Plain extension with selected feature maps; (**c**) Main-Side bi-parted extension with selected feature maps.

Orig.M
Loss Softmax
FC6 ~ FC8
Conv5
Conv4
Conv3
Conv1 ~ Conv2
Input, Label

(**a**)

Orig.16
Loss Softmax
FC6 ~ FC8
Conv5_1, 5_2, 5_3
Conv4_1, 4_2, 4_3
Conv3_1, 3_2, 3_3
Conv1_1, 1_2 ~ Conv2_1, 2_2
Input, Label

(**b**)

New Ext.	
Loss Softmax	
FC6 ~ FC8	
Conv5	Conv5 Ext.N
Conv4	Conv4 Ext.N
Conv3	Conv3 Ext.N
Conv1 ~ Conv2	
Input, Label	

(**c**)

Select Ext.		
Loss Softmax		
FC6 ~ FC8		
Conv5	Conv5 Filt	
Conv4 Conv3	Conv3 Sel.N_1	Conv4 Sel.N_2
Conv1 ~ Conv2		
Input, Label		

(**d**)

New S-Ext.	
Main-Side Loss	
FC6 ~ FC8	FC6 Ext.
Conv5	Conv5 Ext.N
Conv4	Conv4 Ext.N
Conv3	Conv3 Ext.N
Conv1 ~ Conv2	
Input, Label	

(**e**)

Select S-Ext.		
Main-Side Loss		
FC6 ~ FC8	FC6 Ext.	
Conv5	Conv5 Filt	
Conv4 Conv3	Conv3 Sel.N_1	Conv4 Sel.N_2
Conv1 ~ Conv2		
Input, Label		

(**f**)

Figure 13. The five network structures studied in the experimental section. (**a**) The baseline network miniature miniature visual geometry group (VGG-M) (*Orig.M*) and (**b**) 16-layered VGG (*Orig.16*), the comparative extensions with either (**c**,**d**) the Loss of Softmax (*New Ext.*, *Select Ext.*) or (**e**,**f**) the proposed Main-Side Loss (*New S-Ext.*, *Select S-Ext.*).

More specifically, the network structures *Orig.M*, *Orig.16*, *New Ext.*, *Select Ext.*, and *Select S-Ext.* with the three penalization modes are compared and studied in Section 5.2 for a holistic comparison. After that, the structures *New Ext.* and *Select Ext.* are compared in Section 5.3 to showcase the effect of using selected features in the plain softmax loss-based network extension instead of the blank kernels. Finally, the main factors including the coefficient λ, coefficients $\mathbf{B}$, and the three penalization modes are compared based on the *New S-Ext.* structure to analyze the behavior of the Main-Side loss function in Section 5.4.

5.2. Experimental Results

The proposed network extension scheme is verified in this section. The networks chosen for comparison are the baseline VGG-M (*Orig.M*), the 16-layered full-sized VGG (*Orig.16*), extensions based on the softmax loss (*New Ext.* and *Sel. Ext.*), and extensions based on the selected features and Main-Side loss (*Sel. S-Ext.*).

The parameter model file sizes, memory consumption sizes, and their overheads are shown in Table 2, in which the memory consumption is measured with batch size 96. All data are measured based on the Caffe CNN platform. Generally, Main-Side loss-based network extensions have the least overhead compared to those using softmax loss. The parameter file size increments are trivial since the *FC6 Ext.* layer has only five hidden neurons. For the memory consumptions, extra memory space saving is done by reusing existing feature maps from preceding layers. Moreover, there are also implicit computation savings by eliminating the convolutions in layers *Conv3 Ext.* and *Conv4 Ext.*.

Table 2. Trained model file sizes and GPU-memory consumption for batch size of 96.

Net Struct.	*Orig.M*	*Orig.16*	*New Ext.* 128	*New Ext.* 256	*Sel. Ext.* 128	*Sel. Ext.* 256	*Sel. S-Ext.* 128	*Sel. S-Ext.* 256
Model (Mb)	361.7	537.1	439.6	519.8	426.1	460.9	362.2	362.8
Δ Model (Mb)	-	175.4	77.9	158.1	64.4	99.2	**0.5**	**1.1**
Mem (Mb)	1820.3	10547.1	1988.4	2093.0	2018.5	2053.7	1977.4	2004.3
Δ Mem (Mb)	-	8726.8	168.0	272.7	192.7	223.1	**157.1**	**183.9**

Class-wise classification performances measured by accuracies and F1 scores are presented in Tables 3 and 4, based on extensions with $N_{sel} = 128$ and $N_{sel} = 256$ by Algorithm 1. In them, the global, local and batch-wise based penalization modes for *Select S-Ext.* are abbreviated as *Glb.*, *Lcl.*, and *Bat.*. Due to the limitation of page space, the *Select Ext.* is further abbreviated as *Sel. Ext.*. The trailing keyword ReLU indicates the usage of ReLU layer to constrain the Side-Network probabilities. In each column, the first-, second-, and third-highest scores are marked by bold, underline, and double-underline.

From these two tables, several important phenomena need to be taken care of. Firstly, compared to the small version *Orig.M*, the full-sized *Orig.16* is superior in achieving high F1 scores and high accuracy for recognizing negatives, but it is bad for making accurate predictions on the positive classes. This means that the depth-based network extension is more likely to be affected by the class-imbalance. Secondly, the softmax loss-based *Sel Ext.* has more high scores when $N_{sel} = 128$, meaning that selected features are better utilized for minority classes under smaller extension quantity. Thirdly, the Main-Side loss-based selective feature map extensions are stabler at maintaining high performance for the minority classes (Sedan, Van) except for ones too trivial in size (Working Truck). Fourthly, the usage of ReLU slightly decreases the improvement in accuracies while helping with the enhancement of F1 score. Considering the small overhead cost for the *Select S-Ext.* variants, their network extension efficiencies are better than the others.

Table 3. Best averaged F1 score cases of classification performance for 128 feature map extension.

	Negative		Sedan		Station Wagon		Van		Working Truck	
	ACC	F1	ACC	F1	ACC	F1	ACC	F1	ACC	F1
Orig.M	96.83%	0.9791	58.40%	0.6247	81.81%	0.8010	90.11%	0.8422	69.72%	0.5435
Orig.16	**99.68%**	0.9624	57.13%	0.6433	82.04%	**0.8329**	91.64%	**0.8650**	70.05%	**0.5914**
New Ext.	97.20%	0.9822	63.38%	0.6474	82.13%	0.8245	91.92%	0.8459	**74.52%**	0.5666
Sel. Ext.	96.90%	0.9808	63.56%	**0.6487**	**82.25%**	0.8247	**93.00%**	0.8501	68.16%	0.5609
Glb.	97.01%	0.9810	65.73%	0.6438	81.51%	0.8303	92.38%	0.8471	71.29%	0.5377
*Glb.*ReLU	97.22%	0.9820	**65.95%**	0.6410	81.27%	0.8315	92.69%	0.8477	71.25%	0.5406
Lcl.	97.00%	0.9813	65.45%	0.6408	81.65%	0.8310	92.66%	0.8497	69.63%	0.5418
*Lcl.*ReLU	97.27%	**0.9823**	64.61%	0.6404	81.53%	0.8285	92.56%	0.8498	70.70%	0.5375
Bat.	97.12%	0.9814	64.40%	0.6417	81.54%	0.8276	92.89%	0.8471	70.38%	0.5351
*Bat.*ReLU	97.30%	0.9820	63.87%	0.6407	81.67%	0.8273	92.66%	0.8505	72.78%	0.5556

Note: The first, second and third topmost values in each column are marked by **bold**, underline and double-underline. Meanings of abbreviations are: the baseline VGG-M (*Orig.M*), the 16-layered full-sized VGG (*Orig.16*), the softmax loss based extensions (*New Ext.* and *Sel. Ext.*), and Main-Side loss based extensions (*Sel. S-Ext.*).

Table 4. Best averaged F1 score cases of classification performance for 256 feature map extension.

	Negative		Sedan		Station Wagon		Van		Working Truck	
	ACC	F1	ACC	F1	ACC	F1	ACC	F1	ACC	F1
Orig.M	96.83%	0.9791	58.40%	0.6247	81.81%	0.8010	90.11%	0.8422	69.72%	0.5435
Orig.16	**99.68%**	0.9624	57.13%	0.6433	82.04%	**0.8329**	91.64%	**0.8650**	70.05%	**0.5914**
New Ext.	97.23%	**0.9823**	61.97%	0.6431	82.26%	0.8207	92.26%	0.8484	71.97%	0.5804
Sel. Ext.	96.96%	0.9808	61.98%	**0.6453**	**82.39%**	0.8192	91.69%	0.8451	70.85%	0.5439
Glb.	97.12%	0.9816	64.26%	0.6419	81.47%	0.8263	92.83%	0.8453	70.31%	0.5409
*Glb.*ReLU	97.16%	0.9818	64.47%	0.6441	81.80%	0.8290	**92.94%**	0.8505	**73.60%**	0.5472
Lcl.	96.94%	0.9809	**65.91%**	0.6384	81.45%	0.8306	92.46%	0.8507	68.53%	0.5469
*Lcl.*ReLU	97.11%	0.9816	65.11%	0.6439	81.69%	0.8296	92.46%	0.8481	72.05%	0.5564
Bat.	97.01%	0.9809	63.75%	0.6413	81.52%	0.8242	92.92%	0.8457	69.94%	0.5442
*Bat.*ReLU	97.31%	**0.9823**	65.12%	0.6420	81.45%	0.8303	92.70%	0.8482	71.70%	0.5419

Note: The first, second and third topmost values in each column are marked by **bold**, underline and double-underline. Meanings of abbreviations are: the baseline VGG-M (*Orig.M*), the 16-layered full-sized VGG (*Orig.16*), the softmax loss based extensions (*New Ext.* and *Sel. Ext.*), and Main-Side loss based extensions (*Sel. S-Ext.*).

Finally, a brief illustration of the effectiveness of the proposed network extension is given in Figure 14, where *Orig.M* and *Select S-Ext.* with $N_{sel} = 256$ and $\lambda = \exp(-2)$ are chosen for comparison. In Figure 14a, newly recognized images by *Select S-Ext.* are listed in Figure 14a by their types in each row. According to the common characteristics in appearance, three categories can be established in the columns: those with rare structures or confusing appearances (*Rare Instances*), those being blurred by shadows (*Shadowing*), and those being partially covered by trees and buildings (*Covering*). These are the challenging conditions to which the extended network structure is devoted. At last, prediction accuracies of *Orig.M* and *Select S-Net.* on the three sample categories discussed in Section 5.1.2 are illustrated in Figure 14b, abbreviated as *Orig.* and *Imprv.*. Therein, accuracy values of *Imprv.* are marked above its curve markers, and the improvement values *Diff.* are shown as bars. As expected, samples with greater vehicle center distances (*Far Distance*) are better predicted by their recognition easiness. Additionally, consistent with the design pattern of Main-Side Loss, greater improvement happens on positives in the *Centered* category.

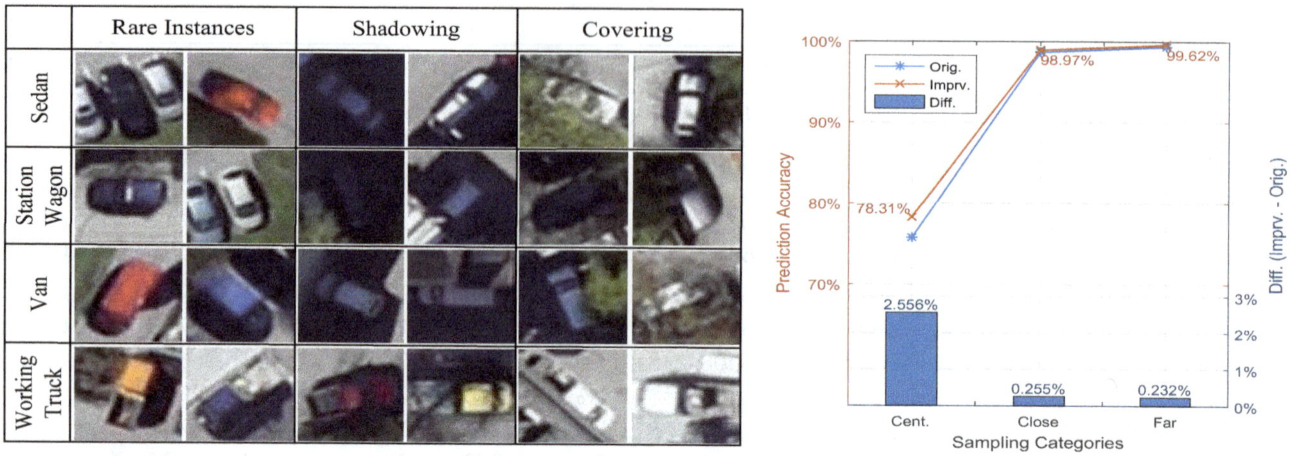

(a) (b)

Figure 14. Network classification performance improvement illustrated by the established classification dataset. (**a**) Newly recognized positives after extension. (**b**) Prediction accuracies and the increments on sample categories: Centered (Cent.), Close Range (Close), and Far Range (Far).

5.3. Network Extension Efficiency by Selected Feature Maps

This sub-section discusses the network extension efficiency in using the selected convolutional feature maps. To justify the comparisons, only the extensions *New Ext.* and *Select Ext.* based on the softmax loss are adopted, so all kernels will be penalized equally over different classes.

Table 5 shows the classification accuracies between the original VGG-M network and its two extended counterparts, *New Ext.* and *Select Ext.*. Seven extension quantities are involved in the comparison, ranging from 64 to 256. Since the feature maps selection scheme described in Algorithm 1 will introduce duplications, the number of feature maps used in the selective extension is always smaller. As can be observed from Table 5, outperforming instances frequently occur on large and medium-sized classes (e.g., Sedan and Station Wagon) which have occupation ratios at 23.06% and 66.96%. For class Van, which has an occupation ratio of 9.10%, only one outperforming is detected. Accuracy differences on class Working Truck fluctuate radically, a class which has the smallest data occupation ratio at 0.88%.

Table 5. Classification accuracies for softmax loss-based extensions *New Ext.* and *Select Ext.*.

Original	**Sedan**		**Station Wagon**		**Van**		**Working Truck**	
	ACC 58.40%		**ACC 81.81%**		**ACC 90.11%**		**ACC 69.72%**	
New Ext. & Sel. Ext.	New	Select	New	Select	New	Select	New	Select
N64/S50	63.02%	**63.19%**	81.85%	**82.27%**	92.58%	91.81%	74.06%	**76.30%**
N96/S71	63.00%	62.78%	82.11%	**82.35%**	92.76%	91.90%	75.84%	67.97%
N128/S89	63.38%	**63.56%**	82.13%	**82.25%**	91.92%	**93.00%**	74.52%	68.16%
N160/S109	63.23%	**63.59%**	82.07%	81.96%	92.62%	92.48%	75.30%	**75.73%**
N192/S130	62.55%	62.30%	81.99%	**82.40%**	92.53%	92.07%	72.54%	**73.62%**
N224/S150	63.38%	63.10%	81.84%	**82.05%**	92.46%	91.95%	69.97%	68.73%
N256/S168	61.97%	**61.98%**	82.26%	**82.39%**	92.26%	91.69%	71.97%	70.85%

Note: For each pair of accuracies given by *New Ext.* and *Select Ext.*, instances where the *Select Ext.* outperforms the *New Ext.* are emphasized by **bold** font.

The phenomenon mentioned above can be regarded as the equal penalization nature of the softmax loss. As most of the feature maps selected by Algorithm 1 have high class-significance for all classes, they are more likely to be assigned to majority classes. For the kernels only efficient on minority classes, fluctuations occur as softmax loss attempts to bias them to the majority ones. As a result, poor overall performance refinement is obtained by *Select Ext.*. As in Figure 15, averaged F1 scores and accuracies are shown for the three networks, with differences between *Select Ext.* (*Select*) and *New Ext.* (*Simple*) displayed as bars *Diff.*. Instances where *Select Ext.* is comparable to *New Ext.* are marked by arrows, and values for *Select Ext.* are listed above the markers. For the aforementioned reasons, these instances are rare, and the superiorities of *Select Ext.* are less significant.

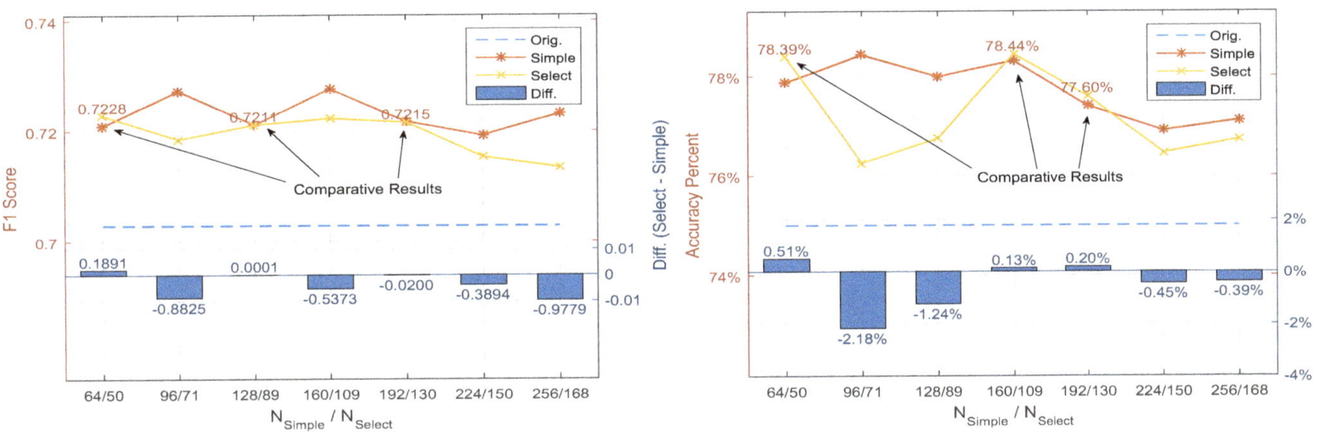

(**a**) F1 Score (**b**) Accuracies

Figure 15. Overall performance comparisons between the *Orig.M*, *New Ext.* and *Select Ext.* under different extension sizes. (**a**) the averaged F1 scores, (**b**) the averaged accuracies. Instances where *Select Ext.* is comparable to *New Ext.* are marked by arrows.

Finally, in Figure 16, a more fair comparison for showing the feature map extension efficiency is performed based on a per-kernel evaluation, where the increase in F1 score for each newly added feature map is calculated by $(F1_{ext} - F1_{orig})\,/N_{ext}$, in which $F1_{orig}$ and $F1_{ext}$ for the baseline and extended network, and N_{ext} is the number of extended feature maps. As can be observed from Figure 16, the selective feature map-based extension is more efficient in medium-sized minority classes (Sedan and Van) for small extension quantity, while dropping more rapidly than the blank kernel-based one since the selected ones lack enough flexibility.

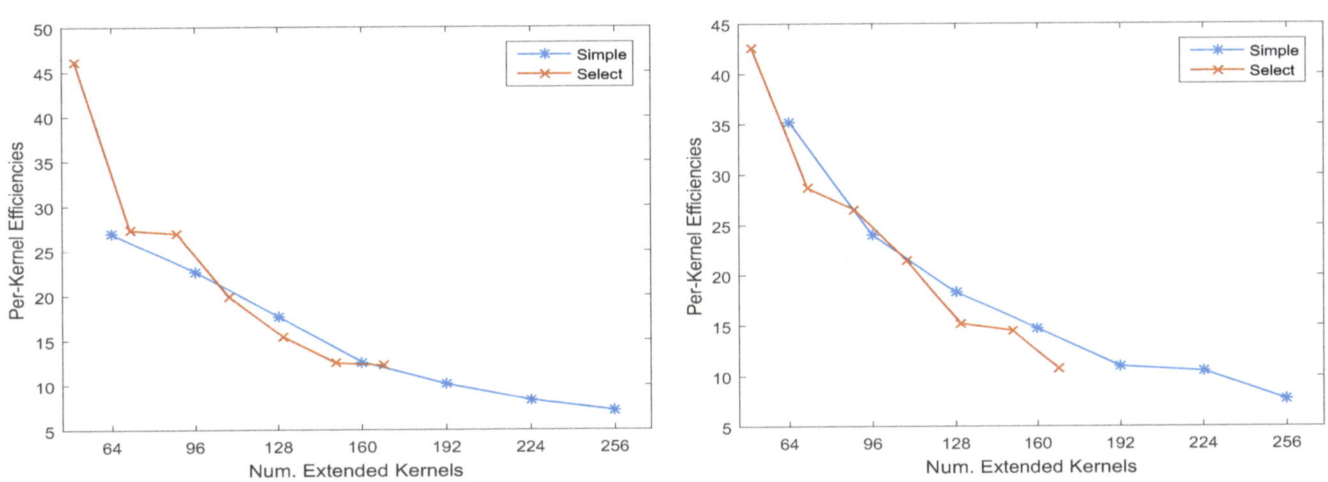

(**a**) Sedan (23.06% * N_{pos}) (**b**) Station Wagon (66.96% * N_{pos})

Figure 16. *Cont.*

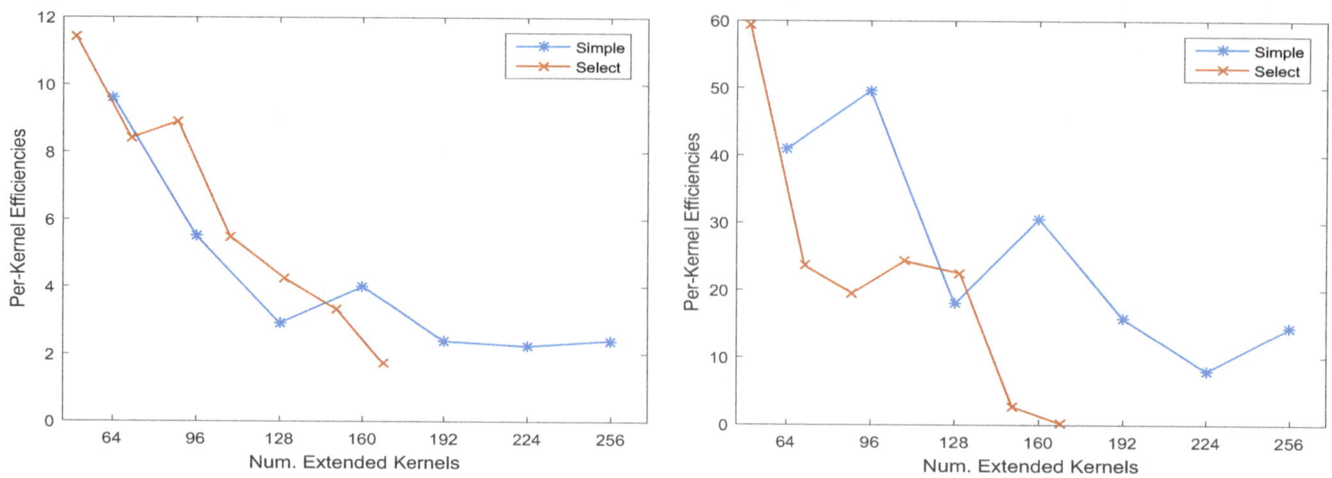

(**c**) Van (9.10% * N_{pos})

(**d**) Working Truck (0.88% * N_{pos})

Figure 16. Efficiency comparison of extended feature maps (kernels). N_{pos} is the quantity of all vehicles. Selected feature maps (kernels) are more effective for small extension and minority classes.

5.4. Main Factors in Main-Side Loss Function-based Fine-Tuning

Three major configurations have to be considered when using the Main-Side loss to fine-tune the bi-parted Main-Side network, which will be analyzed on the *New S-Ext.* extension: the first one is whether to fix the FC layers in the Main Network during the fine-tuning; the second one is the penalization coefficient λ and the positive constraint imposed on the single layered *FC6 Ext.* by an extra ReLU layer; and the third one is the three penalization modes. The reason for choosing the *New S-Ext.* extension for inspection is that it uses feature maps generated by blank kernels, which ensures that all extended kernels are equally flexible, and thus can be useful for an objective illustration of the impact caused by different configurations.

The first configuration—which involves partial fixation or joint optimization—determines whether the FC6 to FC8 layers should be updated during the fine-tuning. This configuration is only examined on the global penalization version of *New S-Ext.*, with the class-wise performances shown in Table 6. It is then obvious from the table that the joint optimization settings outperform the partially fixed ones in almost every class, including the accuracies and F1 scores, except the trivially populated class Working Truck. The superiority in performance for the joint optimization version is caused by the simultaneous adjustment of the estimation accuracies by the Main Network component, which means that kernels in the Main and Side Networks are optimally re-assigned. It is worth note that the existence of the positive constraint by ReLU layer is less significant for the joint optimization versions, in which the scores are almost identical.

Table 6. Categorization accuracies for fixed-Main and joint optimization, best average F1 cases.

	Sedan		Station Wagon		Van		Working Truck	
	ACC	F1	ACC	F1	ACC	F1	ACC	F1
Global No ReLU, Fix-M	62.12%	0.6275	81.10%	0.8212	91.46%	0.8479	72.67%	0.5492
Global ReLU, Fix-M	60.26%	0.6289	81.48%	0.8132	91.03%	0.8463	72.33%	**0.5542**
Global No ReLU, Joint	65.42%	0.6435	81.44%	0.8320	**93.29%**	**0.8497**	**73.77%**	0.5508
Global ReLU, Joint	**65.66%**	**0.6449**	**81.58%**	**0.8325**	92.96%	0.8484	72.06%	0.5490

Note: Fixed and non-fixed optimization settings are abbreviated as *Fix-M* and *Joint*, and the top-2 highest scores are marked as **bold** and underline.

For the second configuration involving the coefficient λ and the positive constraint ReLU, comparing results are shown in Figures 17 and 18. As can be observed from Figure 17a, the influences of λ are more correlated with the prediction accuracies, as they arise when the coefficient decreases. This is because smaller penalization encourages larger likelihood rectifications from the Side Network. This rectification effect is more clear in Figure 18, where medium-sized minority classes (e.g., class Sedan and Van) have greater accuracy improvements compared with majority ones (e.g., Station Wagon). However, the class that is too small (e.g., Working Truck) seems to benefit less from this effect because of the possibility of over-fitting.

In contrast, as seen previously in both figures, the existence of the ReLU layer has little or no influence on the resulting accuracies, while the removal of the ReLU layer seems to help stabilize the fluctuations in accuracies and F1 scores as the penalization λ changes. This is reasonable, since by permitting negative adjustments from the Side Network, they help with pruning the Main Network likelihoods too high to cause over-fitting.

For the third configuration—which involves the comparison between the three penalization modes (Global, Local, and Batch-wise)—experimental results are presented in Table 7. Judging from the scores, there is no apparent winner: Batch-wise penalization is more suitable for improving the small-sized classes (e.g., Working Truck), the Global penalization is more suitable for medium-sized classes (e.g., Sedan and Van), and the Local penalization performs better for the large and medium-sized ones (e.g., Station Wagon and Sedan).

Table 7. Best accuracies and F1s for three modes with or without the ReLU layer on FC6 Ext. layer.

	Sedan		Station Wagon		Van		Working Truck	
	ACC	F1	ACC	F1	ACC	F1	ACC	F1
Global No ReLU, Joint	65.42%	0.6435	81.44%	0.8320	93.29%	0.8497	73.77%	0.5508
Global ReLU, Joint	65.66%	**0.6449**	81.58%	0.8325	**92.96%**	0.8484	72.06%	0.5490
Local No ReLU, Joint	**65.79%**	0.6415	81.63%	**0.8335**	92.91%	0.8494	69.67%	0.5491
Local ReLU, Joint	65.53%	0.6441	81.55%	0.8320	92.88%	0.8468	72.17%	0.5432
Batch-wise No ReLU, Joint	64.79%	0.6442	**81.95%**	0.8308	92.40%	**0.8498**	71.04%	0.5548
Batch-wise ReLU, Joint	65.00%	0.6435	81.72%	0.8307	92.58%	0.8477	**74.92%**	**0.5662**

Note: In each column, the first and second topmost values are emphasized by **bold** and underline. Implementations with and without ReLU layer are marked by 'ReLU' and 'No ReLU'. 'Joint' means non-fixed optimization same as that in Table 6.

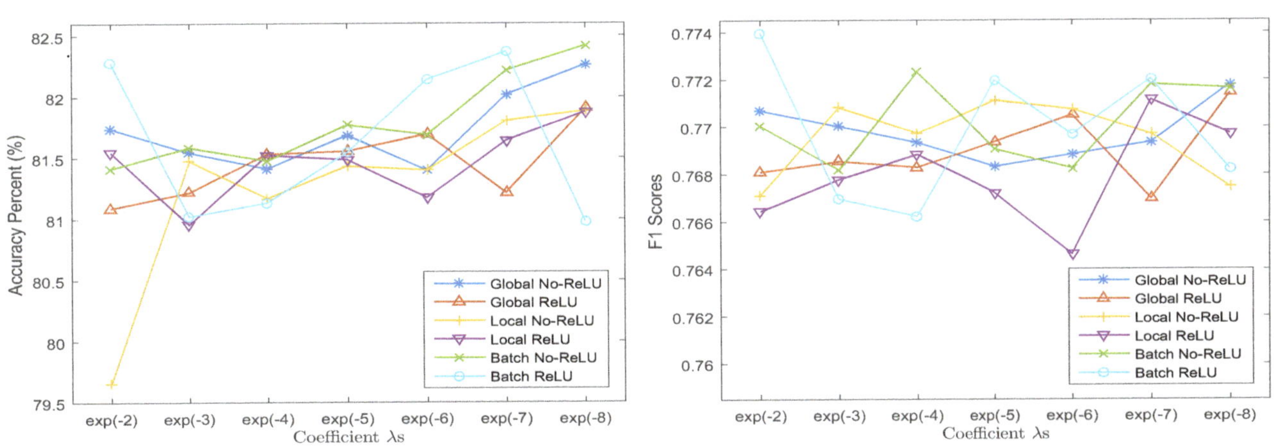

(**a**) Accuracies by λ and ReLU (**b**) F1 Scores by λ and ReLU

Figure 17. Influences of the coefficient λ and ReLU constraint on the overall accuracy and $F1$ score in three modes. (**a**) the averaged accuracies; (**b**) the averaged $F1$ scores.

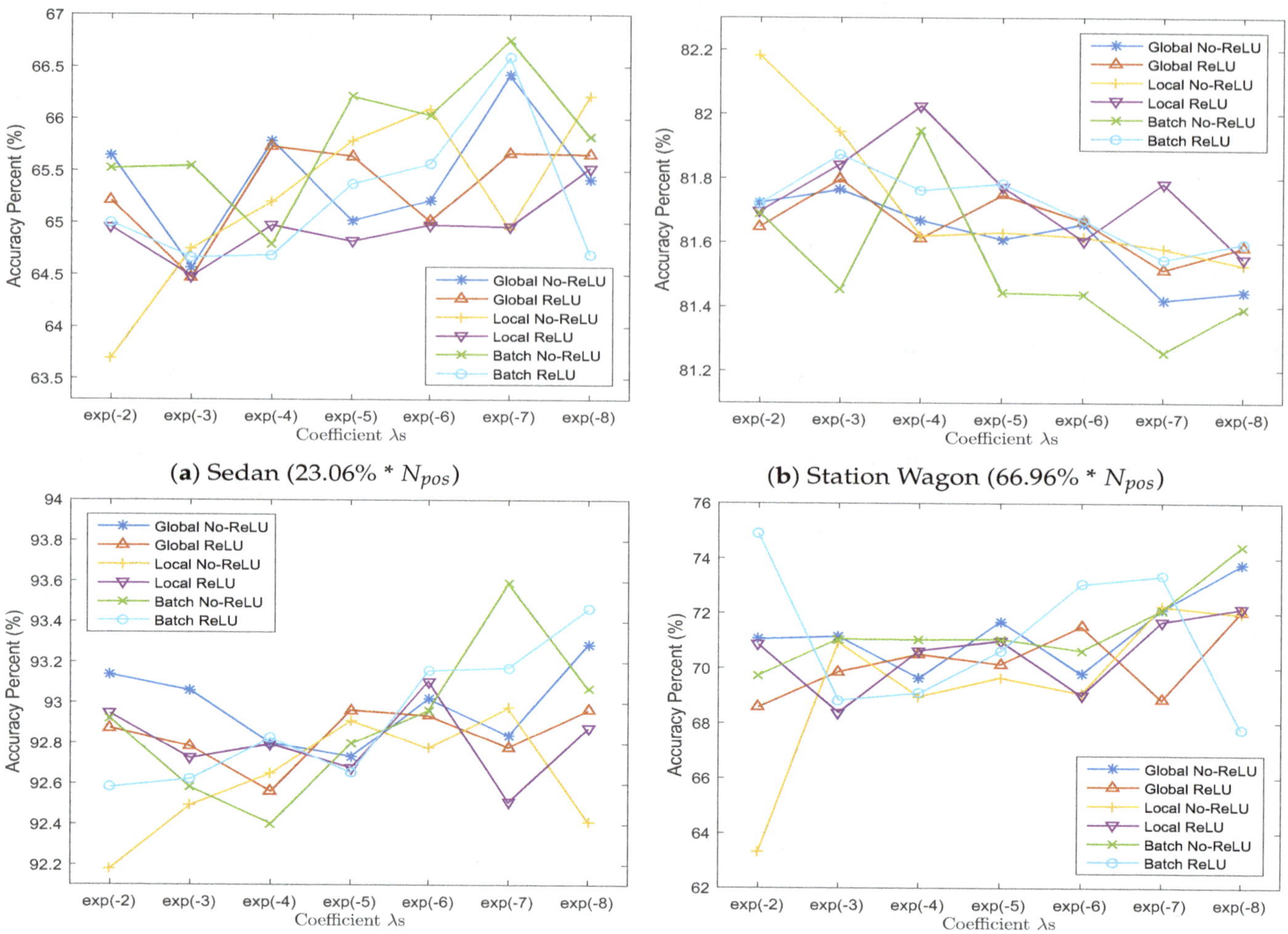

(**a**) Sedan (23.06% * N_{pos}) (**b**) Station Wagon (66.96% * N_{pos})

(**c**) Van (9.10% * N_{pos}) (**d**) Working Truck (0.88% * N_{pos})

Figure 18. Influences of the penalization mode and the coefficient λ on accuracy and $F1$ score for different vehicle types. N_{pos} is the quantity of positives, which is all the vehicles.

6. Discussion

As mentioned in Section 1, few articles have been found to address the class-imbalance issue in high-resolution aerial image-based vehicle localization and categorization using CNN structure extension; this article serves as an exploration of such methods. A principally similar work named hierarchical deep CNN (HD-CNN) [50] has studied the effectiveness of a tree-structured CNN ensemble on general object classification problems involving dozens of classes. However, with only a few classes, it is inconvenient to build multi-level class taxonomy, and appending a near-full-sized CNN structure might not be sufficiently cost-efficient considering the size of the problem.

The effectiveness of the proposed extension scheme is exemplified in a moderately-sized VGG-M network in a self-proven manner. According to the analysis in the previous experimental section, several general conclusions can be drawn on the network extension-based class-imbalance dealing methods, which can be useful for applying the methods on other similar applications:

(a) According to Tables 3 and 4, using wider and deeper network structure with plain extension (blank kernels and softmax loss) will generally improve the classification performances on all classes, while using deeper structure will help more with the generalized performance measured by F1 score.

(b) According to Table 5, the effectiveness of the softmax loss-based plain width extension with either blank kernels or selected feature maps will decrease rapidly as the extension quantity increases. Additionally, the selected feature maps are more effective under small extension quantity, while losing their advantage in large extension, as they lack flexibility.

(c) As can be seen from Tables 3–5, selected feature maps are more helpful for improving the classification accuracies, while they can barely keep up with the blank kernel-based extension in overall F1 score by Figure 15a. To maintain a reasonably high F1 score performance, the penalization mode *Glb.ReLU* and *Bat.ReLU* are preferred, as in Tables 3 and 4.

(d) As seen by Figure 17a, penalization modes without ReLU constraint in the Main-Side loss-related fine-tuning can produce a more significant increment in accuracies as the global penalization λ decreases. The existence of a ReLU layer helps to stabilize the fluctuation in F1 scores when λ changes, as in Figure 17b.

(e) By Figure 18, the class-imbalance-sensitive penalization term $\Omega(\mathbf{z}^*)$ helps to improve the classification accuracies for the medium-sized minority classes (Sedan and Van), but is not so ideal for classes with an absolutely trivial sample quantity (Working Truck).

(f) The sizes of most effective vehicle classes for the three penalization modes are different. Shown by Table 7, the Global penalization mode is effective on medium-sized classes (Sedan and Van), the Local mode is effective for large- and medium-sized classes (Station Wagon and Sedan), while the Batch-wise mode is effective for small-sized classes (Working Truck).

7. Conclusions

Methods for joint vehicle localization and categorization in aerial images helps with important applications such as traffic flow analysis and suspicious vehicle detection. By treating samples who exceed the permitted location deviation as negatives and classifying them along with the other vehicle classes, the problem of cascaded localization error in separated estimation is eliminated. Top-3 accuracy as high as 99% can be achieved when a typical CNN-based classifier is employed (e.g., the 16-layered VGG network), but it still suffers from the class-imbalance issue, which causes poor classification performances on minority classes.

Based on the R-CNN detection structure, a cost-effective network extension scheme is proposed in this paper to address this issue by introducing less computation and memory consumption overhead. Such efficiency is achieved by two means: the feature map selection and bi-partite Main-Side Network extension, which are performed with the help of a feature map class-importance measurement and a class-imbalance-aware loss function newly proposed in this article. The resulting extended network structure is verified along with its similarly-shaped strong counterparts on a 0.13 m GSD aerial image dataset captured over the urban region of Munich. Experimental results show that the selectively extended feature maps are more effective than those produced by randomly initialized new kernels. By applying the Main-Side loss on this bi-partite network, classification performances on medium-sized minority classes can be further improved. The three Main-Side loss penalizing schemes help with this performance improvement differently, showing varied refinement effect on different-sized classes. Generally, by jointly employing the feature map selection and Main-Side loss optimization schemes, comparable vehicle categorization results can be achieved compared to the counterparts with less parameter and memory overheads.

Key contributions of this study are as follows: First, a novel multi-class feature map importance measurement is proposed by extending the existing significance score for binary classification problems. Second, an easy-to-use cost-effective network extension scheme called Main-Side Network is proposed to greatly improve the classification performances on minority classes with small amount of overhead. Third, three penalization modes are proposed for regularizing the Main-Side loss adopted in this extension, which are simple to implement and beneficial for minority classes with different properties.

In future work, the existing classification deficiencies on tiny classes (e.g., the Working Truck class) is planned to be deeply investigated by using stronger models from the one-class classification. Difficult detection conditions involving shadowed and partially sheltered vehicles caused by skyscrapers and street trees will be further analyzed with harder experimental dataset. Behaviors of the three penalization modes for the Main-Side Loss should be further analyzed in detail to enhance the performance. The Main-Side Network extension structure is intended to be replaced by a network splitting method; thus, the convolution and memory consumption overhead can be completely eliminated.

Acknowledgments: This work is jointly supported by the National Science Foundation of China (NSFC) with granting No. 61302154 and No. 61573350, which mainly focus on aerial image and video analysis. The author would like to acknowledge the editor and reviewers by contributing their precious time to view this article. The authors would also like to thank authors Jason Yosinski et al. in [65] for their selflessness in sharing the great visualization tool.

Author Contributions: Feimo Li designed the experiment and prepared the manuscript; both Hongxing Chang, Shuxiao Li and Chengfei Zhu directed the research project and provided conceptual advices. Xiaosong Lan helped in the revision the manuscript.

Abbreviations

The following abbreviations are used in this manuscript:

ADASYN	Adaptive Synthetic Sampling
CBO	Cluster-based Oversampling
CNN	Convolutional Neural Network
DBN	Deep Belief Network
FCN	Fully Convolutional Neural Network
HOG	Histogram of Oriented Gradients
GAN	Generative Adversarial Network
GSD	Ground Sampling Distance
LBP	Local Binary Pattern
R-CNN	Regions with Convolutional Neural Network Features
ROI	Region of Interest
SIFT	Scale Invariant Feature Transform
SMOTE	Synthetic Minority Over-sampling Technique
SVM	Support Vector Machine
t-SNE	t-Distributed Stochastic Neighbor Embedding
VGG	Visual Geometry Group

References

1. Xu, Y.; Yu, G.; Wang, Y.; Wu, X.; Ma, Y. A Hybrid Vehicle Detection Method Based on Viola-Jones and HOG + SVM from UAV Images. *Sensors* **2016**, doi:10.3390/s16081325.
2. Tuermer, S.; Kurz, F.; Reinartz, P.; Stilla, U. Airborne vehicle detection in dense urban areas using HoG features and disparity maps. *IEEE J. Sel. Top. Appl. Earth Obs. Remote Sens.* **2013**, *6*, 2327–2337.
3. Hinz, S.; Schlosser, C.; Reitberger, J. Automatic car detection in high resolution urban scenes based on an adaptive 3D-model. In Proceedings of the 2nd GRSS/ISPRS Joint Workshop on Remote Sensing and Data Fusion over Urban Areas, Berlin, Germany, 22–23 May 2003; pp. 167–171.
4. Qu, T.; Zhang, Q.; Sun, S. Vehicle detection from high-resolution aerial images using spatial pyramid pooling-based deep convolutional neural networks. *Multimedia Tools Appl.* **2016**, doi:10.1007/s11042-016-4043-5.
5. Chen, X.; Xiang, S.; Liu, C.L.; Pan, C.H. Vehicle detection in satellite images by hybrid deep convolutional neural networks. *IEEE Geosci. Remote Sens. Lett.* **2014**, *11*, 1797–1801.
6. Cao, L.; Jiang, Q.; Cheng, M.; Wang, C. Robust vehicle detection by combining deep features with exemplar classification. *Neurocomputing* **2016**, *215*, 225–231.

7. Zhu, H.; Chen, X.; Dai, W.; Fu, K.; Ye, Q.; Jiao, J. Orientation robust object detection in aerial images using deep convolutional neural network. In Proceedings of the 2015 IEEE International Conference on Image Processing (ICIP), Quebec City, QC, Canada, 27–30 September 2015; pp. 3735–3739.
8. Qu, S.; Wang, Y.; Meng, G.; Pan, C. Vehicle Detection in Satellite Images by Incorporating Objectness and Convolutional Neural Network. *J. Ind. Intell. Inf.* **2016**, doi:10.18178/jiii.4.2.158-162.
9. Girshick, R.; Donahue, J.; Darrell, T.; Malik, J. Rich feature hierarchies for accurate object detection and semantic segmentation. In Proceedings of the IEEE Conference on Computer Vision and Pattern Recognition, Columbus, OH, USA, 23–28 June 2014; pp. 580–587.
10. Girshick, R. Fast r-cnn. In Proceedings of the IEEE International Conference on Computer Vision, Los Alamitos, CA, USA, 7–13 December 2015; pp. 1440–1448.
11. Ren, S.; He, K.; Girshick, R.; Sun, J. Faster r-cnn: Towards real-time object detection with region proposal networks. In Proceedings of the Advances in Neural Information Processing Systems, Montréal, QC, Canada, 7–12 December 2015.
12. Liu, W.; Anguelov, D.; Erhan, D.; Szegedy, C.; Reed, S.; Fu, C.Y.; Berg, A.C. SSD: Single shot multibox detector. In *European Conference on Computer Vision*; Springer: New York, NY, USA, 2016; pp. 21–37.
13. Redmon, J.; Divvala, S.; Girshick, R.; Farhadi, A. You only look once: Unified, real-time object detection. In Proceedings of the IEEE Conference on Computer Vision and Pattern Recognition, Stanford, CA, USA, 27–30 June 2016; pp. 779–788.
14. He, K.; Zhang, X.; Ren, S.; Sun, J. Spatial pyramid pooling in deep convolutional networks for visual recognition. In *European Conference on Computer Vision*; Springer: New York, NY, USA, 2014, pp. 346–361.
15. Tang, T.; Zhou, S.; Deng, Z.; Zou, H.; Lei, L. Vehicle Detection in Aerial Images Based on Region Convolutional Neural Networks and Hard Negative Example Mining. *Sensors* **2017**, doi:10.3390/s17020336.
16. Wang, Q.; Lin, J.; Yuan, Y. Salient band selection for hyperspectral image classification via manifold ranking. *IEEE Trans. Neural Netw. Learn. Syst.* **2016**, *27*, 1279–1289.
17. Zhang, F.; Du, B.; Zhang, L. Saliency-guided unsupervised feature learning for scene classification. *IEEE Trans. Geosci. Remote Sens.* **2015**, *53*, 2175–2184.
18. Holt, A.C.; Seto, E.Y.; Rivard, T.; Gong, P. Object-based detection and classification of vehicles from high-resolution aerial photography. *Photogramm. Eng. Remote Sens.* **2009**, *75*, 871–880.
19. Liu, K.; Mattyus, G. Fast multiclass vehicle detection on aerial images. *IEEE Geosci. Remote Sens. Lett.* **2015**, *12*, 1938–1942.
20. Razakarivony, S.; Jurie, F. Vehicle detection in aerial imagery: A small target detection benchmark. *J. Vis. Commun. Image Represent.* **2016**, *34*, 187–203.
21. Bell, S.; Lawrence Zitnick, C.; Bala, K.; Girshick, R. Inside-outside net: Detecting objects in context with skip pooling and recurrent neural networks. In Proceedings of the IEEE Conference on Computer Vision and Pattern Recognition, Stanford, CA, USA, 27–30 June 2016; pp. 2874–2883.
22. Ma, Z.; Yu, L.; Chan, A.B. Small instance detection by integer programming on object density maps. In Proceedings of the IEEE Conference on Computer Vision and Pattern Recognition, Boston, MA, USA, 7–12 June 2015; pp. 3689–3697.
23. Arteta, C.; Lempitsky, V.; Noble, J.A.; Zisserman, A. Interactive object counting. In *European Conference on Computer Vision*; Springer: New York, NY, USA, 2014, pp. 504–518.
24. Yuan, Y.; Wan, J.; Wang, Q. Congested scene classification via efficient unsupervised feature learning and density estimation. *Pattern Recognit.* **2016**, *56*, 159–169.
25. He, H.; Garcia, E.A. Learning from imbalanced data. *IEEE Trans. Knowl. Data Eng.* **2009**, *21*, 1263–1284.
26. Branco, P.; Torgo, L.; Ribeiro, R. A survey of predictive modelling under imbalanced distributions. *arXiv* **2015**, arXiv:1505.01658.
27. He, H.; Ma, Y. *Imbalanced Learning: Foundations, Algorithms, and Applications*; John Wiley & Sons: New York, NY, USA, 2013.
28. Krawczyk, B. Learning from imbalanced data: Open challenges and future directions. *Prog. Artif. Intell.* **2016**, *5*, 221–232.
29. Chawla, N.V.; Bowyer, K.W.; Hall, L.O.; Kegelmeyer, W.P. SMOTE: synthetic minority over-sampling technique. *J. Artif. Intell. Res.* **2002**, *16*, 321–357.

30. He, H.; Bai, Y.; Garcia, E.A.; Li, S. ADASYN: Adaptive synthetic sampling approach for imbalanced learning. In Proceedings of the IEEE International Joint Conference on Neural Networks, Hong Kong, China, 1–8 June 2008; pp. 1322–1328.
31. Jo, T.; Japkowicz, N. Class imbalances versus small disjuncts. *ACM Sigkdd Explor. Newsl.* **2004**, *6*, 40–49.
32. Zhou, Z.H.; Liu, X.Y. On Multi-Class Cost-Sensitive Learning. *Comput. Intell.* **2010**, *26*, 232–257.
33. Ting, K.M. A comparative study of cost-sensitive boosting algorithms. In Proceedings of the 17th International Conference on Machine Learning, Stanford, CA, USA, 29 June–2 July 2000.
34. Krawczyk, B.; Woźniak, M.; Schaefer, G. Cost-sensitive decision tree ensembles for effective imbalanced classification. *Appl. Soft Comput.* **2014**, *14*, 554–562.
35. Huang, C.; Li, Y.; Change Loy, C.; Tang, X. Learning deep representation for imbalanced classification. In Proceedings of the IEEE Conference on Computer Vision and Pattern Recognition, Stanford, CA, USA, 27–30 June 2016; pp. 5375–5384.
36. Jeatrakul, P.; Wong, K.W.; Fung, C.C. Classification of imbalanced data by combining the complementary neural network and SMOTE algorithm. In *International Conference on Neural Information Processing*; Springer: New York, NY, USA, 2010; pp. 152–159.
37. Simpson, A.J. Over-sampling in a deep neural network. *arXiv* **2015**, arXiv:1502.03648.
38. Goodfellow, I.; Pouget-Abadie, J.; Mirza, M.; Xu, B.; Warde-Farley, D.; Ozair, S.; Courville, A.; Bengio, Y. Generative adversarial nets. In Proceedings of the Advances in Neural Information Processing Systems, Montreal, QC, Canada, 8–13 December 2014; pp. 2672–2680.
39. Khan, S.H.; Bennamoun, M.; Sohel, F.; Togneri, R. Cost sensitive learning of deep feature representations from imbalanced data. *arXiv* **2015**, arXiv:1508.03422.
40. Cheng, G.; Zhou, P.; Han, J. Rifd-cnn: Rotation-invariant and fisher discriminative convolutional neural networks for object detection. In Proceedings of the IEEE Conference on Computer Vision and Pattern Recognition, Stanford, CA, USA, 27–30 June 2016; pp. 2884–2893.
41. Bertasius, G.; Shi, J.; Torresani, L. Deepedge: A multi-scale bifurcated deep network for top-down contour detection. In Proceedings of the IEEE Conference on Computer Vision and Pattern Recognition, Boston, MA, USA, 7–12 June 2015; pp. 4380–4389.
42. Shen, W.; Wang, X.; Wang, Y.; Bai, X.; Zhang, Z. Deepcontour: A deep convolutional feature learned by positive-sharing loss for contour detection. In Proceedings of the IEEE Conference on Computer Vision and Pattern Recognition, Boston, MA, USA, 7–12 June 2015; pp. 3982–3991.
43. Robinson, J.P.; Shao, M.; Wu, Y.; Fu, Y. Families in the Wild (FIW): Large-Scale Kinship Image Database and Benchmarks. In Proceedings of the 2016 ACM on Multimedia Conference, Amsterdam, The Netherlands, 15–19 October 2016; pp. 242–246.
44. Santos, C.N.d.; Xiang, B.; Zhou, B. Classifying relations by ranking with convolutional neural networks. *arXiv* **2015**, arXiv:1504.06580.
45. Schroff, F.; Kalenichenko, D.; Philbin, J. Facenet: A unified embedding for face recognition and clustering. In Proceedings of the IEEE Conference on Computer Vision and Pattern Recognition, Boston, MA, USA, 7–12 June 2015; pp. 815–823.
46. Oh Song, H.; Xiang, Y.; Jegelka, S.; Savarese, S. Deep metric learning via lifted structured feature embedding. In Proceedings of the IEEE Conference on Computer Vision and Pattern Recognition, Stanford, CA, USA, 27–30 June 2016; pp. 4004–4012.
47. Wen, Y.; Zhang, K.; Li, Z.; Qiao, Y., A Discriminative Feature Learning Approach for Deep Face Recognition. In *Computer Vision—ECCV 2016*; Springer International Publishing: Cham, The Netherlands, 2016; pp. 499–515.
48. Chu, J.L.; Krzyak, A. Analysis of feature maps selection in supervised learning using convolutional neural networks. In *Canadian Conference on Artificial Intelligence*; Springer: New York, NY, USA, 1994; pp. 59–70.
49. Marcu, A.; Leordeanu, M. Dual Local-Global Contextual Pathways for Recognition in Aerial Imagery. *arXiv* **2016**, arXiv:1605.05462.
50. Yan, Z.; Zhang, H.; Piramuthu, R.; Jagadeesh, V.; DeCoste, D.; Di, W.; Yu, Y. HD-CNN: hierarchical deep convolutional neural networks for large scale visual recognition. In Proceedings of the IEEE International Conference on Computer Vision, Los Alamitos, CA, USA, 7–13 December 2015; pp. 2740–2748.

51. Jaderberg, M.; Simonyan, K.; Zisserman, A.; Kavukcuoglu, K. Spatial transformer networks. In Proceedings of the Advances in Neural Information Processing Systems, Montréal, QC, Canada, 7–12 December 2015.
52. Wang, N.; Li, S.; Gupta, A.; Yeung, D.Y. Transferring rich feature hierarchies for robust visual tracking. *arXiv* **2015**, arXiv:1501.04587.
53. Ng, W.W.; Zeng, G.; Zhang, J.; Yeung, D.S.; Pedrycz, W. Dual autoencoders features for imbalance classification problem. *Pattern Recognit.* **2016**, *60*, 875–889.
54. Guyon, I.; Gunn, S.; Nikravesh, M.; Zadeh, L.A. *Feature Extraction: Foundations and Applications*; Springer: New York, NY, USA, 2008.
55. Bar, Y.; Diamant, I.; Wolf, L.; Lieberman, S.; Konen, E.; Greenspan, H. Chest pathology identification using deep feature selection with non-medical training. *Comput. Methods Biomech. Biomed. Eng. Imaging Vis.* **2016**, doi:10.1080/21681163.2016.1138324.
56. Matsugu, M.; Cardon, P. Unsupervised feature selection for multi-class object detection using convolutional neural networks. In *Advances in Neural Networks—ISNN 2004*; Springer: New York, NY, USA, 2004; pp. 864–869.
57. Zou, Q.; Ni, L.; Zhang, T.; Wang, Q. Deep Learning Based Feature Selection for Remote Sensing Scene Classification. *IEEE Geosci. Remote Sens. Lett.* **2015**, *12*, 2321–2325.
58. Wang, L.; Ouyang, W.; Wang, X.; Lu, H. Visual tracking with fully convolutional networks. In Proceedings of the IEEE International Conference on Computer Vision, Los Alamitos, CA, USA, 7–13 December 2015; pp. 3119–3127.
59. Liu, D.R.; Li, H.L.; Wang, D. Feature selection and feature learning for high-dimensional batch reinforcement learning: A survey. *Int. J. Autom. Comput.* **2015**, *12*, 229–242.
60. Yang, B.; Yan, J.; Lei, Z.; Li, S.Z. Convolutional channel features. In Proceedings of the IEEE International Conference on Computer Vision, Los Alamitos, CA, USA, 7–13 December 2015; pp. 82–90.
61. Zhong, B.; Zhang, J.; Wang, P.; Du, J.; Chen, D. Jointly Feature Learning and Selection for Robust Tracking via a Gating Mechanism. *PLoS ONE* **2016**, *11*, e0161808.
62. Goodfellow, I.; Bengio, Y.; Courville, A. *Deep Learning*; MIT Press: Cambridge, MA, USA, 2016; pp. 166–373.
63. Jarrett, K.; Kavukcuoglu, K.; LeCun, Y.; Ranzato, M. What is the best multi-stage architecture for object recognition? In Proceedings of the 2009 IEEE 12th International Conference on Computer Vision, Kyoto, Japan, 29 September–2 October 2009; pp. 2146–2153.
64. Nair, V.; Hinton, G.E. Rectified linear units improve restricted boltzmann machines. In Proceedings of the 27th International Conference on Machine Learning (ICML-10), Haifa, Israel, 21–25 June 2010; pp. 807–814.
65. Yosinski, J.; Clune, J.; Nguyen, A.; Fuchs, T.; Lipson, H. Understanding neural networks through deep visualization. *arXiv* **2015**, arXiv:1506.06579.
66. Simonyan, K.; Vedaldi, A.; Zisserman, A. Deep inside convolutional networks: Visualising image classification models and saliency maps. *arXiv* **2013**, arXiv:1312.6034.
67. Nguyen, A.; Yosinski, J.; Clune, J. Deep neural networks are easily fooled: High confidence predictions for unrecognizable images. In Proceedings of the IEEE Conference on Computer Vision and Pattern Recognition, Boston, MA, USA, 7–12 June 2015; pp. 427–436.
68. Szegedy, C.; Zaremba, W.; Sutskever, I.; Bruna, J.; Erhan, D.; Goodfellow, I.; Fergus, R. Intriguing properties of neural networks. *arXiv* **2013**, arXiv:1312.6199.
69. Zeiler, M.D.; Fergus, R. Visualizing and understanding convolutional networks. In *European Conference on Computer Vision*; Springer: New York, NY, USA, 2014; pp. 818–833.
70. Erhan, D.; Bengio, Y.; Courville, A.; Vincent, P. Visualizing Higher-Layer Features of a Deep Network. *Univ. Montr.* **2009**, *1341*, 3 .
71. Krizhevsky, A.; Sutskever, I.; Hinton, G.E. Imagenet classification with deep convolutional neural networks. In Proceedings of the Advances in Neural Information Processing Systems, Lake Tahoe, ND, USA, 3–6 December 2012; pp. 1097–1105.
72. LeCun, Y.; Denker, J.S.; Solla, S.A.; Howard, R.E.; Jackel, L.D. *Optimal Brain Damage*; NIPs: Tokyo, Japan, 1989; Volume 2, pp. 598–605.
73. Maaten, L.v.d.; Hinton, G. Visualizing data using t-SNE. *J. Mach. Learn. Res.* **2008**, *9*, 2579–2605.

74. Chatfield, K.; Simonyan, K.; Vedaldi, A.; Zisserman, A. Return of the devil in the details: Delving deep into convolutional nets. *arXiv* **2014**, arXiv:1405.3531.
75. Simonyan, K.; Zisserman, A. Very deep convolutional networks for large-scale image recognition. *arXiv* **2014**, arXiv:1409.1556.
76. He, K.; Zhang, X.; Ren, S.; Sun, J. Deep residual learning for image recognition. In Proceedings of the IEEE Conference on Computer Vision and Pattern Recognition, Stanford, CA, USA, 27–30 June 2016; pp. 770–778.
77. Huang, G.; Liu, Z.; Weinberger, K.Q.; van der Maaten, L. Densely connected convolutional networks. *arXiv* **2016**, arXiv:1608.06993.
78. Targ, S.; Almeida, D.; Lyman, K. Resnet in Resnet: Generalizing residual architectures. *arXiv* **2016**, arXiv:1603.08029.
79. Szegedy, C.; Liu, W.; Jia, Y.; Sermanet, P.; Reed, S.; Anguelov, D.; Erhan, D.; Vanhoucke, V.; Rabinovich, A. Going deeper with convolutions. In Proceedings of the IEEE Conference on Computer Vision and Pattern Recognition, Boston, MA, USA, 7–12 June 2015; pp. 1–9.

Permissions

All chapters in this book were first published in MDPI; hereby published with permission under the Creative Commons Attribution License or equivalent. Every chapter published in this book has been scrutinized by our experts. Their significance has been extensively debated. The topics covered herein carry significant findings which will fuel the growth of the discipline. They may even be implemented as practical applications or may be referred to as a beginning point for another development.

The contributors of this book come from diverse backgrounds, making this book a truly international effort. This book will bring forth new frontiers with its revolutionizing research information and detailed analysis of the nascent developments around the world.

We would like to thank all the contributing authors for lending their expertise to make the book truly unique. They have played a crucial role in the development of this book. Without their invaluable contributions this book wouldn't have been possible. They have made vital efforts to compile up to date information on the varied aspects of this subject to make this book a valuable addition to the collection of many professionals and students.

This book was conceptualized with the vision of imparting up-to-date information and advanced data in this field. To ensure the same, a matchless editorial board was set up. Every individual on the board went through rigorous rounds of assessment to prove their worth. After which they invested a large part of their time researching and compiling the most relevant data for our readers.

The editorial board has been involved in producing this book since its inception. They have spent rigorous hours researching and exploring the diverse topics which have resulted in the successful publishing of this book. They have passed on their knowledge of decades through this book. To expedite this challenging task, the publisher supported the team at every step. A small team of assistant editors was also appointed to further simplify the editing procedure and attain best results for the readers.

Apart from the editorial board, the designing team has also invested a significant amount of their time in understanding the subject and creating the most relevant covers. They scrutinized every image to scout for the most suitable representation of the subject and create an appropriate cover for the book.

The publishing team has been an ardent support to the editorial, designing and production team. Their endless efforts to recruit the best for this project, has resulted in the accomplishment of this book. They are a veteran in the field of academics and their pool of knowledge is as vast as their experience in printing. Their expertise and guidance has proved useful at every step. Their uncompromising quality standards have made this book an exceptional effort. Their encouragement from time to time has been an inspiration for everyone.

The publisher and the editorial board hope that this book will prove to be a valuable piece of knowledge for researchers, students, practitioners and scholars across the globe.

List of Contributors

Edgaras Janusonis and Romualdas Bausys
Department of Graphical Systems, Vilnius Gediminas Technical University, Sauletekio al. 11, LT-10223 Vilnius, Lithuania

Giruta Kazakeviciute-Januskeviciene
Department of Graphical Systems, Vilnius Gediminas Technical University, Sauletekio al. 11, LT-10223 Vilnius, Lithuania
Faculty of Public Governance and Business, Mykolas Romeris University, Ateities str. 20, LT-08303 Vilnius, Lithuania

Tadas Limba and Mindaugas Kiskis
Faculty of Public Governance and Business, Mykolas Romeris University, Ateities str. 20, LT-08303 Vilnius, Lithuania

Kyungil Lee, Hyun Chan Sung, Youngjae Yoo, Yoonji Kim and Seong Woo Jeon
Division of Environmental Science & Ecological Engineering, Korea University, 02841 145 Anam-ro, Seongbuk-gu, Seoul 02841, Korea

Joung-Young Seo
Division of Environmental Science & Ecological Engineering, Korea University, 02841 145 Anam-ro, Seongbuk-gu, Seoul 02841, Korea

Jung Hyun Kook
Eco, Environment, Education, 583, Yangcheon-ro, Gangseo-gu, Seoul 07547, Korea

Yubao Sun, Sujuan Wang, Qingshan Liu, Renlong Hang and Guangcan Liu
Jiangsu Key Laboratory of Big Data Analysis Technology, Jiangsu Collaborative Innovation Center on Atmospheric Environment and Equipment Technology, Nanjing University of Information Science and Technology, Nanjing 210044, China

Li Shen, Linmei Wu, Yanshuai Dai, Wenfan Qiao and Ying Wang
State-Province Joint Engineering Laboratory of Spatial Information Technology for High-Speed Railway Safety, Southwest Jiaotong University, Chengdu 611756, China
Faculty of Geosciences and Environmental Engineering, Southwest Jiaotong University, Chengdu 611756, China

Xinran Lv and Xu Liang
The College of Information Science and Technology, Beijing Normal University, Beijing 100875, China

Libao Zhang
The College of Information Science and Technology, Beijing Normal University, Beijing 100875, China
The State Key Laboratory of Remote Sensing Science, Beijing Normal University, Beijing 100875, China

Milad Janalipour and Ali Mohammadzadeh
Faculty of Geodesy & Geomatics Engineering, K. N. Toosi University of Technology, Tehran 1996715433, Iran

Lei Pan, Heng-Chao Li, Yang-Jun Deng and Xiang-Dong Chen
School of Information Science & Technology, Southwest Jiaotong University, Chengdu 610031, China

Fan Zhang
College of Information Science & Technology, Beijing University of Chemical Technology, Beijing 100029, China

Qian Du
Department of Electrical & Computer Engineering, Mississippi State University, Starkville, MS 39762, USA

Yuxiang Zhang, Ke Wu and Xiangyun Hu
Hubei Subsurface Multi-Scale Imaging Key Laboratory, Institute of Geophysics and Geomatics, China University of Geosciences, Wuhan 430074, China

Bo Du
School of Computer, Wuhan University, Wuhan 430079, China

Liangpei Zhang
State Key Laboratory of Information Engineering in Surveying, Mapping, and Remote Sensing, Wuhan University, Wuhan 430079, China

Daniel Guidici
Department of Engineering Science, Sonoma State University, 1801 E Cotati Ave, Rohnert Park, CA 94928, USA

Matthew L. Clark
Center for Interdisciplinary Geospatial Analysis (CIGA), Department of Geography, Environment and Planning, Sonoma State University, 1801 E Cotati Ave, Rohnert Park, CA 94928, USA

Chengyi Wang, Jingbo Chen and Jianglin Ma
Institute of Remote Sensing and Digital Earth, Chinese Academy of Sciences, Datun Road North 20A, Beijing 100101, China

Lu Li
Institute of Remote Sensing and Digital Earth, Chinese Academy of Sciences, Datun Road North 20A, Beijing 100101, China
The University of Chinese Academy of Sciences, Beijing 100049, China

Feimo Li, Shuxiao Li, Chengfei Zhu, Xiaosong Lan and Hongxing Chang
Institute of Automation Chinese Academy of Sciences, Beijing 100190, China
University of Chinese Academy of Science, Beijing 100049, China

Index

www.ingramcontent.com/pod-product-compliance
Lightning Source LLC
LaVergne TN
LVHW060548210926
848485LV00005B/1280